Tropical Plant Collecting

From the Field to the Internet

Edited by
Scott A. Mori
Amy Berkov
Carol A. Gracie
Edmund F. Hecklau

TECC Editora
Florianópolis, Brazil

This book is dedicated to
John Daniel Mitchell
in recognition of his support of the
Institute of Systematic Botany and the
LuEsther T. Mertz Library of
The New York Botanical Garden

Published by
TECC Editora
Florianópolis, Brazil
www.tecceditora.com - info@tecceditora.com

Distributed by Itasca Books

Cover painting by Michael Rothman

Contents

Preface

The diversity of plants is described in two important kinds of publications—Floras and monographs (Box 1-1). A Flora describes all plants in a given geographic area, and a monograph treats all species of a particular group of plants throughout its geographic range.

Monographs, which include the information needed for identifying plants as well as a summary of what is known about the group studied, provide the baseline information needed for other studies of plants, such as cytology, molecular biology, ecology, environmental assessments, conservation biology, and economic botany. Mares (1991) has pointed out that modern research on mammals in the United States is "built upon a firm foundation of basic research" and without this foundation "the exciting questions now being asked would be difficult to formulate." Likewise, well-done and complete Floras are the most useful tools for determining the names of plants in specific areas, and also facilitate biological studies that would be impossible without this important baseline information. An outstanding example is Croat's (1978) classic *Flora of Barro Colorado Island*, which has been followed by studies of the island's forest ecology (Leigh et al., 1982, 1996), as well as by studies of the insects, birds, mammals, and all other organisms that depend upon the plants described in the Flora. In other words, Floras and monographs serve as the basis for studies of evolution, ecology, ethnobotany, and conservation biology because they allow researchers to identify the plants they study.

Published monographs and Floras provide the foundation needed to design studies to test ecological and evolutionary hypotheses. We do not mean to imply, however, that the botanists who produce monographs and Floras do not test hypotheses. Contrary to the claims of some critics of systematic botany, who maintain that systematics is not a science, but merely "stamp collecting" or an enumeration of facts about the living world, hypothesis testing is the very foundation of systematic botany. The overwhelming amount of descriptive work in systematics masks the hypothesis testing, but every taxon (e.g., family, genus, species) defined by a systematist is a hypothesis that is open to testing by the addition of new collections and the accumulation of new data. In short, there are as many hypotheses in systematic biology as there are taxa on the planet! In addition, the establishment of species boundaries paves the way for testing hypotheses about the evolutionary and ecological relationships between plants and animals, as well as the relationships that plants have with their abiotic environment.

Mori et al.'s (2005) discussion demonstrates how the availability of both a published Flora and a bat Fauna facilitated a study of plant/bat interactions in the Neotropics (Lobova et al, 2009). Bats play an important role in the pollination and dispersal of tropical plants; however, details of these interactions in lowland neotropical forests are not thoroughly documented because there are few areas in the Neotropics with published inventories of both plants and bats. An exception is found in central French Guiana, where the existence of both a Flora (Mori et al. 1997, 2002a) and a bat Fauna (Charles-Dominique et al. 2001) made it possible to identify the regional species. This, in turn, facilitated determining the identities of plants that specific bats depend upon for at least part of their nutrition, and the identities of bats that specific plants depend on for seed dispersal (Lobova et al. 2009). Lobova et al. (2009) addressed the hypothesis that the relationships between bats and plants have resulted in morphological adaptations in both groups. They also demonstrated that bats play a key role in maintaining biodiversity in tropical lowland forests.

This book describes the field work involved in making plant collections and gathering the data used in producing monographs and Floras. Chapters 1 and 2 deal with Mori's career as a tropical botanist and Berkov's adaptation to the rigors of a year's field work in French Guiana. The third chapter discusses the adventures and misadventures inherent to tropical field studies. Chapters 4, 5, and 6 provide insights into how botanical specimens are collected, processed, and integrated into herbaria, and how information about them is disseminated via the Internet. The final chapter reviews the future for tropical forests in a world of increasing human population and consumption.

Chapter 1 (My Career as a Tropical Botanist) provides an overview of Mori's career as a tropical botanist. Although this career (which started in 1965) began before the advent of many of the technologies now available (most significantly the use of computers and computerized databases, the development of software used in cladistics and other methods of phlyogenetic inference, and automated DNA sequencing), the goals of tropical botany (to inventory plants, understand the relationships among them, and define the relationships between them and their environment) have remained essentially the same. He describes the impacts that his mentors, other botanists, and botanical institutions have had on his career.

The paths that lead students to study tropical biology are many. In Chapter 2 (Amy's Year in the Rain Forest: Romance Versus Arduous Reality), Amy Berkov relates how, after changing from a career as a graphic artist working in New York City to that of a tropical biologist, she managed to survive an exciting, but harrowing, year studying interactions between wood boring cerambycid beetles and trees in the Brazil nut family (Lecythidaceae) in central French Guiana. This chapter was initially written as a long "thank you" letter to the many people who helped Amy realize her dream of spending a year living alone in the middle of the rain forest. We hope that it will help future tropical biologists prepare for the challenges they may face in adapting to a very novel environment.

Chapter 3 (Tips for Tropical Biologists) presents the dangers and difficulties associated with tropical field work. Content ranges from exposure to tropical nuisances, such as parasitic botflies, to encounters with genuinely dangerous organisms (for instance pathogens associated with tropical diseases like leishmaniasis), to basic tropical housekeeping, including how to set-up and sleep in a hammock. This chapter includes short essays written by field biologists about such misadventures as inflicting a serious wound with a machete; getting lost and spending an uncomfortable night in the forest; interacting with a tarantula hawk wasp; encountering a herd of white-lipped peccaries in Amazonian Peru; and narrowly escaping capture by guerilleros in Colombia. The goal of the chapter is to provide information that will make the work of tropical biologists safer, easier, and more comfortable.

The next three chapters provide suggestions on how plant collection and herbarium management can be refined, and include recommendations on taking advantage of database and Internet technology to manage and present biodiversity data. The major theme in Chapter 4 (From the Field) is that collections must be well-prepared and informative to justify their being archived in herbaria; and that trees and the epiphytic plants that grow in them can be adequately inventoried only if collectors climb into trees to gather specimens. Chapter 5 (Into the Herbarium) describes herbarium practices, and emphasizes that herbarium administrators should develop mission statements for their herbaria. These statements should articulate the goals of the herbarium and set guidelines for specimens that will and will not be archived. Chapter 6 (Onto the Internet) discusses the contemporary computer technologies that are available for managing herbarium collections, and for making data and images available on the Internet. Programs are now available that facilitate the production of electronic monographs and Floras. These user-friendly technologies make it possible for those without advanced computer expertise to make information about the plants they study available immediately and almost anywhere in the world, which speeds up the process of biodiversity inventory.

Chapter 7 develops the idea that tropical forests are both fragile and resilient. They are fragile because the co-evolutionary processes that have, over long periods of time, influenced plants, animals, and the habitats in which they live, can be easily torn asunder. They are resilient because even large disruptions, resulting in the destruction of old-growth trees, create forest openings called gaps that can be subsequently repaired by natural regenerative processes. When human population levels are modest and land use technology appropriate, human forest management mimics natural gap processes. If, however, human population levels increase beyond a certain threshold and human activity causes too much forest destruction, natural gap formation and regenerative processes no longer function. The gaps become too large and begin to coalesce, and the time intervals between human-induced gap formations become too short to allow regeneration of mature forest. Although humans can live in harmony in tropical forests if they maintain relatively low populations and do not consume at high levels, they almost always have a negative impact on biodiversity... and the greater the increase in human density and consumption the greater that negative impact will be. Although humans have fostered the generation of plant and animal diversity through the creation of domesticates that benefit humans directly, they have had little to do with the evolution of biodiversity in tropical forests. The suggestion that humans are essential to the functioning and diversity of tropical forests is based on an anthropocentric world view that does not adequately account for evolutionary processes; most of the finely tuned interactions among plants and animals inhabiting tropical forests arose long before humans were on the scene. Finally, tropical forests provide ecosystem services to the planet, and the entire world must be willing to pay for those services as part of the cost of maintaining stable climates, stable hydrological cycles, and safe levels of atmospheric gases. By safeguarding the ecosystem services provided by tropical forests, it may also be possible to safeguard the forests themselves—these incredible repositories of life, with their "stranger-than-fiction" tales of evolution and co-evolution; stories both well known and as-yet-undiscovered.

The book also includes four appendices. Appendix A describes the "Adopt-a-Forest" strategy that has oriented Mori, along with many other tropical biologists, in his career trajectory. Tropical research, including field and laboratory studies, as well as subsequent specimen archival, is expensive and requires external funding. Acquiring this funding can be much more daunting than actually conducting research so Appendix B discusses the diverse funding sources available for supporting studies of tropical biology. The final two appendices are lists of personal items needed for field work (Appendix C) and the basic equipment needed to make botanical collections (Appendix D).

We hope that this book will be of use to those planning to carry out biological studies in the tropics (especially in the New World) that involve plant collection. The first step in learning about plants, and the animals and other organisms that depend on them, is to learn their names, which in turn, often depends on collecting the specimens needed to facilitate their identification.

The Editors

Scott A. Mori
Amy Berkov
Carol A. Gracie
Edmund F. Hecklau

Acknowledgments

We are grateful to the funding organizations and individuals that provided the financial support which made the research upon which this book is based possible. The funders are listed by project as follows: Flora Neotropica Monographs on Lecythidaceae (National Science Foundation [BMS 75-03724 and DEB-8020920]; *Guide to the Vascular Plants of Central French Guiana* (Beneficia Foundation, Oliver S. and Jennie R. Donaldson Charitable Trust, Dorothy Salant and the G. A. G. Charitable Cooperation, Eppley Foundation for Research, Harriet Ford Dickinson Foundation, National Geographic Society, National Science Foundation [BSR-9024530], Andrew W. Mellon Foundation, Rhulen Family Foundation, and the Rockefeller Foundation); French Guiana e-Flora (Centre National de la Recherche Scientifique of France); *Seeds Dispersed by Bats in the Neotropics* (Bat Conservation International, Beneficia Foundation, and the National Science Foundation [NSG-NATO 0309534 and DEB 0414098]; Flowering Plants of the Neotropics (Beneficia Foundation and the Samuel Friedman Charitable Trust); the Plants and Lichens of Saba project (Conservation International); and the Vascular Plants of the Osa Peninsula (Chris Davidson and the Blue Moon Fund). We also thank the following individuals for making significant financial contributions to our work: Sheila and Thane Asch, Sharon Christoph, Chris Davidson, Susan Fredericks, Gerhard Hass, Anne Hubbard, Katie Lee, Naomi Pitcairn (including a generous subsidy for publishing this book), and Hazel Tuttle. From 1991 through 2008, Carol Gracie and the senior editor organized ecotours in which approximately 350 people (excluding those on cruise ships) travelled with us on 34 trips to destinations in Brazil, Costa Rica, Ecuador, France, French Guiana, Greece, Hawaii, Italy, Spain, Trinidad, and Venezuela. We thank The New York Botanical Garden for facilitating these trips and those we explored the botanical world with us for electing to join the trips. We also thank the cruise ship passengers with whom we interacted when we lectured and served as guides on cruises to the Amazon, Carribean, and several other destinations in tropical America.

We appreciate the effort made by the scientists who contributed to our multi-author projects such as the *Guide to the Vascular Plants of Central French Guiana* (Mori et al., 1997, 2002a); for pictures of all 80 of them, see the second volume of the Flora, as well as to the 150 botanists who contributed to the *Flowering Plants of the Neotropics* (Smith et al., 2004).

We thank Melissa Tulig and Tony Kirchgessner for helping us develop the databases upon which the results of several projects discussed in this book are published, Nestor Pérez-Moliere for help with imaging of specimens, and Arthur Fairwheather and Patrick Maraj for resolving IT problems. We express special gratitude to Melissa Tulig for designing and maintaining our project web pages and to the Directors of the Herbarium, Barbara Thiers and Patricia Holmgren before her, for facilitating our herbarium studies. In addition, we thank Barbara Thiers and the IT Department of NYBG for making it possible to place data associated with herbarium specimens in the KE Emu database and then making it available on the Internet. We thank John Brown and Edmund F. Hecklau for their volunteer service which led to a number of publications and The New York Botanical Garden's Volunteer Program, especially Jackie Martinez, for making the participation of top-notch volunteers possible. The bibliophiles, William R. Buck, John D. Mitchell, and Tom Zanoni, have brought literature of interest to the attention of the senior editor over many years and he is thankful to them for taking the time to do this. In addition, the senior editor is grateful to Daniel Atha, Xavier Cornejo, Hugh Iltis, Thomas Croat, Doug Daly, Jim Lutyen, Michael Nee, Ghillean T. Prance, Rob Naczi, and Tom Zanoni for their contributions to his knowledge of tropical plant collecting.

We are grateful to R. H. J. Erkens and P. J. M. Maas for their review of our comments on the integration of the Utrecht Herbarium into the collection of the Netherlands Centre for Biodiversity in the Leiden, Alex Popovkin for reading the part of the manuscript about him, and Damon Little for sharing his knowledge of the use of computer programs in Systematic Botany. We thank Thomas Couvreur, Hugh Iltis, Melissa Tulig, Jacquelyn Kallunki, Linde Ostro, Rob Naczi, Paola Pedraza, Don Stone, Allen Young, and Tom Zanoni for reviewing parts of the manuscript. Their suggestions greatly improved the book, but the ideas presented herein do not necessarily reflect their views. We are grateful to Nathan P. Smith for copy editing, designing, and producing this book.

We thank Naomi Pitcairn for preparing the drawings of the capuchin monkeys, the tarantula hawk wasp, and the peccaries; Bobbi Angell for the drawings of the mosquito net and hammock net knots (Chapter 3); Carmen Galdames for the field images of *Lecythis tuyrana* used in Fig.

5-1; Sheranza Alli and Cindy Hirsch for help in the preparation of Fig. 5-1; and Michael Rothman for the paintings of the understory and the canopy of French Guianan rain forest (Chapter 1), the cover painting, and most of the line drawings in Chapter 3. We thank Jean-Jacques de Granville, David Campbell, Nathan Smith, Naomi Pitcairn, and Michael Nee for writing about some of their experiences in the boxes associated with Chapter 3.

We are thankful to The New York Botanical Garden for providing the senior editor with a curatorial position and the access to the herbarium, laboratory, and library facilities needed to carry out research on tropical plants. In addition, we are grateful to the administrators of The New York Botanical Garden during the time that the senior editor worked at NYBG, Jim Hestor and Gregory Long (the Presidents); Ghillean Prance, Michael Balick, Brian Boom, Dennis Stevenson, and Jim Miller (Vice Presidents for Science), and Richard Howard, Enrique Forero (former Directors of the Institute of Systematic Botany) and Douglas Daly (the current Director) for their support of much of the research that has gone into this book.

Amy Berkov gratefully acknowledges the PSC-CUNY Research Foundation and the Fund for Neotropical Plant Research of The New York Botanical Garden for funding her year of field research in French Guiana. She is also profoundly grateful to thesis co-advisors Scott Mori and Barbara Meurer-Grimes who, along with committee members David Grimaldi and Lee Herman, took on a high-risk investment. She thanks Gérard Tavakilian for his painstaking work rearing and identifying cerambycids that stimulated her to study plant/beetle interactions. She would not have been able to carry out research on canopy insects without the expertise and generosity of arborists Bob Weber and Chris Roddick, and might never have withstood a year of splendid isolation without the support of Hugette and Gerald Dumas, friends, fellow students, and family.

Scott Mori is grateful to his Post-docs: Tatyana Lobova and John Janovec and his research assistants, Xavier Cornejo, Carol Gracie, Scott Heald, and Nathan Smith, for working so hard and enthusiastically on the projects we did together. He thanks his students Pedro Acevedo-Rodriguez , Amy Berkov, Brian Boom, Cullen Geiselman, Vanessa Hecquet, Ya-Yi Huang, Maria Lúcia Kawasaki, Samuel Kisseadoo, Amy Litt, John Pipoly, Carolina Potascheff, and Chi-Hua Tsou for adding to his knowledge of tropical biology.

The senior editor is also grateful to all those who facilitated his field work, provided field data, and hosted him during herbarium visits; among them are Pedro Acevedo-Rodriguez, Reinaldo Aguilar, Andy Allinckx, Ivan Allinckx, the rest of the Allinckx family, Lucien Aboukrat,

Patti Anderson, Bobbi Angell, Paulo A. Cost Lima Assunção, Catherine Bainbridge, Michael Balick, Peter Becker, Julio Betancur, Amy Berkov, Frédèric Blanchard, Alan Bolten, Brian Boom, Michel Boudrie, William R. Buck, Romeu Cardoso, Pierre Charles-Dominique, Patrick Chatelet, Georges Cremers, Thomas B. Croat, Françoise Croizier, Cullen Geiselman, Douglas C. Daly, Robert Dressler, James Duke, Chris Davidson, Christian Feuillet, Beat Fischer, Adrian Forysth, Moacir Fortes, Maria A. de Freitas, Vicki Funk, Claude Gascon, Carol A. Gracie, Philippe Gauchet, Ara Görts-van Rijn, Eric Gouda, Jean-Jacques de Granville, Vanessa Hecquet, Edmund F. Hecklau, Bernard Hermier, Scott V. Heald, Leslie Holdridge, Noel Holmgren, Ya-Yi Huang, Roger Hutchings, Hugh Iltis, Venise Isidore, John Janovec, Brian Keeley, Nadja Lepsch-Cunha, Horace Lofton, Jim Luteyn, Alexandre A. de Oliveira, Ghillean T. Prance, Amy Litt, Tatyana Lobova, Paul and Hiltje Maas, Luis Alberto Mattos Silva, J. F. Menezes, John D. Mitchell, Michel Modde, Joep Moonen and his wife Marijke and son Bernie, Michael Nee, Heather Peckham Griscom, Terence D. Pennington, Everaldo da Costa Pereira, John Pipoly, Naomi Pitcairn, Alex Popovkin, Carolina Potascheff, Benedito Rabelo, Lauren Raz, Helen Richard, Daniel Sabatier, Talmon Soares dos Santos, Nancy Simmons, Nathan Smith, Marie-Françoise Prévost, Peter Raven, C. F. da Silva, Gerard Tavakilian, Adrian Tejedor, Merlin Tuttle, Chi-Hua Tsou, Robert Voss, Eileen Whalen, Marina Wong, and many others.

Finally, we express our gratitude to John Daniel Mitchell for his intellectual and financial support of research at the Institute of Systematic Botany. His knowledge of the literature, facilitation of connections with people with whom we subsequently developed projects, financial support for numerous projects, support of the library, and deep commitment to conservation have enabled us to accomplish a great deal more than would have been possible without his generosity. In recognition of what he has done for systematic botany and conservation, we dedicate this book to John.

John Daniel Mitchell. Photo by Carol A. Gracie (1995).

Chapter 1
My Career as a Tropical Botanist

by Scott A. Mori

This chapter provides an overview of my career as a tropical botanist, which began in 1965 when I first became familiar with lowland tropical forests on a field trip to Mexico. During my 46 years as a botanist, tropical biology became a field of study, and biology itself has been transformed by new technology. The most important innovations influencing the study of systematic botany during this period have been the use of computers for recording, storing, searching and analyzing data. Today we have software for word processing, literature searches, and phylogenetic analyses that help us better understand evolutionary relationships; automated DNA-sequencing; and we can disseminate research results worldwide using the Internet. Studying plants in the field and herbarium; and illustrating, photographing, and describing them in Floras and monographs (Box 1-1) have changed less dramatically, but have been made easier because of advances in technology.

In 1980, I read the report Research Priorities in Tropical Biology (Committee on Research Priorities in Tropical Biology, 1980) which suggested that the best way to understand tropical biology would be to establish a series of research areas throughout the tropics, inventory the regional plants and animals, and then conduct detailed studies of ecological relationships in these areas. By that time I had already decided to dedicate part of my research to a study of the Brazil nut family, but this report stimulated me to make a botanical inventory and ecological study of the forests of central French Guiana, then to expand that idea to other areas. This chapter tells the story of how the Research Priorities in Tropical Biology report (see Appendix A), the projects I have participated in, and other people have influenced my career in tropical botany.

Box 1-1. The Difference between Monographs and Floras
by Scott A. Mori

Monographs and Floras are still among the most important products of plant systematics. Curators of The New York Botanical Garden's Institute of Systematic Botany are selected based on their expertise in a given plant family or families, and their interest in the plants of a given part of the world. For example, Paola Pedraza is a specialist in the blueberry family and plants of the Andean region. Her interest in the Andes is based, to a large extent, on the fact that most species of the blueberry family are found in the Andes of South America and the mountains that continue into Central America.

A monograph is a treatment of a given plant group (= taxon) throughout its entire geographic distribution. That taxon could be a genus, a division of a family, or even an entire family. For example, our treatments of the Lecythidaceae (a plant family restricted to tropical areas) subfamily Lecythidoideae comprise a monograph of all New World (= Neotropics) species of this family. The Lecythidaceae are also found in the Old World (= Paleotropics), but the species occurring there belong to different subfamilies (Mori et al., 2007). Flora Neotropica (see http://www.nybg.org/botany/ofn/OFN.html) is a series of monographs that ultimately aims to include all New World species of plants (including mosses, liverworts, hornworts, ferns and their allies, gymnosperms, and flowering plants) and fungi. A list of published Flora Neotropica monographs is available at the Organization for Flora Neotropica Website at http://www.nybg.org/botany/ofn/Monograph%20List.htm. In contrast, a Flora is a treatment of all of the species in a given geographic area. For example, the Vascular Plants of Central French Guiana (Mori et al., 1997, 2002) treats only species of vascular plants from a restricted geographic area, but includes many different taxa. In this case, the word Flora is capitalized to distinguish it from flora, which means all of the plants of a given area. A checklist enumerates all of the species of a given area, i.e., the flora, and a checklist is the first step in producing a Flora. A list of Floras published for the Neotropics, as well as for the rest of the world, can be found in Frodin (2001).

In some cases, a monograph is a hybrid between a monograph and a Flora; for example, genera in the family may be found in both the New and Old Worlds, but only species of one region are treated in the monograph. Although a monograph usually includes considerably more information than a Flora, some Floras are monographic in nature, e.g., the Intermountain Flora published by The New York Botanical Garden.

> Modern monographs almost always include a phylogenetic analysis of the taxon treated based on anatomical, morphological, and molecular data, something that is not possible for a flora because a phylogeny based on a restricted number of species from a limited geographic area can not provide a phylogeny of a group also represented in other areas. On the other hand, DNA barcoding the species in a flora is useful because it facilitates the identification of sterile specimens at all stages of a plant's life cycle in the area of the Flora.

Initiation to Botany

By the time I reached high school, I had already developed an interest in natural history because of my experiences camping with the Boy Scouts and hunting for rabbits and pheasants with my father, my "Uncle" Charley (my grandmother's sister's husband), and a neighbor, Jack Garnett. I lived in the village of Milton (then with a population of about 1500) in fertile farming country in southeastern Wisconsin. There were hunting preserves, woodlands, lakes, and marshes scattered among the corn fields, but a wildlife refuge called Storr's Lake (Fig. 1-1, top) was especially important because it put nature within a mile of my back door. Even before I had my driver's license, I would hoist my canoe onto my back and portage it to the lake where I hunted, fished, and ran a trap line to catch muskrats and an occasional mink. Spending so much time at Storr's Lake also turned me into a bird watcher, and stimulated me to learn about the plants that birds and other animals depend upon for food and shelter. This was the start of my fascination with plant/animal interactions, which is reflected in my interests in the pollination and dispersal of tropical plants. While in high school, I also developed an awareness of conservation as the result of the influence of our local game warden, Royce Dallman, and through my participation in the Milton Union High School Conservation Club.

After graduating from high school in 1960, I enrolled at the University of Wisconsin at Stevens Point to pursue an undergraduate degree in biology and conservation. Although I had a small scholarship to pay tuition, I worked as a janitor in the student union, washing and waxing floors and cleaning classrooms, to pay other expenses. I made a paltry 85 cents an hour but my expenses were low; for example, tuition was approximately $65 a semester. Although I was a somewhat above average student in high school, I was worried about my academic ability because I had not developed good study habits. Consequently, I studied a great deal my first weeks and the results of my first test, by coincidence in botany, gave me great satisfaction when I received the highest grade in the class. My new, improved study habits carried over for my entire undergraduate career at Stevens Point, and I finished with a high grade point average and a B.S. degree with a major in biology and a minor in conservation. Several of my biology teachers encouraged me to go on to graduate school, so I applied to the University of Wisconsin at Madison (UW) to pursue an M.S. in biology.

That first semester I enrolled in the plant geography class taught by botanist Hugh Iltis (Fig. 1-1, bottom), and my career aspirations soon took a different course. Iltis, a fiery teacher and avid conservationist, inspired me to change my major to botany and I then set my goals on becoming a university professor. Iltis, now a professor emeritus at the University of Wisconsin, became the mentor of my M.S. and Ph.D. research and the in-

Fig. 1-1. Early influences. Top: The author at the Storrs Lake sign in 2000, the place where he first became interested in natural history while in high school. Bottom: Hugh Iltis and the author at the University of Wisconsin in 1986. Photos by Carol A. Gracie.

spiration for my conviction that systematists should play a major role in the conservation of the plants and animals they study.

Iltis was an avid plant collector famous for collecting large numbers of well-prepared plant specimens, a process he called "baling hay." In December 1962, Iltis and his Ph.D. student Don Ugent (later a professor at Southern Illinois University in Carbondale) travelled to Peru to study species of wild potatoes as part of Ugent's thesis research (Iltis, 1988). One of their discoveries was a weedy tomato, with small, sweet, green-and-white berries slightly smaller than a cherry, which they collected and numbered as *Iltis and Ugent 832*. Iltis squeezed out the seeds from the fruits and put them in a packet. When he returned to Madison he sent them to Charles Rick, a renowned tomato geneticist at the University of California, Davis who published the collection as the new species *Lycopersicon chmielewskii* C.M. Rick named in honor of a Polish tomato expert (H. H. Iltis, pers. comm., May 2010).

In 1980, Iltis learned that Rick had crossed the progeny of 832 with a commercial tomato in an effort to improve the desirability of the commercial variety. Their progeny produced larger fruit with increased pigmentation and, most important, higher soluble sugar content. Rick informed Iltis that each 0.5% increase in sugar content was worth about 8 million dollars a year (in 1987 dollars), an astonishing return on a modest $21,000 NSF grant. And this does not include the other results from the project, especially with regard to potatoes, which were the focal organisms of the grant in the first place!

Iltis inspired students to appreciate the natural world in less tangible ways. In my case, he completely transformed my career goals: I initially wanted to be a high school biology teacher and track and wrestling coach, and after witnessing Itlis's zeal for nature I decided to obtain a Ph.D. and pursue a career in systematic botany. The following quote from Iltis (1988) captures the message that still guides the way I look at human interactions with the natural world.

> "Mankind depends on plants for food, fiber, drugs—and a livable world. But more than that, our children will want nature to experience while growing up—to explore, love, and enjoy its beauty and diversity. Corn and cows, concrete and cars are not enough to sustain and empower a human psyche that until only a few generations ago lived in daily contact with a variety of plants and animals, a psyche that, winnowed and sifted by natural selection, is genetically programmed to respond positively to nature and its patterns. By destroying so much of the natural environment, we humans are now destroying crucial parts of our own psychological as well as physical habitat. For those in the know, it is a gloomy picture indeed."

Iltis was among the first modern biologists, along with Paul Erlich (1968), to emphasize the negative impact that continued human population growth was having on the natural world. He is a dynamic and inspirational speaker who either filled those in his audiences with admiration for what they also felt was happening in the world, or hatred for the heretical ideas they believed he was preaching (Iltis, 2002).

Hugh was an early advocate of nature protection in Wisconsin, and worked tirelessly alongside the Nature Conservancy to establish local reserves. In 2007, David and Shelly Hamel honored him by naming a 110 acre reserve (harboring sandy prairie, savanna, and a bog) the Iltis Prairie. His studies of plants have led to habitat conservation nearly everywhere he has done fieldwork. For example, the discovery and description of a wild relative of corn, *Zea diploperennis* H. H. Iltis, Dobley & R. Guzmán, inspired Mexican authorities in the state of Jalisco to establish the Las Joyas Biological Field Station of the Universidad de Guadalajara, and the 135,000-hectare (350,000-acre) Reserva Biosfera de la Sierra de Manantlán, under the auspices of the state of Jalisco and the UNESCO Man in the Biosphere (MAB) program (Vázquez G. et al., 1995).

I entered graduate school in the Botany Department at Madison without a fellowship, so the summer before I started, I worked double shifts at a canning factory to earn money for my tuition and other expenses for the upcoming school year. The fact that I paid most of my expenses motivated me to get the most for my money by studying more than ever. As a result, I viewed my courses as an opportunity to learn as much as I could, never just as a means to get the credits needed to graduate. With ongoing encouragement from Iltis, I applied for support from the Department of Botany for the following year and was awarded a teaching assistantship. I no longer needed to work during the summer to pay for my next year at the university, and this made it possible for me to participate in my first adventure in Latin America!

Becoming a Tropical Botanist

It is easy to date the beginning of my fascination with tropical plants—it began in the summer of 1965 during a trip to Mexico with Keith Roe and his wife Eunice (Fig. 1-2). At the time, Keith was an advanced graduate student and I a beginning student, with both of us under the direction of Hugh Iltis. The previous year Keith had been to Mexico collecting species of *Solanum* section *Brevantherum* (Fig. 1-3, left), the topic of his Ph. D. dissertation, as well as specimens of other plant families. Iltis encouraged Keith to return to Mexico to continue making collections of *Solanum* and to gather additional general collections (Fig. 1-3, right) as part of Iltis's effort to enhance the status of the University of Wisconsin Herbarium as a repository of tropical plants. He also wanted Keith to gather duplicate collections for major Mexican herbaria to increase the botanical specimens available to Mexican botanists. With this in mind, Iltis and Keith wrote a successful proposal to finance the expedition as well as to pay Keith's tuition and student stipend. Iltis had observed my enthusiasm for plant collecting on his Easter vacation plant geography class field trip to the southeastern United States, and asked me to accompany Keith and Eunice. In exchange for the trip, he provided me with a small salary and made it clear

Fig. 1-2. Roe & Mori expedition to Mexico in 1965. Top: Travel by Jeep was not always easy, in this case the vehicle stalled in the middle of the Río Verde in Oaxaca when the engine got wet. By luck, a person who saw it happen dried off the spark plugs and removed the fan belt, allowing us to start the car and drive to the other side of the river. Bottom. Puerto Escondido, Oaxaca, our most beautiful campsite. Most of the time camps were set along the side of the road at any place with enough room. Photos by Eunice Roe.

that he expected me to help them gather a record number of specimens!

My lifelong fascination with the beauty of lowland tropical forests began on the 24th of August, 1965 on a day when I collected in the vicinity of Cárdenas in the state of Tabasco. My journal reads "This morning I stayed in camp at kilometer 40 and collected. Keith and Eunice collected further towards Mal Paso. I collected about 30 numbers of some of the fin-

Fig. 1-3. Collecting plants, especially species of Solanum sect. Brevantherum, was the primary goal of the Mexican trip. Left: Solanum rugosum Dunal is a member of this group of Solanum. Photo by Scott A. Mori. Right: Because we made so many collections, we often had to stack our presses two deep on the dryer. Photo by Eunice Roe.

est material of the trip and was rewarded by seeing the most fantastic area of the whole trip—virgin tropical rain forest with a beautiful stream running through it. The stream has crystal clear water and a series of rapids and waterfalls. It is one of the nicest spots I have ever been in my life."

When we returned to the UW at the end of the summer, Hugh was ecstatic with our 1823 collections. Specimens of *Solanum* Sect. *Brevantherum* dominated the collection, and five of them were subsequently designated as the types of new species for science described by Keith (Roe, 1967). There were also many collections from other families, and collecting them with Keith and Eunice was my introduction to the beauty and diversity of tropical plants. This 18,000 mile trip opened up the world of tropical botany for me!

This adventure put me in the field with a student experienced in making high quality specimens who communicated his enthusiasm for plants and Latin America to me, taught me that successful fieldwork is based on writing grants to support the trips, and demonstrated that much can be accomplished by three people working very hard day after day to achieve common goals. This experience, coupled with my realization that new species of plants could still be discovered, convinced me to become a tropical botanist.

To get more tropical experience, Iltis next recommended that I travel to Costa Rica to participate in a summer field course offered by the recently established Organization for Tropical Studies (OTS). Thus Roger Anderson, an ecology graduate student of J. T. Curtis at UW, and I enrolled in Richard Pohl's course on tropical grasses and, in the summer of 1966, travelled to Costa Rica several weeks before its start.

We arrived early so that we might undertake a study of some aspect of tropical botany before the OTS course started. We didn't know exactly what we wanted to study, but decided to collect plants in the lowlands of northeastern Costa Rica near the sleepy village of Puerto Viejo, and hoped that we would discover something interesting to investigate. Our first day out, we hiked to the Sarapiquí River and were exploring along its banks when we found an amazing plant that neither of us had ever seen before. As we examined the white and red flowers clustered along its trunk, a boat loaded with bananas pulled up to the shore and the crew began unloading the fruit into a pick-up truck. Roger and I struck up a conversation with the driver and casually asked if he knew the name of our mystery plant. "Certamente," he said, "it is, the chocolate tree *Theobroma cacao* L." Much to our astonishment, our informant turned out to be world-renowned tropical ecologist Leslie Holdridge (Fig. 1-4, top), famous for the *Holdridge Life Zone* system of vegetation classification (Holdridge, 1967). Learning that we were enthusiastic (but poorly informed) botanists, he took us under his wing and helped us on both the current and my subsequent trip to Costa Rica.

The next day Roger and I explored a swamp dominated by the palm *Raphia taedigera* (Mart.) Mart., and were so impressed by this habitat, completely different from anything we had ever seen in Wisconsin, that we decided to describe the vegetation using the methods Roger had learned in his ecology courses (Cottam & Curtis, 1962). For several reasons, this experience had a profound impact on my future. In the first place, it taught me how difficult it can be to study tropical vegetation. The swamp was so wet that by the end of the day our feet were shriveled beyond recognition, and we were stung by so many mosquitoes that our faces were soon covered with ugly bumps. We accused one another of being "the ugliest human... ever seen," but we persisted with the research and, based on the data we collected in the *Raphia* swamp, I became the co-author of my first botanical publication (Anderson & Mori, 1967). This study demonstrated that while there is still a great deal to learn about tropical forests, reaching an understanding of their composition and ecology involves a lot of hard and sometimes very difficult fieldwork.

On the way to the swamp, Roger and I passed by a towering tree of the Brazil nut family (*Lecythis ampla* Miers). Woody, capsular fruits the size of a baby's head were scattered on the ground under its crown (Fig. 1-4, bottom). I gathered up a number of them as souvenirs, not then realizing that these bizarre objects, mistaken by a custom's official upon my entry into the States as Amerindian pottery, would play a significant role in my career. When I returned to classes the next fall and proudly presented Iltis with one of these giant, pottery-like fruits and the idea to work

Fig. 1-4. Student years in Costa Rica. Top: Leslie R. Holdridge near Puerto Viejo in 1967. Photo by Scott A. Mori. Bottom: The author at the age of 26 holding a monkey pot fruit (Lecythis ampla) somewhere in the lowlands of northeastern Costa Rica. Photographer unknown.

on Lecythidaceae for my thesis, he told me that he had discussed possible research projects with legendary plant explorer and melastome specialist John Wurdack at the Smithsonian Institution who suggested the Brazil nut family as being in dire need of taxonomic revision. When I learned that the tree producing those spectacular fruits belonged to a group of plants in need of taxonomic and ecological attention, I decided to make Lecythi-daceae the topic of my future research, an interest that has persisted for my entire career.

To gather information for my M.S. thesis on *Lecythis* (monkey pots) in Central America, Iltis suggested that I travel to Costa Rica, Pana-ma, and Colombia to observe plants in the field and examine collections in local herbaria. My goal, he said, would be to find as many trees of Lecythis as possible, gather specimens for herbaria, and make observations about the taxonomy, ecology, and economic botany of this poorly known genus of giant trees. This trip would differ from my previous field experiences because it would be the first time that I travelled alone to the tropics.

After using herbarium collections to map the known localities of *Lecythis*, I worked out an itinerary and departed for San José, Costa Rica on 15 August 1967. I didn't return to Madison until mid-January and, al-though I collected only 113 numbers, I gained invaluable knowledge about the Brazil nut family. At that time, permits to make botanical collections were not necessary, and I was able to move from one country to the next and collect without having to deal with the difficult-to-navigate, some-times impossible, bureaucracies associated with contemporary plant col-lecting. Fieldwork could also be done on a shoestring budget; for example, I reported in a letter to Iltis (21 October 1965) that I paid less than $4 per day for room-and-board in San José. There were few research stations that could be used as a base of operations, but a young botanist travelling on his own was an object of curiosity and people frequently invited me to stay with them so they could find out what a botanist does. I also learned about the importance of local colleagues. Not only did local botanists allow me to accompany them on trips, at a fraction of the cost, but they also directed me to sites where I would have the best chance of seeing monkey pots in the field. Although many people assisted me along the way, Leslie Hold-ridge, in Costa Rica, and Robert Dressler, Jim Duke, and Horace Loftin, in Panama, were especially generous with their assistance.

When I arrived in San José, I reported to Dr. Holdridge at the Tropical Science Center, and he provided me with temporary office space, with access to his personal library, and helped me plan collecting trips. The first was an expedition into the lowlands of northeastern Costa Rica to revisit the forests Roger Anderson and I had seen during the previous summer. I wanted to study fruit variation in *Lecythis ampla*, and knew that trees could be easily located in the vicinity of Puerto Viejo. In particular, I wanted to determine if fruit size was as variable as Dugand (1947) had documented for *L. minor* Jacq. At the time of my trip undisturbed forest was within easy walking distance of Puerto Viejo, and a particularly fine patch of forest with monkey pots could be seen on the Holdridge farm on the banks of the Sarapaqui River. The farm subsequently became the La Selva Biological Station of the Organization for Tropical Studies, where many past and present students have been initiated into the stimulating world of tropical biology.

From Puerto Viejo, I traveled down the Sarapiquí River to a small Costa Rican customs post at La Trinidad, situated at the confluence of the Sarapiqui and San Juan rivers. I spent a week studying fruits from different trees of *L. ampla*, and learned that individual trees displayed considerable fruit variation. With fruit and leaf collections in my pack, I then headed down the San Juan River on a local freight boat, with Nicaragua on my left and Costa Rica on my right, until I reached the village of Barra del Colorado on the Caribbean coast. There I added more monkey pot fruits to my load and, with my equipment and botanical treasures on my back, proceeded to walk the 17 miles on the soft, black sand beach to what is now known as the John H. Phipps Biological Field Station near Tortuguero. When I came to rivers emptying into the Caribbean Sea, I was sometimes able to pay someone a small fee to ferry me across. I otherwise tied my pack onto a makeshift log raft and swam with it across the river; it wasn't until I reached the research station that I learned how dangerous swimming in those rivers could be. When I yelled across the river, hoping to get someone to come over and ferry me across, and no one responded, I stripped to my underpants, left my equipment behind, and swam to the station. As I pulled myself from the water none other than Archie Carr, the founder of the station, greeted his unexpected visitor by saying "do you realize that you have just crossed a river loaded with sharks?" He calmly walked out on the dock, pulled in a line with a large hook baited with what had been a foot-long fish, and now supported shark-mangled remains.

Plant collectors can also have unnerving encounters with "domestic" animals. At that time of the year green sea turtles were arriving at the Tortuguero beach to lay their eggs, and feral dogs had gathered to feed on the eggs. While I was hiking along the coast close to Tortuguero, a pack of dogs appeared in the distance and as they got closer and closer, I felt threatened and pulled out my machete. Suddenly, one of the dogs lunged at me, grabbed the leg of my pants near my ankle and ripped a piece of it off. I responded by wildly swinging my machete at the dogs while letting out loud, terrified screams, and the pack scurried away faster than it had approached! If I had not had that machete within easy reach, perhaps I would not be telling this story now.

Upon arriving in Panama, I visited Dr. Robert Dressler at the Smithsonian Tropical Research Institute. At the time Bob was studying orchids, euglossine bees, and anything else that struck his fancy. Nevertheless, he took the time to arrange visits for me to Barro Colorado Island and Summit Gardens, and introduced me to James Duke who was in Panama making a study of the biological impact of a proposed sea level canal in the Darién Province. My first impression of Duke, as recorded in my notebook, was that he "looks as much like an explorer as he is. He is tall, slender, in good shape, good looking, and deeply tanned. He knows the Darién better than any American." Duke made arrangements for me to go to Cerro Pirre and the Chucunaque River with his guide Narcisco Bristan for two weeks for a fraction of the total cost of the trip. This opportunity permitted me to get into one of the most remote areas of the world and provided the collections and information needed to describe *Lecythis mesophylla* S. A. Mori, my first new species (Mori, 1971).

Data that I gathered on this trip served as the basis for my M.S. thesis (Mori, 1970); moreover, the experience convinced me that I wanted to pursue further studies of the Brazil nut family under the guidance of Iltis at the University of Wisconsin. For my Ph.D. thesis I decided to monograph *Gustavia* (Mori, 1976), a genus of understory trees, because it was so difficult to collect specimens from large individuals of *Lecythis*. Thus my next trip was an expedition to southern Central America and northern South America to collect *Gustavia* and other Lecythidaceae. I was fortunate to enlist fellow graduate student and inveterate plant collector Michael Nee to accompany me on the expedition. In exchange for his participation, the general collections were numbered on his series.

The journey with Mike started on 26 January 1971 and ended with my return in mid July. We traveled by bus, hitch-hiking, rented car, boat, airplane, and on foot to collect plants in Costa Rica, Panama, Colombia, Venezuela, Suriname, and Brazil. Our goals were to see as many species of *Gustavia* as possible in the field, study collections in local herbaria, learn as much as possible about tropical plants in general, and to experience the inevitable adventures associated with botanizing in the tropics.

When we arrived in Panama, Dressler informed us that a new species of *Gustavia* had been overlooked because of its superficial similarity to the common *G. superba* (Kunth) O. Berg. This species had first been recognized as distinct by R. E. Woodson in 1938 when he, P. H. Allen, and R. J. Siebert collected it in the Canal Zone. A manuscript name was written on their collection (*Woodson et al. 1582*), but a description was never published. Dressler rediscovered it in 1970, and notified Thomas Croat, then stationed in Panama and working on the *Flora of Barro Colorado*

Fig. 1-5. Comparison of two species of Gustavia. Top: Mike Nee holding a branch of Gustavia superba in the right hand and a branch of G. grandibracteata in the left hand. Bottom left: Ramiflorous inflorescence of G. superba. Bottom right: Terminal inflorescence of G. grandibracteata, the later was a new species discovered by Robert Dressler and Tom Croat and described by Croat and the author. Photos by Scott A. Mori.

Island. Dressler and Croat told us that individuals of the new species (Fig. 1-5) could be observed along the trans-isthmian railroad, especially just to the northwest of Gamboa, where it could easily be compared with numerous sympatric individuals of *G. superba* (Fig. 1-5). Excited that even in this well-explored area we might participate in the description of that new species of *Gustavia*, Mike and I traveled by bus to Gamboa, and then hiked along the railroad. We found many plants of the undescribed species, with its spectacular flowers hidden above a terminal rosette of leaves, growing next to *G. superba*, a gorgeous plant in its own right with cauliflorous flowers arising from the stem below the leaves. We took copious notes and photographs, and in 1974 Croat and I published the new species as *G. grandibracteata* Croat & S. A. Mori (Croat & Mori, 1974).

While studying the types of *Gustavia* for my dissertation, I soon realized that John Dwyer had described *G. grandibracteata* as *G. superba* var. *puberula* Dwyer (Dwyer, 1965). I examined the holotype and realized that Dwyer's variety was based on a mixed collection: the herbarium sheet of the holotype includes *G. grandibracteata* leaves on the sheet, but also contains a packet with *Eschweilera* flowers! In addition, the isotype fruit deposited in the herbarium of the Missouri Botanical Garden has a fruiting rachis of *G. nana* Pittier along with the fruits of *G. grandibracteata*. Based on the mixed fruit collection, Croat and I concluded that the fruits of *G. superba* var. *puberula* arose from below the leaves and decided that the name did not apply to our *G. grandibracteata*. Study of duplicates from other herbaria, however, revealed that Dwyer's plant did indeed produce fruit at the apex of the terminal rosette, making it necessary to treat the variety as a synonym of *G. grandibracteata*. The *International Code of Botanical Nomenclature* (McNeill et al., 2006) states that a plant name has priority only at the rank at which it is published, and therefore *G. grandibracteata* stands as the valid epithet for the species because *puberula* was published as a variety, not as a species.

Mike and I also discovered a new *Gustavia* from Colombia on this trip! Our examination of specimens at the Herbario Nacional Colombiano in Bogotá, and discussions with Colombian botanist H. Garcia-Barriga, suggested that another undescribed species was growing in the rapidly disappearing forests of the Magdalena Valley. Mike and I promptly made arrangements to visit the German/Colombian forestry project at Campo Capote to search for the new species, but our efforts were unexpectedly interrupted. My journal entry on the 26th of March states "a band of *guerilleros* kidnapped four Germans and four Colombians. The kidnapping took place at around 8:00 a.m. We traveled the same road at around 2:00 p.m. It appears that the rest of the Germans will fly out by helicopter this afternoon...." The next day adds "We have been confined to camp by the army general. Rumor has it that we are to fly to Barrancabermeja this afternoon." (see Box 3-5 for Mike's description of this adventure). Consequently, our efforts to find the new *Gustavia* had to be put on hold until we returned to Campo Capote on July 8th. Although I then note "The Germans have not returned and *guerillero* activity is said to be worse than ever" we were able to make three collections of the new *Gustavia* which was subsequently published as *G. romeroi* S. A. Mori & Garr.Barr. (Mori & García-Barriga, 1975). The species was named to honor Rafael Romero-Castañeda for his many contributions to Colombian botany, and because he was one of the first botanists to collect the species. We subsequently learned that the Germans and Colombians taken hostage at the time of our first trip to Campo Capote were rescued with only a few minor injuries, but prior to our second trip, at least one soldier was killed by the *guerilleros*.

From Colombia, Mike and I traveled via Venezuela and Suriname to Belém, at the mouth of the Amazon. In Belém, we boarded the large passenger boat called the Lauro Soudré and ascended the river to Manaus, where we planned to join a botanical expedition led by Ghillean (Iain) Tolmie Prance. Unfortunately, by the time we arrived in Manaus, Prance and his team had already departed for Pôrto Velho, in what is now the southwestern Amazonian state of Rondônia. We were, however, invited to stay at the Prance home with Iain's wife Anne and their daughters Rachel and Sarah. Anne facilitated our botanical explorations in the vicinity of Manaus and, as will be described later, Iain and Anne Prance were to have a lasting impact on my career as a tropical botanist.

After this expedition, I returned to my job as an Instructor at the University of Wisconsin at Marshfield and, working between classes, finished my dissertation entitled "Taxonomic and Anatomic Studies of *Gustavia* (Lecythidaceae)" in 1974. Although I enjoyed teaching, by then I also knew that I wanted to be a tropical field botanist and, in particular, to dedicate my career to study the Brazil nut family (Lecythidaceae).

Monograph of the Brazil Nut Family

Because species of the Brazil nut family, for the most part, are very large trees it is essential to be able to climb them to gather specimens and to study pollination and dispersal biology. See Chapter 4 for a detailed discussion of the methodology I have used for gathering specimens from trees and other plants, such as epiphytes and lianas, found in the crowns of trees.

The Prance and Mori monographic studies of the Brazil nut family, which established background data for collaborations with Amy Berkov, Xavier Cornejo, Ya-Yi Huang, Nate Smith, and Chih-Hua Tsou, have also led to most of my other research topics. For example, I initially became interested in the floras of French Guiana and the Osa Peninsula of Costa Rica when I visited them to collect species of Lecythidaceae. After studying the Lecythidaceae of French Guiana, I wanted to learn the species of the family in another species-rich area, and chose the Lecythidaceae study at the Biological Dynamics of Forests Fragments in Brazil. I first became captivated by pollination and seed dispersal mechanisms because the flowers and fruits of Lecythidaceae show such amazing variation in response to different pollinators and dispersal agents. After learning that the seeds of *Lecythis pisonis* Cambess. were dispersed by bats, I suspected that most of the zygomorphic-flowered species were also dispersed by them. As a result, my colleagues and I developed a general interest in the roles bats play in both seed dispersal and pollination of tropical plants, and that led to a book on bat seed dispersal in the Neotropics (Lobova et al., 2009). My only research project without some relationship to Lecythidaceae was our inventory of the plants and lichens of Saba, where no Lecythidaceae occur (Mori et al., 2007a). For me, the Lecythidaceae have been a window through which I have been able to see the entire neotropical flora.

For those interested in learning more about New World Lecythidaceae visit the Lecythidaceae Pages (Mori et al., 2010).

A Year in Panama

In early 1974 Mike Nee, who had just completed a year as the Missouri Botanical Garden's (MO) plant collector in Panama, informed me that MO was looking for someone to fill the position. This opportunity came at a perfect time because I had just received my Ph.D. at the UW and was looking for a job. I applied for the low-paying but exciting position and when the director of the project, Tom Croat, offered it to me, I jumped at the opportunity. Fellow botanist Jacquelyn Kallunki (another graduate of the UW who studied under Iltis for her M.S. and Robert Kowel for her Ph.D., Fig. 1-6, top right) and I (Fig. 1-6, top left) departed for Panama in September 1974, where I assumed curatorship of the Summit Herbarium (Fig. 1-7). My goals for the year were to collect interesting plants and generally learn more about tropical botany. Although I had worked as a biology instructor at the University of Wisconsin – Marshfield/Wood county campus from 1969 to 1974, and my teaching assignments included general botany, this would be the first time that I would actually be paid to spend time in the field collecting plants!

We lived on the grounds of Summit Gardens, which had been established by the Panama Canal Company in 1923 as the Canal Zone Plant Introduction Gardens. Since their inception these gardens have been unofficially known as Summit Gardens, a name referring to the Continental Divide located a short distance to the south. The original garden staff not

Fig. 1-6. The author and some of the botanists he collected with in Panama and Guyana. Top left: The author on the El Llano-Carti Road in Panama in 1975. Photo by Jacquelyn Kallunki. Top right: Jacquelyn Kallunki in the Darién of Panama in 1975. Bottom left: Alan Bolten in Guyana in 1975. Bottom right: Al Gentry on a Flora Neotropica field trip in Brazil in 1987. Photos B–D by Scott A. Mori.

only landscaped the grounds, but also introduced ornamentals, fruit trees, forest trees, forage crops, fiber plants, grain crops, medicinal plants, beverage plants, and rubber-producing plants. It is estimated that nearly 15,000 species were introduced into Summit Gardens (Croat, 1971). Today Summit Gardens, which includes a small zoo of Panamanian animals, is managed by the government of Panama City and is used mostly for public recreation and education.

The Summit Herbarium and Library, incorporating the U. S. Army Tropic Test Center herbarium, was established in 1971 (Dwyer, 1967). One of my first jobs as curator was to organize the herbarium in its new quarters in a building previously occupied by the Hotel Tivoli. This formerly luxurious hotel had been opened by the Canal Zone authority in 1907 (Collins, 2009) to house employees not yet assigned to living quarters, for Canal Zone business associates, and as a social center for Canal Zone employees. At its peak there were 220 rooms and dining facilities for up to 700 people. I had enjoyed the hotel on my two previous trips to Panama, but by the time I arrived to start my curatorship it was no longer functioning. The herbarium had just been relocated to a dilapidated room at the Tivoli from the battery-recharging shed of the Dredging Division of the Panama Canal Company (M. Nee, pers. comm. 2009). The herbarium in the Tivoli consisted of a large area for drying herbarium material and herbarium cases to store the specimens. Today, the Summit Herbarium is a modern facility and part of the Smithsonian Tropical Research Institute (STRI) Herbarium, under the direction of Dr. Mireya Correa.

Our collection strategy in Panama was to visit several areas in as many different months of the year as possible. This made it possible to collect plants that flowered and fruited in different seasons, facilitating the collection of specimens representing different stages of a plant's life cycle. For example, if a plant was collected in flower on one trip, we looked for it in fruit on subsequent trips and *vice versa*. The El Llano-Carti Road in Panamá Province and the Cerro Colorado mine area in Veraguas Province were areas that were visited repeatedly. In addition, we made less frequent trips to other parts of Panama including a trip to one of the most exciting places to collect, the Darién in the extreme eastern part of the country. We wanted to visit these poorly collected localities as often as possible, and to maximize the number of collections made on each trip. This composite collection strategy yielded 6000 collections between September 1974 and August 1975 (Dwyer, 1980), including at least 108 type specimens of new taxa (Missouri Botanical Garden, 2009).

My most exciting fieldwork that year was a National Geographic Society sponsored expedition organized by Alwyn Gentry (Fig. 1-6, bottom right) to explore Cerro Tacarcuna in the Darién Gap, the road-free area between Panama and the Chocó of Colombia. All of the vascular plants collected on the expedition were numbered on Al's collection series and I collected the mosses and lichens on my series. The Darién was one of the most remote and least botanically explored areas in Panama; perhaps in all of Central America. This trip gave me the opportunity to be in the field with Al, considered by many to be the greatest botanical explorer of our time. His many accomplishments are documented in a tribute that Miller et al. (1996) paid to Gentry in the Annals of the Missouri Botanical Garden shortly after he perished, along with the renowned ornithologist Theodore A. Parker, in a plane crash in western Ecuador (on the 3rd of August in 1993).

I first met Al in June 1967 when he came to the University of Wisconsin as a prospective graduate student. Hugh Iltis was introduced to Gentry by Al's undergraduate mentor Ted Barkley after Iltis gave an invited lecture entitled "A requiem for the prairie" at the University of Kansas in Manhattan. Impressed by Gentry, Iltis asked him to consider graduate school at the UW. When Al arrived at UW, I showed him the campus, gave him a tour of Birge Hall and the graduate program's facilities, and explained to him what it was like to study under Iltis. Al completed an M.S. degree at UW on Bignoniaceae and then moved on to the Missouri Botanical Garden where he finished his Ph.D. under the tutelage of Walter Lewis. From the time Al was at the UW, he and I corresponded about our mutual interest in plants, and I sent him my Bignoniaceae collections for determination, and he sent me his Lecythidaceae collections. To show my appreciation for the special effort he made to send me collections and images of Lecythidaceae, I named *Gustavia gentryi* S. A. Mori in his honor, and he likewise named *Parmentiera morii* A. H. Gentry after me. However, it was not until the Tacarcuna trip that we, both in our 30's, collected plants together.

We first flew to El Real, and then travelled by boat to Yaviza to pick up our supplies and hire boats to take us to the foothills of the Serranía del Darién. We departed from Yaviza in two dug-out canoes on January 11, 1975 for the Cuna Indian village of Pucuro. From there we hiked into the mountains on the ridge between the Pucuro and Tapalisa Rivers (Gentry, 1975a, 1975b, 1975c, 1977). We walked several days to our first base camp and proceeded to make excursions until we had collected all the fertile plants that we could find in the area. Subsequently, we established other base camps closer to Cerro Tacarcuna. From our final base camp, we walked to what we thought was the elfin forest covering the summit of Cerro Tacarcuna at 1875 meters on January 28th. On a subsequent trip to

Fig. 1-7. Project locations. A = Panama. B = Bahia. C = Central French Guiana. D = Biological Dynamics of Forest Fragments Project in central Amazonian Brazil. E = Nouragues. F = Saba. G = Osa Peninsula. The projects are lettered in the order they were initiated by the author.

this area, Al discovered that there was another slightly higher peak a short distance away (Gentry, 1977).

We had hired several Panamanians to carry gear to the base camp and stay with us for the entire trip, and contracted Cuna Indians in Pucuro to carry supplies from the boats to the first base camp at Cerro Mali. But when we left Pucuro Al and I were carrying our personal gear, as well as a 10-gallon supply of formaldehyde used to preserve our collections until we could dry them back at the Summit Herbarium, and clipper poles, so that we could make collections along the way. Toward the end of the first afternoon, we were so exhausted that whoever was carrying the formaldehyde could barely stagger a few hundred meters before collapsing. Then we would exchange formaldehyde for clipper poles, and repeat the process. At dusk, we cut a small clearing on the ridge, had dinner, and crawled into our sleeping bags. I slept on an air mattress on a platform constructed above the ground, but Al crawled into his hammock where he was safe from terrestrial insects and the roving bands of white-lipped peccaries that anyone sleeping near the ground feared most (See Box 3-4). From that expedition on, I have been a converted hammock sleeper.

Shortly after Al retired to his hammock I heard his teeth chattering, went over to his hammock, and found him shaking and sweating uncontrollably. The attacks lasted for the next several days and I was convinced that Al had contracted malaria. For his safety, I suggested that he return to Pucuro with the porters, and from there travel back to Panama City. Al adamantly refused to abandon the expedition, so I eventually stopped asking and hoped for the best—and fortunately, after several days he returned to normal. There are many similar stories about Al, and all illustrate his determination to accomplish his mission no matter how extreme the discomfort. Al would collect all day without even thinking about eating unless someone else happened to bring lunch; when a pair of pruning shears fell from a tree and hit him on the head he simply let the blood clot and continued working; and after he got lost in Peru with Camilo Díaz and didn't eat for three days, he returned to collecting the very same day he was rescued (Stap, 1990). Miller (Miller et al., 1996) states, "Gentry's stamina and enthusiasm on these expeditions and his near obliviousness to physical discomfort were legendary."

Al loved the physical challenges associated with plant collecting. When he considered attending the University of Wisconsin in 1966, he wrote Iltis about his "… desire for physical adventure with tropical plants, tropical field trips, and the chance to be away from civilization exploring little-known jungles and discovering exotic new species…" (Iltis pp. 446–449 in Miller et al., 1996). On the Tacarcuna trip Al found a medium-sized tree of *Quararibea* that he did not recognize and desperately wanted to collect. We tried to cut it down—something that I would never do today—but it wedged in another tree as it fell. Al proceeded to climb up the trunk even as its branches and those of the support trees cracked, and the *Quarariba* periodically dropped short distances under his weight. Thinking that the tree would fall the rest of the way and that Al would be injured when it crashed to the ground, I pleaded with him to forget about the collection. My pleas were ignored and he made the collection (Gentry & Mori 14000), which turned out to be the relatively common species *Quararibea asterolepis* Pittier.

We brought 27 fifty-pound loads of provisions to Pucuro with us at the start of the expedition and were shocked, after only a couple of weeks, to learn that we were running out of provisions. We thought we had plenty of food for 13 men, but we'd either made a serious miscalculation on the amount of food we would need, or the porters had lightened their loads on the way in. Although we suspected the latter, we couldn't be sure because we'd failed to make complete inventories of what the porters were carrying at the start and end of the hike. We first tried to supplement our provisions by hunting and fishing, but soon realized that we would have to send some of the men to the Colombian village of Unguia for more supplies. A few days later, the men returned without any supplies because they had spent all the money on booze and women in the village. Al and I then realized that we would have to accompany a small group of men on a second trip into Colombia to buy provisions. We departed at daybreak, hiked as fast as we could, reached the village late in the afternoon, and purchased our supplies. Al didn't want to lose an extra day of collecting, and tried to convince our exhausted party to return to the base camp early the next morning. They, however, wanted to rest a day in the village before returning, so Al decided that he would go by himself. I knew it would be dangerous for him to return alone and decided to hike back with him.

We left at dawn because walking into the mountains would take even longer than it had taken us to descend. Late in the afternoon, we came to a fork in the trail. We mistakenly took the trail to the right, and were still walking long after we should have reached our base camp. At dusk, we were soaked by a thunderstorm, but soon after the storm stopped we stumbled across a hut. The only person there was a Chocó Indian woman who invited us to spend the night in her house. She first made us a meal of porridge, and then indicated that we could sleep on the split palm floor. This turned into a night of horrors that even Al had a hard time tolerating! Cockroaches kept crawling across our faces, and then, because we were still damp from the rain, we started to shiver with cold. Al opened the plant press and we covered ourselves with the newspapers, but soon learned that they are no substitute for blankets! I reluctantly suggested that we sleep closer together in an effort to warm up, but between the cold, the insects, and our own fetid aromas, sleep was nearly impossible. The next morning we thanked our hostess and got an early start for the base camp, but we didn't arrive until dusk, at almost the same time as our men who had spent an extra day relaxing in the village.

Living in the botanical paradise of Summit Garden and serving as a curator of the nearby herbarium, taking weekly field trips to all ends of Panama, learning from visiting botanists such as Al Gentry, working for

Thomas Croat, and contributing to the botanical inventory of the Panamanian flora made my year in Panama one of the most exciting of my career!

NYBG Post-Doc

After completing my year in Panama, I joined the staff of The New York Botanical Garden (NYBG) in the fall of 1975 as a post-doctoral fellow under the supervision of Ghillean T. Prance (Fig.1-8, top left). Iain had developed an interest in plants and animals by exploring the English and Scottish countryside where he collected natural history objects and displayed them in his own museum. In the autumn of 1957 he enrolled as an undergraduate at Oxford University to study botany and, after graduation, was the leader of a student botanical expedition to Turkey. He returned to Oxford University to pursue an advanced degree under the direction of Frank White. His dissertation, entitled "A Taxonomic Study of Chrysobalanaceae" and published many years later (Prance & White, 1988), led to Iain's D. Phil. in Forest Botany in 1963.

Bassett Maguire invited Iain to study the collection of Chrysobalanaceae at NYBG, and soon after receiving his doctorate he joined its staff. At New York, he ascended the ranks and eventually became Senior Vice President for Science in 1981, leaving in 1988 to become Director of the Royal Botanic Gardens, Kew. During his careers at NYBG and Kew, Prance became a legendary Amazonian plant collector; prolific publisher and lecturer; teacher of botany at all levels; and originator of new programs in botany at NYBG, Kew, and the Instituto de Pesquisas da Amazônia in Manaus, Brazil (Langmead, 1995; Mori, 1999). Although retired from Kew since 1999, he has pursued his interest in botany in many ways, including serving as the McBryde Fellow at the National Tropical Botanical Garden of the United States; Director of Science of the Eden Project in Cornwall, England, and Principle Investigator of a Darwin Initiative grant for the management of the Yabotí Biosphere Reserve in Misiones, Argentina (Laurance, 2008). During his "retirement," Iain has published a Flora Neotropica monograph of Proteaceae (Prance et al., 2007), a revision of *Foetidia* (Lecythidaceae) (Prance, 2008), and is currently preparing a monograph of *Barringtonia* (Lecythidaceae).

My post-doctoral stay at NYBG was possible because Iain had been awarded a grant from the National Science Foundation to produce a monograph of neotropical Lecythidaceae for Flora Neotropica. The Organization for Flora Neotropica (OFN) was established in 1964 with the aim of publishing monographs of the plants and fungi of the Neotropics

Fig. 1-8. The author and some botanists he collected with in Amazonian Brazil and French Guiana. Top left: Ghillean T. Prance driving a Zodiac in the Amazon in 1977. Top right: Jean-Jacques de Granville at Pic Matécho in French Guiana in 2000. Bottom left: George Cremers at the base of Mt. Galbao in French Guiana in 1976. Photos by Scott A. Mori. Bottom right: The author and Carol Gracie at Km 41, a reserve of the Biological Dynamics of Forest Fragments Project in central Amazonian Brazil in ca. 1988. Photographer unknown.

(Forero & Mori, 1995). Since its inception, the goals of the OFN have also been part of the mission of scientific research at NYBG as evidenced by NYBG serving as the base of operations for OFN, the Executive Directors all being NYBG staff members, and publication of nearly all of the monographs through The New York Botanical Garden Press. Iain served as the Executive Director of OFN from 1975 to 1988 and was dedicated to the study of the neotropical flora, producing monographs himself (Mori & Prance, 1990; Prance, 1972a, 1972b, 1972c, 1989; Prance & Freitas da Silva, 1973; Prance & Mori, 1979; Prance et al., 2007), and promoting the studies of others through his leadership.

As part of the fieldwork supported by the NSF grant, I collected Lecythidaceae in Guyana, Suriname, and French Guiana in 1976. Alan Bolten (Fig.1-6, bottom left), an entomologist, joined me on trips to remote rain forest sites in Suriname and French Guiana. One of our most spectacular trips was to the Wilhelmina Mountains where we used the trails the government had opened for mineral exploration. Although the survey team was no longer in the area, it had been there so recently that the trails were easy to follow and their shelter frames were still in good condition. Using their trails and camps, Alan and I botanized along kilometers of transects through undisturbed rain forest with little chance of getting lost. Southern Suriname still harbors some of the most pristine tropical areas in the entire world, and we saw monkeys, agoutis, peccaries, toucans, and other animals daily and collected spectacular Lecythidaceae and other plants. Alan and I visited many similar areas on that trip, but central French Guiana was where we encountered the greatest diversity of Lecythidaceae, collecting a total of 18 species on a two-week trip!

This first phase of my career at NYBG inspired me in several ways. In the first place it gave me the opportunity to work with Prance whose productive fieldwork—he is usually the first person up in the morning, participates in all of the work of collecting and maintaining the camps, and exhibits a contagious enthusiasm for collecting— and success as a leader, fund raiser, and author of scientific papers provided a model I attempted to follow. In the second place, I was in the field with an outstanding entomologist whose outgoing personality facilitated making the contacts needed to explore the forests of the Wilhelmina and Lely Mountains in Suriname and the area surrounding Saül in central French Guiana where we collected and made observations on the pollination of Lecythidaceae (Mori et al., 1978). And finally, exposure to such vast expanses of undisturbed rain forest in the Guianas convinced me that the Guianas were ideal for studying the botany of undisturbed rain forest, especially the ecology of Lecythidaceae that Prance and I were planning for the future.

Bahia, Brazil

After completing my post-doc at NYBG, I was hired in 1978 by Dr. Paulo de Tarso Alvim as the Curator of the Herbarium of the Centro de Pesquisas de Cacau (CEPEC) in Bahia, Brazil (Fig. 1-7). Alvim had received his doctorate at Cornell University in 1948, and was among the first Ph.D. plant physiologists in Brazil. His research interests focused on the ecological and physiological factors that determine plant productivity, especially of coffee (*Coffea arabica* L.), chocolate (*Theobroma cacao* L.), forest trees, and the plants of tropical savannas. In 1963, he worked with the Comissão Executiva do Plano da Lavoura Cacaueira (CEPLAC) to establish CEPEC and served as its scientific director until 1988. He was among the founders of the Sociedade Botânica do Brasil (in 1949) and, in recognition of his contributions to science, was a member of the Academia Brasileira de Ciências (Academia Brasilieira de Ciências, 2009).

In addition to Alvim, another important individual that contributed to my work in Bahia was Luiz Alberto Mattos Silva. We made many field trips together and he was a coauthor and editor of the papers our group produced during that period (Mori & Mattos Silva, 1979a, 1979b, 1979c; Mori et al., 1980a, 1980b, 1985, 1989), and was critically important for those published in Portuguese. Although I can speak Portuguese fairly well, I could never have published in Portuguese without the careful editing by Luiz.

At the 15th Brazilian Botanical Congress held in Pôrto Alegre in 1964, Alvim enlisted the expertise of Dr. João Murça Pires to plan an inventory of the plants of the cocoa-producing regions of the states of Bahia and Espírito Santo. The project began in January 1965, and benefited from the input of Drs. Alberto Castellanos of the Herbarium Bradeanum and Graziela Maciel Barroso of the Rio de Janeiro Botanical Garden. In 1966, Sérgio Guimarães da Vinha became the first curator of the CEPEC Herbarium, a position that he held until he took a leave of absence to study ecology at the University College Wales, UK in 1975 (Mori & Mattos Silva, 1979a).

CEPEC is located in the panhandle of southern Bahia in the Brazilian Atlantic Forest, which extends for the length of most of coastal Brazil from the states of Rio Grande do Norte in the north to Rio Grande do Sul in the south. It is estimated that the coastal forest originally covered some 1,300,000 km^2 before Brazil was discovered by the Portuguese explorer Pedro Alvares Cabral in 1500 when he landed near Santa Cruz de Cabrália in southern Bahia. Today 100 million Brazilians live in more than 3000 cities scattered throughout the Atlantic Forest, including São Paulo and Rio de Janeiro, and that forest has been reduced to approximately 98,800 km²—only 7.6% of its original extent (Galindo-Leal & Gusmão Câmara, 2003; Morellato & Haddad, 2000). More detailed descriptions of the Atlantic Forest are found in Mori et al. (1983a, 1983b), Oliveiro-Filho & Fontes (2000), and Thomas (2008).

My first collection made as curator of the CEPEC Herbarium (*Mori & Kallunki 9241*) was gathered on 10 February 1978 and the last (Mori & Benton 13627) on 26 March 1980. Although forests (Fig. 1-9, top right) or their remnants dominate the landscape in the cocoa-producing region of Bahia, there is an amazing diversity of other vegetation types found in and adjacent to the region. My botanical excursions (Fig. 1-9, bottom) took us to these major vegetation types: mangroves; beach communities; a low shrub-dominated community between the beach and inland wet forest (*restinga*); periodically flooded forest along rivers (*várzea*); marsh communities; tropical wet forest; tropical mesophytic forest; liana forest; and a dry thorn-scrub community (*caatinga*; Fig. 1-9, top left) (Mori & Mattos Silva, 1979a). Further inland and outside of the cocoa region, I also visited low elevation savanna (*cerrado*; Fig. 1-10, top) and high elevation rocky savanna (*campo rupestre*; Fig. 1-10, bottom) (Mori, 1989). The plants were spectacular, many had been collected infrequently, and others had never been collected at all, and represented new species. Bahia was, and still is, a botanist's paradise!

Having recently explored the essentially untouched forests of the Guianas, it was a shock to see how fast natural vegetation in Bahia was being destroyed—goats grazed *caatinga*; vacation homes replaced *restinga*; coffee plantations encroached upon liana forest; and logging, firewood extraction, charcoal production, cattle ranching, agriculture, and the construction of cities altered all forest types, but was especially hard on the band of mesophytic forest found slightly inland from the coast (Morellato & Haddad, 2000; Mori, 1989; Mori & Mattos Silva, 1979a). Our group followed the same collecting strategy as in Panama: visiting the same localities throughout the year. In Brazil, I was frequently dismayed to find that what had previously been a good forest collecting site was covered by tree stumps on a subsequent visit. A major factor in this destruction was the opening of the Brazilian coastal highway (BR 101) in southern Bahia in 1973, which facilitated the cutting and transportation of timber from the once extensive forests of the cocoa-producing region of southern Bahia.

To point out the impact that this destruction was having on plant diversity, we analyzed the geographic distributions of 127 tree species found in the wet and mesophytic forests of eastern Brazil, and concluded that 53.5% of them were endemic to that region (Mori et al., 1981). In ad-

Fig. 1-9. Scenes from Bahia, Brazil. Top left: Cavanillesia arborea K. Schum (Malvaceae) growing in caatinga, a kind of thorn scrub vegetation of dry areas. Top right: Undisturbed lowland moist forest is found in a narrow band along most of the Atlantic coast of Brazil. This area is a biodiversity hotspot that has been reduced to approximately 7% of its original extent. Photos by Scott A. Mori. Bottom: The author and Terry Pennington, specialist in Elaeocarpaceae, Inga (Fabaceae), Meliaceae, and Sapotaceae, pausing for lunch on a collecting trip. Visits by specialists advance the knowledge of the families of plants they study by collecting additional species and by identifying collections in local herbaria. Photo by Jackie Kallunki.

Fig. 1-10. Two other vegetation types found in Bahia. The variation of vegetation types makes Bahia one of the most species rich areas in the world. Top: Cerrado, a general name for savanna in Brazil. Bottom: Campo rupestre, a type of rocky savanna found at higher elevations than cerrado. The mountain peak in the background is Pico das Almas and the ruble of rocks in the foreground is the result of mining for gold and diamonds. Photos by Scott A. Mori.

dition, we analyzed the distributions of several herbaceous families (e.g., the Bromeliaceae) and demonstrated that 37.5% to 77.4% of their species were found only in eastern Brazil. Studies of birds (Haffer, 1974), reptiles (Müller, 1973), primates (Kinzey, 1981), and butterflies (Brown, 1979) also support the hypothesis that the Atlantic Coastal Forest of Brazil is a center of endemism for many plant and animal species. This high level of endemism, coupled with the extensive human modification of natural vegetation, has made many species susceptible to extinction. These coastal forests have now been designated as one of the world's biodiversity hotspots (Galindo-Leal & Gusmão Câmara, 2003; Mittermeir et al., 2005; Myers et al., 2000), and Brazil has achieved some success in protecting the remaining forests.

Although I had been interested in conservation since my days as a graduate student, during my two years in Bahia I witnessed first-hand how vulnerable natural habitats are to the destructive forces of humans. This experience convinced me that natural areas will eventually disappear from most places in the world unless they are protected in biological reserves, and that one of the most important roles of plant and animal systematists is to alert governmental and non-governmental agencies, and the public, when the organisms they study are endangered by human activities.

NYBG Curator

At the end of my two-year contract with CEPEC in 1980, I returned to NYBG as an Associate Curator. Although I was tempted to stay in Bahia because there were still so many botanical discoveries to be made, I decided to return to NYBG because it would allow me to continue my research on the Brazil nut family, and to initiate floristic work in the Guayana Shield of northeastern South America, a region that still harbors one of the last large tropical wilderness areas of the world. At that time, curators at NYBG were encouraged to become experts in plant families as well as carry out botanical exploration in specific areas, mostly in North America and the Neotropics. Since then, the research program at the Garden has expanded to include research on economic botany, molecular biology, and genomics, and in other parts of the world, but curators in the Institute of Systematic Botany still study their focal plant or fungal groups in depth, and sometimes specialize in the floras and mycota of given parts of the world.

During my post-doc at NYBG, Prance and I had completed our monograph of the actinomorphic-flowered Lecythidaceae (Prance & Mori, 1979). Our next goal was to publish a monograph of the zygomorphic-flowered species; thus upon returning to New York, my first job was collaborating with Prance on an NSF proposal for project support. We decided to follow a strategy outlined in the report *Research Priorities in Tropical Biology* (Committee on Research Priorities in Tropical Biology, 1980) for our study of Lecythidaceae. Peter Raven, and other leading tropical biologists, suggested that the best way to understand tropical biodiversity would be to conduct detailed studies of selected areas throughout the tropics. This strategy has recently been called the "adopt-a-forest" strategy (Appendix A) by Bill Laurance (2008) and is referred to as such in this book. We used Tom Croat's *Flora of Barro Colorado Island* (Croat, 1978) and the studies of ecology and evolution made possible because of it, (many summarized in Leigh et al., 1996), as a specific model of what we wanted to accomplish with our monograph of Lecythidaceae.

Prance and I chose central French Guiana (Fig. 1-7), an area I had first visited in 1976, as our study site because of its high diversity of Lecythidaceae, including an extraordinary richness of zygomorphic-flowered species (of the 29 species of Lecythidaceae now known to grow there only *Gustavia augusta* L. and *G. hexapetala* (Aubl.) Sm. possess actinomorphic flowers). We had included a plan to study the ecology of Lecythidaceae in our proposal, and thought that this Overseas Department of France in northeastern South America would be a perfect site because it was still covered by vast expanses of relatively undisturbed forest, and because of the ease of working there.

French Guiana. French Guiana, Guyana, and Suriname are collectively known as the Guianas. The westernmost, Guyana, covers 215,000 km² (Lindeman & Mori, 1989); the middle, Suriname (also spelled Surinam) is 165,942 km² (Lindeman & Mori, 1989); and the easternmost, French Guiana, is approximately 84,000 km² (Barret, 2001). Guyana, formerly British Guiana, achieved its independence in 1966 from Great Britain; Suriname, formerly Dutch Guiana, became independent from the Netherlands in 1975; and French Guiana is still an Overseas Department of France (Lindeman & Mori, 1989). These are some of the least populated areas of the world, with Guyana having an estimated population of 772,298 (in 2008, U. S. Department of State, 2010a), Suriname 492,829 (in 2004, U.S. Department of State, 2010b), and French Guiana somewhere around 200,000 (in 2009).

While my students Brian Boom, John Pipoly, and I were studying Guianan species of the Brazil nut family as part of the NSF project, we also made collections of other plants. Based on these preliminary data I next submitted an NSF proposal, awarded in 1991, to prepare a Flora of the vascular plants of central French Guiana. In all, between 1976 and 2000 we made 23 expeditions to central French Guiana, and after that carried on with expeditions to other parts of the Department. The main products of the floristic inventory are the two-volume *Guide to the Vascular Plants of Central French Guiana* (Mori et al., 1997, 2002a), a guide to the mosses (Buck, 2003), and another to the liverworts and hornworts (Gradstein & Ilkiu-Borges, 2009), making this area one of the few places in the tropics where preliminary inventories for all plant groups have been published.

Our work in French Guiana was facilitated by excellent logistical support from the herbarium of the French Office de la Recherche Scientifique et Technique Outre-Mer (ORSTOM) now known as the Institut de Recherche pour le Développement (IRD). The herbarium was established in 1965 by Roelof A. A. Oldeman, and a vigorous program of botanical exploration has been carried out ever since. During most of our expeditions, Jean-Jacques de Granville (Fig. 1-8, top right) was the Director of the botanical program and Georges Cremers (Fig. 1-8, bottom left) was the curator of the Herbier de Guyane (CAY) (IRD, 2009). Under their direction, thousands of collections were gathered, duplicate specimens were sent to specialists, the specialist determinations were attached to the specimens, and other collections were identified by matching them to the correctly identified specimens. As a result of this long-term effort, the herbarium served as a valuable resource when we were identifying our collections, and provided data derived from collections made by other botanists. Michel Hoff, IRD computer specialist and botanist, came to NYBG and worked with Carol Gracie and NYBG information managers to facilitate the import of French Guiana records in the CAY database (now called AUBLET2) into NYBG's Virtual Herbarium.

Our work was made much easier by the presence of French botanists who knew the local flora; if we gathered a specimen of a plant that we did not recognize, de Granville, Cremers, Françoise Crozier, Christian Feuillet, Michel Hoff, Denis Loubry, Marie-Françoise Prévost, Bernard Riera, Daniel Sabatier—or some other botanist working in the herbarium—often knew its name. Although we never had a formal arrangement with IRD, the *Guide to the Vascular Plants of Central French Guiana* represented collaboration between that institution and NYBG, and Cremers, de Granville, and Hoff are co-authors of the Flora.

From 1993 onward, Joep Moonen and his family (Fig. 1-11) also facilitated our expeditions. Joep met us at the airport, we stayed at Moonen's Emerald Jungle Village ecotour lodge, he helped us buy supplies, took us to the airport or the helicopter pad to send us off on our expeditions, and was waiting for us when we returned. If our collecting took us to a place that could be reached by roads, Joep would drive us there and return at the appointed time to pick us up after we had finished our collecting. In addition, we often took day-long excursions using Emerald Jungle Village as our base of operations. For long-term projects, these kinds of relationships are invaluable.

The French Guiana expeditions were streamlined because plant collecting permits are not required. This liberal policy is possible because it is relatively easy to keep track of the activities of foreign botanists in a political entity only 84,000 square kilometers in size. In contrast, stricter control is needed in larger countries, such as Brazil, where it is nearly impossible to control the borders. During most of the time that we worked in French Guiana, visas were not even required; thus to make botanical collections all we had to do was inform the curator of the Herbier de Guyane when and where we would be collecting. The main reasons that I was able to dedicate such a large part of my career to the botanical exploration of French Guiana were the complete absence of bureaucratic obstacles, and the hospitality and collaboration received from French botanists.

French Guiana has other

Fig. 1-11. The family of Joep Moonen. Top: In 1994. Left to right: Joep, Marijke, and Bernie. Photo by Carol A. Gracie. Bottom: In 2010. Left to right: Joep, Bernie, and Marijke. This family facilitated many of our trips to French Guiana. Photo by Joep Moonen.

advantages as a field destination: there are comfortable lodgings and good restaurants in Cayenne, it is easy to purchase provisions for field trips, and there are numerous research stations and eco-lodges that can be used as bases for collecting trips. In addition, in the earlier years of my expeditions there was little or no crime in Cayenne or in the countryside. This, however, has changed somewhat, and one has to be more careful walking around Cayenne at night, and in selecting campsites. Nevertheless, it is still one of the safest and easiest places in Latin America to do fieldwork. As the dollar has lost ground against the euro, the increased expense has been one disadvantage for American scientists working in French Guiana.

Central French Guiana. The village of Saül (Fig. 1-7) is situated approximately 300 kilometers southwest of Cayenne in the geographic center of French Guiana. The village was founded around 1935 by the priest Père Didier, who named it after a gold miner named Saül. At his time there were 2000 to 3000 gold prospectors scattered throughout the region, and these inhabitants made up the priest's congregation (Modde, 1981). During our studies, Air Guyane had several regularly scheduled 45-minute flights, so getting to and from the village was relatively easy. Although the official population of Saül was listed as 100, it was probably closer to 50.

The area surrounding Saül is covered mostly by undisturbed, non-flooded forest between 200 m and 400 m elevation (Mori & Boom 1987; Oldeman 1974). The most obvious deviations from the predominant vegetation pattern are found in poorly drained areas, often dominated by *Euterpe oleracea* Mart. (Arecaceae); on outcroppings of large, granitic rocks, i.e. inselbergs (Granville & Sastre 1991), which occur as steep slopes surrounded by forests of varying height (as at Pic Matécho); and small areas of secondary vegetation found near the village and at scattered homestead sites. In addition, elevations above 500 m, for instance the highest peak on Mont Galbao (762 m), are covered by cloud forest. There are no large lakes or rivers, but small streams are found everywhere. A small lake caused by impeded stream drainage near Pic Matécho is home to species not yet found elsewhere in the area, e.g., *Mayaca sellowiana* Kunth (Mayacaceae), *Nymphaea glandulifera* Rothschied (Nymphaeaceae), and *Tabebuia insignis* (Miq.) Sandwith (Bignoniaceae). There are no large areas of white sand and seasonally flooded ground; hence the plants associated with these habitats are, for the most part, lacking in central French Guiana. Average annual rainfall is 2413 mm, with a well-defined dry season beginning in July and extending through November, and a less pronounced and less reliable dry period for several weeks sometime in February or March. The wettest months of the year are from March through June.

From 1976 to 1989, we used the ORSTOM house (Fig. 1-12, top) in the village as our headquarters. On our first trip we shipped food to Saül and prepared our own meals. From 1982 to 1988, Michel Modde (Fig. 1-12, bottom), who came to French Guiana in 1978 to establish an agricultural cooperative, provided our meals. When his farming attempt failed, he organized a hostel that catered to tourists interested in natural history and to scientists, like us, carrying out research in tropical forests. Michel provided us with breakfast, field lunches, and excellent dinners. We ate family-style, and meals were coupled with language lessons: Michel did not speak English, and only those willing to speak French were permitted to sit next to him. I was one of his students and learned a great deal of French from him. In recognition of his contributions to our expeditions, Terry Pennington and I named *Guarea michelmoddei* T. D. Penn. & S. A. Mori in his honor (Pennington & Mori, 1993).

After Michel died in 1988, we stayed in Saül for one more year, and then moved our base of operations to Eaux Claires (Fig. 1-13). We had visited Eaux Claires in 1989, and the Allinckx family had invited us to move our collecting base to their homestead. The family is French but had lived in Algeria until it became independent, then— in sequence—Belgium, Canada (too cold), Guadeloupe (too developed), and France, before finally settling in French Guiana.

Fig. 1-12. Our base of operations in the village of Saül. Top: The ORSTOM house in Saül in 1982. This is where we stayed on our expeditons from 1982 to 1989 when we moved to Eaux Claires after the death of Michel Modde. Photo by Scott A. Mori. Bottom: The author and Michel Modde in Modde's house where we ate breakfast and dinner. Lunch was always taken in the field. Photo by Carol A. Gracie.

Here the Allinckx family was experiencing the dream of Ghislaine, the matriarch, to live with her family in a tropical wilderness. She had lost her husband to cancer the previous year and two of her five children were no longer in French Guiana, so when we came on the scene she was managing Eaux Claires with two of her children, Betty and Yvan, her daughter-in-law Marie-Claude, and the intermittent help of another son, Andy. The family initially subsisted on produce from a small agricultural plot, and by hunting and fishing for personal consumption. They ran an ecotourism business, and rented hammock space in shelters, ran a restaurant featuring outstanding meals based on local foods (including peccary, tapir, and a delicious fish called *aimara*), and guided tourists on adventure tours into the rain forest. Our new base at Eaux Claires was surrounded by rain forest, and the beautiful Eaux Claires creek, the source of our drinking water

Fig. 1-13. Our base of operations at Eaux Claires from 1989 onwards. Top: The Allinckx family homestead in 1989. Bottom: The Allinckx family in 1993 from left to right: Andy (son), Jean-Yves (friend), Betti (daughter), Ghislaine (mother), Michel (future husband of Ghislaine), Mandy (oldest granddaughter), Yvan (son), Marie-Claude (wife of Yvan), and Cindy (youngest granddaughter). Photos by Carol A. Gracie.

Fig. 1-14. Eaux Claires in central French Guiana. Top: Approximately 200 years of plant collecting in the tropics is represented by the botanists in this image. Sitting on the steps of Carbet Carol at Eaux Claires in 1995 are: Bottom: Bottom row: Carol Gracie, Bernard Jardin, Frieda Billet, Jean-Jacques de Granville; middle row: Tom Croat, Hiltje Maas, the author, George Cremers; top row: Paul Maas, David Read, Terry Pennington. Bottom. Brian Boom, the author's first Ph.D. student, and the author arranging the plant presses on the drying frames at Eaux Claires. Photos by Carol A. Gracie.

Fig. 1-15. Botanical artists associated with the author's career. Top: Bobbi Angell, creator of nearly all of the botanical line art appearing in the author's publications. In this book she prepared some line drawings as indicated. Photo by Carol A. Gracie. Bottom: Michael Rothman, artist responsible for the cover, most line art images, and the paintings in this chapter. Photo by Scott A. Mori.

and where we bathed, ran right through the homestead. In all ways this was a tropical paradise, but one with many of the comforts of home provided by the Allinckx family.

Our expeditions would have been much more difficult to plan and execute without the help of Michel Modde and the Allinckx family. In addition to preparing meals, they helped us obtain supplies and transport specimens from Saül to Cayenne and on to New York, and gave us a place to store our equipment between expeditions. Bottled gas, used to dry plant specimens, is an especially difficult item to get into remote field stations because it is prohibited on passenger flights, but we never had to worry about this because our hosts always made sure that the gas was carried in on special cargo flights and waiting for us at the start of each expedition.

Although we personally had few problems during our expeditions, the small village of Saül was periodically plagued by contentious interactions among its inhabitants, including the shooting of a settler over a land dispute (he survived and the culprit was jailed for only a short time), the stabbing of a local gold miner in a lover's quarrel (he survived and forgave her), the theft of a tourist's camera equipment because the tourist had offended the thief (nobody was punished), the burning of the mayor's home in the middle of the night (it was never determined if it was arson or an accident but most villagers

Fig. 1-16. Looking SW across the canopy of old growth lowland forest from near Saül, French Guiana toward Mt. Galbao, the highest point in the area at 762 meters and nearly the highest peak in all of French Guiana. Cloud forest is found on the top of this mountain. The biggest threats to this area are large gold mining operations and the possibility of the construction of a road from the coast to Saül. The latter would open the area to logging and unsustainable agriculture. Photo by Carol A. Gracie.

suspected the former), and the defiant machete-cutting of the power lines leading to Michel Modde's home, in front of our eyes as we were eating dinner (nothing was done about it). Although our groups were dedicated to our work and usually managed to avoid problems, I once became involved in a serious altercation with a villager. This confrontation took place on our 1989 expedition, which was modeled after Earth Watch trips and designed to finance the trip as well as to get extra hands to help with research.

On that trip, we had six volunteers helping us collect plants and gather data for our studies. I first become aware of a problem when the fe-

Fig. 1-17. A painting representing the rain forest canopy in central French Guiana. The largest flower is of Clusia grandiflora Splitg., flowers in the upper left are of Couratari stellata A. C. Sm., the large, brown fruits are those of the sapucaia (Lecythis zabucajo), the mammals on the large limbs are South American coatis (Nasua nasua Linnaeus, 1766), the bird with a Clusia seed in its beak is a purple honeycreeper (Cyanerpes caeruleus Linnaeus, 1766), the snake is an Amazon tree viper (Bothrops bilineatus Wagler, 1830), and the hummingbird below the limb is a crimson topaz (Topaza pella Linnaeus, 1758). Painting by Michael Rothman.

male volunteers complained that the town drunk, whom I will fictitiously call Jacques, was hanging around their house and annoying them. They had also noted that small items, such as flashlights, were disappearing. The problem grew worse when Jacques also began to spend time at the OR-STOM house, where we held evening meetings to discuss the work plan for the following day and where some of us lived. The tension erupted when Jacques disrupted one of our meetings and I told him that he should leave. When he refused, I became so incensed that I enlisted several volunteers, and we subdued Jacques and physically threw him out of the house. In the ensuing heated exchange, I used my limited French to tell Jacques that he was "a bad boy," and Jacques responded that he was going to get his shotgun and come back to kill me! He then left the house, and apparently was on his way back to carry out his threat when the villagers intercepted him and took away his shotgun. Unfortunately they neglected to inform us that they had disarmed him and had him under control… as a result, those sleeping in the ORSTOM house had restless nights, fearing that they might suddenly be blown out of their hammocks by a shotgun blast intended for me! The next day, I visited the most influential villagers and told them that I would report this incident to the *gendarmerie* in Cayenne if there were further confrontations with Jacques—fortunately, I never saw him again on that or on any other expedition. Needless to say, at the end of the trip most of us were elated to get onto the Air Guyane flight and return to the safety of Cayenne.

Eighty botanists (Fig. 1-14) worked over the course of 11 years to complete this project, and throughout the preparation of the *Guide to the Vascular Plants of Central French Guiana*, we were fortunate to have the assistance of a photographer, Carol Gracie (Fig. 1-8, bottom right); a botanical artist, Bobbi Angell (Fig. 1-15, top); and a landscape painter, Michael Rothman (Fig 1-15, bottom) working with us. Carol participated in most of the expeditions and photographed all plants that were collected in flower or fruit; thus, the rest of us were able to concentrate on making the collections. Several times Bobbi joined us to make field sketches of the plants that she later illustrated back in her studio from dried specimens. When Bobbi was not with us, we made a special effort to take more complete photographs and pickle flowers and fruits so she could make more accurate illustrations. Finally, Michael Rothman joined several expeditions to capture the landscapes on canvas. His paintings of the pristine forests surrounding Saül (Fig. 1-16), such as the one of the canopy (1-17), bring to life the beauty of the wilderness still to be found in French Guiana.

Our final expedition to central French Guiana took place in 2000, when the last volume of the *Guide to the Vascular Plants of Central French Guiana* (Mori et al., 2002a) was in press, and there was less incentive for me to collect in the region, so I looked for new French Guianan forests to explore. In addition, at that time regular airline service was disrupted for several years. Most important, the Allinckx family sold their homestead, Ghislaine returned to France, and Yvan established another homestead at Crique Limonade (on the other side of Saül), where he dedicated himself more to agriculture and less to ecotourism. We had been hearing intriguing reports of a research station called Nouragues, located about 70 km

NNE of Saül, and decided to continue our botanical explorations there, which I will describe later in this chapter.

Biological Dynamics of Forest Fragments Project. The Biological Dynamics of Forest Fragments Project, located north of Manaus in Central Amazonian Brazil (Fig. 1-18) and initially known as the Minimum Critical Size of Ecosystems Project, was initiated by Thomas E. Lovejoy in 1979 (Fig. 1-18, bottom left). As stated by E. O. Wilson (2001): "The primary purpose was to assess the effect of reduction in rainforest area on biological diversity and particularly on the number of species of plants and animals in remnant patches." When the project was founded, it was a joint effort between the World Wildlife Fund (WWF) and the Instituto Nacional de Pesquisas da Amazônia (INPA). Following Lovejoy's move from the WWF to the Smithsonian in 1989 it became a joint project of the Smithsonian Institution and INPA.

I became active in the BDDFP in 1988 at the invitation of an ecologist who was searching for a specialist in a family of Amazonian trees that could be used to address ecological hypotheses. My jobs were to identify the trees, collect vouchers, describe new species, and help collect ecological data for sophisticated statistical analyses that were outside my field of expertise, but to be conducted by the ecologist. I was eager to participate because Iain Prance and I wanted to expand our studies of Lecythidaceae ecology to other parts of its range in the Neotropics. We had already conducted basic ecological studies in French Guiana, and the results had just been published (Mori & collaborators, 1987). Central Amazonian Brazil was the perfect place to develop a Lecythidaceae project because some of the French Guianan species reach the western limits of their ranges, and a number of western Amazonian species reach the eastern limits of their distributions there; there are also species endemic to the region. The prospect of joining the BDDFP fit our long-term goal of studying the systematics and ecology of Lecythidaceae in different parts of its range. We were especially anxious to study biotic interactions, to determine the extent to which pollinators and dispersal agents drive speciation in Lecythidaceae.

The Brazil nut family (Mori et al., 2010) was also a perfect model for the ecologist's studies. Because the floras of different regions intermingle in Central Amazonian forests, Lecythidaceae is unusually species-rich in the BDFFP plots; furthermore some species occur at very high densities. The ecologist planned to investigate the influence of abiotic factors, such as soil nutrients and water, on competition among tropical tree species; the preferences of different tree species for different topographies (i.e., do some species preferentially grow in lower areas and others in well-drained

Fig. 1-18. *The Biological Dynamics of Forest Fragments Project (BDFFP). Top: Forest at Km 41 (Reserve 1501) where the 100 ha Lecythidaceae plot is located. This is the control reserve for the BDFFP and is surrounded by more-or-less continuous forest. Photo by Carol A. Gracie. Bottom left: Claude Gascon (left) and Thomas E. Lovejoy (right) in 1993, the former is one of the early field managers and the latter the founder of the project under the name of the Minimum Critical Size of Ecosystems Project. Photo by Scott A. Mori. Bottom right: Pasture which surrounds all of the experimental reserves, one of which is seen in the background. Photo by Carol A. Gracie.*

habitats?); the influence of fires, revealed by the presence of charcoal in the soil, on species distributions and diversity; the impact of past human activity, inferred from the presence of phytoliths (derived from cultivated plants) in the soil, on tree diversity and richness; and many other ecological questions. We thought this collaboration between systematists and an ecologist had great potential to add to our knowledge of the classification and ecology of the Brazil nut family, and would enable us to test hypotheses that could also be applied to other dominant Amazonian tree families.

Professional surveyors first established a 100-hectare plot in the 1000-hectare control reserve of the BDFFP (Reserve 1501 but also known as Km 41) (Fig.1-19). This reserve had been established so that researchers could collect data in an area that had not been subject to major disturbance from humans, at least in recent times. The 100-hectare plot was divided into 20 × 20 m quadrats, with stakes at each of their corners and the southwestern stake of each plot marked with the quadrat's coordinates. A topographic map was made for the entire plot, and soil samples were collected from each of the quadrats within the eastern half of the study area. Experienced woodsmen, working under the direction of the ecologist, located and mapped 7791 Lecythidaceae trees with a DBH of at least 10 cm. I then spent a total of six months, on four different expeditions, identifying the trees and collecting voucher specimens for those that could not be identified with certainty from the ground. Over five years a succession of student research assistants managed the field team (Fig. 1-19, bottom) of two woodsmen. They collected throughout the year, and ultimately gathered flowers and fruits of nearly all of the species. In addition, they made phenological observations every two weeks on selected individuals of each species. This research team helped to resolve the questionable preliminary identifications of many species by collecting fertile specimens (Mori et al., 2001).

We faced numerous problems in reaching satisfactory determinations for individuals of some species. For example, on our first pass through the plot we identified 364 individuals as *Eschweilera collina* Eyma, but when this species started to bloom and then set fruit the woodsmen reported that the flowers and fruits of some individuals were different than those of other individuals. Study of their specimens confirmed their suggestion that we had included two species in our determinations of *E. collina*. From the ground, even with the aid of binoculars, it was impossible to distinguish sterile individuals of the two species because both the bark and leaves could not be distinguished. However, once we had specimens, even when they were sterile, we could tell the difference because the dried leaves of *E. collina* possesed salient higher order venation on both surfaces whereas those of the other species did not. In order to correct the determinations, we were forced to revisit all 364 trees and make voucher specimens from each. Our study of the specimens revealed that only 55 of the individuals represented *E. collina*, whereas 309 belonged to a new species, which was named *Eschweilera romeu-cardosoi* S. A. Mori in honor of Romeu Cardoso, our most experienced woodsman.

We identified a total of 38 species of Lecythidaceae within the 100-hectare plot, and only one additional species, *Lecythis poiteaui* Poit., outside of it (indicating that nearly all of the species in the reserve occurred in our plot). The mean number of species per hectare was 17.3 ± 2.6, with a range of 11–24 species. Five species (*Eschweilera atropetiolata* S. A. Mori, *E. coriacea* [DC.] S. A. Mori, *E. grandiflora* [Aubl.] Sandwith, *E. truncata* A. C. Sm., and *E. wachenheimii* [Benoist] Sandwith) were found in more than 90% of the hectares. The Brazil nut tree (*Bertholletia excelsa* Bonpl.) was the rarest species in our study; it was represented by a single very large tree that subsequently died. The mean density of individuals per hectare was 77.9 ± 19.9, with a range of 45–149 individuals and the mean basal area per hectare was 3.76 ± 0.86 m³, with a range of 2.13-6.56 cubic meters. The total basal area for all Lecythidaceae was 376.35m. These results reveal that central Amazonian Brazil harbors, per hectare, the greatest number of individuals and species of this iconic Amazonian family heretofore reported in the literature.

We hypothesize that high species diversity of central Amazonian Lecythidaceae is the result of western Amazonian species reaching their easternmost distribution, Guayanan species reaching their southwestern-most distribution, the presence of a high number of endemics, and the occurrence of widespread species (Mori & Lepsch Cunha, 1995; Mori et al., 2001). We suggest that other dominant Amazonian families, such as Burseraceae, Fabaceae *sensu lato*, and Sapotaceae, may also reach their greatest species richness in this area. If so, this is a compelling argument for maintaining large tracts of intact forest in central Amazonia.

Although fire has had an impact within our study plot (Bassini & Becker, 1990), these fires probably took place 6000 to 400 years before the Pleistocene, and, therefore, the species of the Brazil nut family that we identified are characteristic of old growth forests; with none of them commonly found in secondary habitats. Moreover, Piperno and Becker (1996) did not find soil phytoliths characteristic of crop plants, such as manioc and corn, and concluded that the 100-hectare plot has been under continuous forest cover since at least 4500 BP, and never been cleared for swidden agriculture. This evidence does not support the hypothesis that most Amazonian forests have been modified by humans, as claimed by some anthropologists (Heckenberger et al., 2008). Even if an area had once been inhabited, that would have been more than 400 years ago, and signs of significant human impact on biodiversity would no longer remain.

Our studies of central Amazonian Lecythidaceae suggest that

Fig. 1-19. The camp at Km 41 (Reserve 1501). Top: An open-sided sleeping shelter at the camp in 1987. Photo by Scott A. Mori. Bottom: The Lecythidaceae field team in 1989. Left to right: Romeu Cardoso, Nadja Lepsch-Cunha, Paulo Costa Lima Assunção, and Everaldo da Costa Pereira. Photo by Carol A. Gracie.

sampling very large plots is not particularly efficient in terms of cost and labor. The species curves, not surprisingly, demonstrated that species richness increases rapidly for small sample sizes, and slowly for large sample sizes. Therefore, most common species of trees in a given area can be determined by sampling relatively few hectares. Species area curves also demonstrate that it is very difficult to sample all species present, even with very large sample sizes. Our one-hectare sample yielded ca. 17 spp./ha, the two-hectare 23 spp./ha, the five-hectare 30 spp./ha, the 20-hectare 36 spp./ha, and the 50-hectare 37 spp./ha (Mori et al., 2001). The single individual of the Brazil nut tree would only be documented with certainty in a 100-hectare sample, but by the time 20 hectares had been sampled, 94% of the species were documented. Therefore, in designing ecological and biodiversity studies, it is important to first establish what hypotheses will be tested, and then determine how large a plot is needed to address them. Otherwise, the expenditure of excessive amounts of labor, money, and research time will seldom justify the small amount of additional data obtained.

Unfortunately, our association with the ecologist was interrupted when he took a job in another country. As a result, most of the ecological questions we had wanted to answer have not been addressed. For example, the phenological records so carefully recorded by our field team have, to the best of my knowledge, been lost and will never be published. Although I sometimes ask if all the work was worth it, my answer is yes, because I have learned two important lessons about tropical field work.

First, the only way to make a complete taxonomic inventory of species in a given area is to sample all trees of the group or groups of interest. This is, of course, impossible for very large areas, but establishing study plots in different habitats increases the possibility of recording different species. A more practical and less expensive inventory system is to walk trails throughout the area every two weeks, at all times of the year, and collect all species in flower and/or fruit over the course of at least five years. Using an experienced field crew to look for the flowers and fruits of species that have not yet been well-collected is a very effective way to describe all life cycle stages of tropical trees. We discovered that 30% of the species of Lecythidaceae growing in the 100-hectare plot had been described within the last 40 years, and our inventory facilitated the description of many of them. With random general collection instead of focused collecting, these species would have taken much longer to discover.

Second, it is difficult to predict if collaborations will be productive. Thus, before making long-term commitments to a project in which you must depend on a critical collaborator, you should ensure that you will be able to collect enough data to make the project worthwhile even if your collaborator drops out. Although our collaboration worked well at the start, it stalled, and then stopped altogether once the collaborator left Brazil—taking most of the data with him. That, however, did not stop the work that we had initiated with student interns, two of whom went on to receive their doctorates in ecology. Helping to train students and learning about the systematics and ecology of central Amazonian Lecythidaceae, made six months in the field and many additional months in the herbarium worth the effort expended. We ultimately published a reasonable number of papers (Bierregaard et al., 2001b, Lepsch da Cunha & Mori, 1999; Lepsch da Cunha et al., 1999; Mori, 1992b, 1995c; Mori & Lepsch-Cunha, 1995; Mori et al., 2001; Oliveira & Mori, 1999), but of course, if the ecologist had remained with the project to its end, we would have learned a great deal more about the ecology of neotropical Lecythidaceae.

Nouragues. The Nouragues Nature Park (Figs. 1-7, 1-20), situated in northeastern French Guiana, was established to promote scientific research in a tropical rain forest setting remote from human activity, protect tropical habitats, and provide environmental education for Guyanese and foreign visitors. The park, created in 1995, encompasses 100,000 hectares and is the largest nature reserve in France (Nugent, 2001). Within the park there are three camps, two (Saut Pararé and Camp Inselberg) serving as centers for scientific study and the third (Camp Arataye) intended to function as a center for ecotourists and students to learn about natural environments. Unfortunately, gold miners robbed Camp Arataye in May 2006, and in the process shot to death two staff members. Although special arrangements can be made to use the facilities of Camp Arataye for research, it has not been open to the public since that tragic incident.

The Saut Pararé Camp is situated on the left bank of the Arataye River approximately 20 kilometers from its confluence with the Approuague River. When the water is high, the camp can be reached from the village of Regina in four to five hours by boat. During low water, it can take as much as three hours longer. This camp was established in 1978 by the Museum National d'Histoire Naturelle and managed by the museum until 1985, when it came under the auspices of the French Centre National de la Recherche Scientifique (CNRS). Saut Pararé is the site of the Canopy Operating Permanent Access System (COPAS), which was established to facilitate research in a 1.5-hectare area of forest canopy. However, as of October 2009 the COPAS system was not operational due to technical difficulties.

The Saut Pararé Camp, surrounded in all directions by what appears to be undisturbed rain forest, was not ideal for ecological studies because hunters could easily reach it by boat, and when researchers were not present they used it as a base camp. Because the large mammals and birds they killed play a vital ecological role in the ecology of the forest, a more remote area was needed to conduct ecological studies in an area not influenced by human activity. In 1986, ecologist Pierre Charles-Dominique led a search for a camp less compromised by hunting. He and his team studied maps of the region, and decided to explore a site about eight kilometers inland with a large granitic dome surrounded by forest. After exploring the area, the team decided that the inselberg was so difficult to reach that negative impact by hunting would be minimal, so it was selected as the new research site and named Camp Inselberg (Fig. 1-20).

The Camp Inselberg area is located at the base of the Nouragues Inselberg (Fig. 1-21) and is dominated by a series of small hills rarely exceeding elevations of 120 meters; however, the Balenfois Ridge, with its highest point at 460 meters, crosses the reserve and the peak of the Nouragues Inselberg reaches 430 meters. The camp is surrounded mostly by lowland rain forest that has been forested for at least 3000 years, although there is evidence of periodic disturbance by Amerindians. In addition, there are smaller tracts of liana and bamboo forests, *Euterpe* palm-dominated forests along streams, savanna-like vegetation associated with inselbergs (Granville & Sastre, 1991), and low, Myrtaceae-dominated forest at the top of the inselberg. Thus, a wide variety of vegetation types harboring a rich flora can be reached within easy walking distance of the camp. The mean

Fig.1-20. The Nouragues Nature Park. Top: Kitchen, dining room, and library at Camp Inselberg. Photo by Scott A., Mori. Bottom: The 2002 expedition consisting of from left to right: the author, Tatyana Lobova, Pierre Charles-Dominique (one of the founders of the station), and Frédèric Blanchard. Photo by Patrick Chatelet.

Fig. 1-21. *The Nouragues Nature Park. The Nouragues Inselberg has vegetation very different from that of the surrounding lowland forest. Photo by Scott A. Mori.*

annual rainfall is 3000 millimeters, with a two-month dry season in September and October and a short dry season usually in March (CNRS, 2009). The camp can be reached by a 35 minute helicopter ride from Cayenne or by a five to seven hour boat trip and an eight kilometer hike north from Saut Pararé (which takes three to five hours, depending on the load and physical condition of the hiker).

Camp Inselberg consists of open-sided shelters called "carbets" (Fig. 1-20, top) that can accommodate a total of about 15 scientists. There is a clean water supply throughout the year, electricity generated by solar panels and a small hydroelectric plant, satellite phone and Internet access, and reasonable prices for food and lodging (researchers take turns preparing meals). A small grants program, open to scientists from throughout world, is designed to encourage the widest possible use of the Nouragues Nature Park for scientific research (see CNRS, 2009 for a list of funded projects). The Bridge Project (Chave, 2006) selected Camp Inselberg as one of its study sites for testing ecological theories about tropical tree diversity. A checklist of the plants has been published (Belbenoit et al., 2001) and online checklists of the plants, birds, fish, mammals, frogs, and reptiles are available (CNRS, 2009). In addition, an online site provides a preliminary Flora of the area, including line drawings and field images (Mori et al., 2007c). This combination of features ranks Nouragues as one of the best places to study the ecology of pristine vegetation in the Guianas, and numerous studies have already generated publications, many of which are cited on the station's web site (CNRS, 2009) and described in Bongers et al. (2001).

In 1975, Jean-Jacques de Granville established a plan to set aside biological reserves throughout much of French Guiana (Granville, 1975)—the areas he proposed are mapped in Lindeman and Mori (1989). This initial effort has led to the nature reserves found in French Guiana today, the most spectacular of which is the Parc amazonien de Guyane, officially established on 27 February 2007 and covering three and one-third million hectares in the southern half of the Department (Le Parc amazonien de Guyane, 2010). The park is divided into four zones: 1) a core zone designed to protect nature and human cultures in which only local inhabitants can hunt, fish, practice slash-and-burn agriculture, and establish villages; 2) a peripheral zone in which regulated development projects can take place; 3) a zone of controlled access in the southern third of the Department in which anyone not native to the area must have authorization from the authorities to enter; and 4) a collective zone along the major rivers in which hunting and fishing can take place in a band within five kilometers of the rivers. Unfortunately, all of these areas permit human manipulation of the natural environment and, thus, with increasing human population and consumption there is no guarantee that any part of the park will be protected from the environmental damage caused by human "development" projects—more appropriately called human "destruction" projects.

Once our goal of establishing the framework for an e-Flora of French Guiana was accomplished, my work in French Guiana was, for the most part, completed. Fortunately, de Granville (Granville, 1975), the French government, and local French Guianan authorities have made considerable progress in preserving a part of the Guianan rain forest, one of the largest pristine rain forests of the world.

Saba. Saba is a five square mile island located off the coast of St. Maarten in the Netherlands Antilles (Fig. 1-7, Fig. 1-22). There are about 1200 permanent inhabitants. English is the official language of the island, but Dutch is also taught in schools and is spoken by many Sabans. (Do not confuse this small island with the country of Sabah, a much larger country located on the northeastern corner of Borneo.) In February 2006, Conservation International, in conjunction with the Saba Conservation Foundation, embarked on a project to survey Saba's biodiversity from beneath the sea to the summit of its highest peak, Mt. Scenery (Fig. 1-22), at 877 meters (2,877 feet). As part of this effort, marine botanists from the Department of Botany of the Smithsonian Institution surveyed the marine algae.

Conservation International invited NYBG to join the project and, in 2006 and 2007, our mission was to prepare checklists of Saba's terrestrial plants and lichens, as well as its marine algae (Mori et al., 2007a). Although the island is small in size, its rugged topography is covered by several vegetation types, ranging from near-desert to cloud forest (Fig. 1-22). Our goal was to gather specimens of non-vascular plants, vascular plants, and lichens from these different habitats, and then summarize the data in an online illustrated checklist. Our group included specialists on lichens (H. Sipman and R. Harris), bryophytes (W. R. Buck), ferns (R. C.

Fig. 1-22. The vegetation of Saba can be very wet and very dry over short distances. Top: Mt. Scenery, at the top of the image, is covered by cloud forest. The town in the lower right side is Windwardside. Photo by Carol A. Gracie.

Moran), and flowering plants (S. A. Mori and C. Gracie) to participate on the expeditions and prepare the checklists. Gracie documented almost all of the flowering plants with digital images, and D. S. Littler, M. M. Littler, and B. L. Brooks photographed the collections. The digital images represented by the collections were then attached to the database records of the collections. This method is called "vouchering" (Fig. 1-23) the images (see Chapter 4), a system that allows others to verify the identity of the images by checking specimens in the herbaria of the Botanischer Garten und Botanisches Museum Berlin-Dahlem (B, most of the lichens), NYBG (the flowering plants, byrophytes, and some of the lichens), and the United States National Herbarium (US, marine algae).

One goal of the flowering plant section of the website was to encourage other botanists to prepare revisions of the families that occur on Saba. As a model for other botanists to follow M. L. Kawasaki, of the Field Museum of Natural History (F), wrote the family, generic, and species descriptions for the Myrtaceae, and prepared an electronic key to facilitate their identification. Each species page includes a description of the taxon; common names; its distribution; comments on its ecology, pollination, and dispersal biology; taxonomic notes as needed to clarify problems with species delimitations; and etymology (the meaning of the species name). In addition, a species page includes a distribution map based on all georeferenced collections in the database; selected images; a list of all specimens that have been entered in the database; and a list of all images available in the database.

Other features of the flowering plant Flora are a key to the families of flowering plants known from Saba; a glossary to help with understanding botanical terminology; a means of identifying plants based on an online list of common names associated with their scientific names; and a method to search for all combinations of data. For example, if a hiker sees a red flower on the Mt. Scenery trail, it is possible to search for all specimens that have been collected from that trail, and refine the search to yield only those species with red flowers. The user can then identify the flower by comparing it to the images in the database.

Saba is one of the first islands in the world to have a preliminary online inventory of all its known plants and lichens. This is, however, only the first step toward completing an online Flora. Unfortunately, funding was not available to complete the next phase of the project, the production of species pages for all of the species of the island. The loss of funding prior to finishing a complete Flora can stop plant systematics projects dead in their tracks—even if major progress such as the creation of specimen-based checklists—have been accomplished. Project support is sometimes acquired from conservation organizations when an individual within the organization has an interest in protecting specific habitats, geographic regions, or groups

Fig. 1-23. *Pressing and drying plants on Saba is much more comfortable than in the more humid tropical rain forests where the author generally explores for plants. Top: The author pressing in the field using a field press and taking notes as they are collected. Bottom: The author placing the plant press on the drying rack. The heat source is a hot plate. Photos by Carol A. Gracie.*

of plants or animals, and thus promotes that particular interest within the organization, and this is what happened to the Saba Project. Funds were raised to get the project started, but with the departure of the interested individual at CI the money needed to carry the project further was lost and the project ended. I moved on to work on a project with major conservation implications, a vascular plant inventory of the plants of the Osa Peninsula in Costa Rica.

Osa Peninsula, Costa Rica. The Osa Peninsula is located in the province of Puntarenas, in the southwestern corner of Costa Rica. To the north and west it is bordered by the Pacific Ocean, to the east by the Golfo Dulce, and to the northeast by the Costa Rican mainland (Fig. 1-24, top). Altitude on the peninsula ranges from sea level to 745 meters at Mt. Rincón (Grayum et al., 2004). The climate of the Osa Peninsula is hot and humid, with an average temperature of 25°C, an annual average rainfall of 6000 mm, a rainy season from August to December, and four months of reduced precipitation from January to April (Herrera, 1985; Carrillo et al., 2000). According to Holdridge (1967), the Osa Peninsula has moist, wet, and pluvial life zones in tropical to premontane altitudinal belts. Among them, the wet forest life zone is predominant and encompasses 92.48 % (= 1,445.28 km^2) of the Osa project area.

I have been interested in the Osa since 1971 when Michael Nee and I collected *Grias cauliflora* L. there while we traveling through Central America. My interest in the Osa grew when I became aware that it harbored many species of plant families found primarily in South America. For the Lecythidaceae, the Osa seemed to be the most northern limit, at least in the Pacific lowlands, where a number of species occur. *Eschweilera mexicana* T. Wendt, S. A. Mori & Prance, which occurs in Mexico, is the absolute northernmost species, but Lecythidaceae species richness declines progressively from the Osa northwards.

When Adrian Forsyth asked me to develop a web site for the vascular plants of the Osa, I jumped at the chance. Our team consisted of Reinaldo Aguilar (Fig. 1-24, bottom left), originally a mechanic who was subsequently trained as a botanist at the Instituto Nacional de Biodiversidad (INBio); his wife Catherine Bainbridge (Fig. 1-24, bottom left), a trained horticulturist and botanical artist; Xavier Cornejo (who served an internship at UW under Iltis, Fig. 1-24, bottom right), with an B.S. in Biology from the University of Guayaquil, Ecuador; Hannah Stevens and Francine Douwes, GIS experts, and me. Our goals were to inventory all of the vascular plants of the Osa, illustrate them with images, apply what we learned to conservation, teach others how to identify the plants, and educate visitors and local people about their importance.

As part of a phytogeographic analysis of the trees of the Osa, we created a checklist of the flowering plants of the Osa, which included 2200 species. We selected 821 tree species from the checklist, and described the overall distribution patterns of the 455 species for which we found distribution data. In our sample, 4.6% of the trees were endemic to the Osa and adjacent mainland, 6.2% to Osa/Costa Rica, 14% to Osa/Costa Rica/adjacent countries, 13.2% were found throughout Mesoamerica, 14.9% had distributions extending into NW South America, 1.8% extended into NW South America and the West Indies, and 45.3% of the trees were widely distributed in tropical America. This study demonstrated the importance of protecting forest on the Osa Peninsula, because it will insure the survival of endemics found only on the peninsula and the adjacent mainland. In addition, nearly 25% of Osa plants are regional endemics (i.e., occurring on the Osa and adjacent mainland, other parts of Costa Rica, Nicaragua, or Panama). Protection of these forests will also safeguard many widespread species whose only remaining Mesoamerican populations are on the Osa.

In spite of conservation efforts on the Osa Peninsula, forest cover there declined from 97% in 1979 to 89% by 1997 (Sánchez-Azofeifa et al., 2002). Hammel et al. (2004) commented "What we witness today is the steady deforestation of the Osa Peninsula forest reserves and, basically, all of the other forest reserve areas that buffer the national parks." Sánchez-Azofeifa et al. (2002) also noted that only 44% of the forest remaining on the Osa is old growth, and that deforestation occurring less than one kilometer away negatively impacts the borders of mature forest in the Corcovado National Park. These authors emphasize that 1500 km^2 are required to maintain a viable reproducing population of jaguars—which are crucial to protecting overall biodiversity on the Osa—but that only 977 km^2, including both the park and unprotected areas outside its borders, remain in forest. Loss of forest cover also reduces ecosystem services (Foley et al., 2007), such as water supply and purity, carbon sequestration, climate stabilization, providing the beauty and biodiversity that support tourism, and protecting genetic diversity, including that of plants with medicinal or other direct economic potential.

If the slow but steady reduction of forest cover on the Osa continues, the biodiversity that Osa forests harbor and the ecosystem services they provide will continue to be compromised. Because the Osa includes the last large tract of lowland rain forest on the Pacific coast of Central America, it is especially critical to protect its remaining forest cover. This would ensure that its regionally endemic tree species would have at least some populations out of harm's way. In addition, maintaining forest cover on the Osa will provide much-needed protection of some Central American populations of widespread species, even if lowland rain forests outside of the Osa become so fragmented that they can no longer sustain viable populations of old-growth plants and animals.

Fig. 1-24. Vascular Plants of the Osa Peninsula Project. Top: Location of the Osa Peninsula. Map prepared by Hannah Stevens and Francine Douwes. Bottom left: Reinaldo Aguilar, Field Manager of the project, and Catherine Bainbridge, the latter prepared orchid descriptions, rapid field guides, and did the accounting for the field work. Bottom right: Xavier Cornejo, NYBG Project Manager. Photographers not known.

Conclusions

The goal of this chapter and this book in general is to allow others to learn from my experiences during 46 years as a tropical botanist. One of the most important things I was reminded of by writing this chapter is how important other botanists have been to the development of my career. There have been, however, a few times when I have spent a great deal of my time and resources working on projects that give little return for my efforts; thus, it is important to select those with whom one collaborates with care. The work of systematic botany costs money so before undertaking a career in this field one has to accept the fact that considerable time will be spent raising funds. This topic is not discussed in this chapter, but it is covered in Appendix B. Having a plan or a mission for one's career facilitates the process of soliciting funds for research. My missions became clear to me after reading the Research Priorities in Tropical Biology report which suggested that the best way to understand tropical biology is to establish a series of research areas throughout the tropics, inventory the regional plants and animals, and then conduct detailed studies of ecological relationships in these areas. This topic is elaborated on in Appendix A. In addition, I extended this philosophy to monographic work by selecting a specific group of tropical plants, the Brazil nut family (Mori et al., 2010), to study over a very long period of time. Thus, the missions of my career were to learn as much as I could about the floras of specific areas (e.g., central French Guiana and the Osa Peninsula) and the systematics, evolution, and ecology of Lecythidaceae. I was fortunate to have selected Lecythidaceae to study and believe there are few persons interested in natural history who would not be as impressed as I am by the relationships between the flowers and fruits of this family with their pollinators and dispersal agents. Modern technology has given biologists the tools to understand evolution better than ever before; thus, I recommend that students learn as much as they can about computer programs, statistical analysis, and molecular biology to prepare them for careers in systematics. Finally, my mentor, Hugh Iltis, taught me that every biologist should be dedicated to the preservation of ecosystems and the organisms that comprise them.

Chapter 2

Amy's Year in the Rain Forest:
Romance versus Arduous Reality

by Amy Berkov

When I precipitously decided to return to school at the ripe age of 36, there were two things I knew: I wanted to climb large tropical trees, and I wanted to spend a year living in the middle of a rain forest. This is the story of my improbable transformation from emphatically non-academic urban artist into... something else! A change from art to science is fairly dramatic; I grew up in a humanities-, language-oriented family, and I think it's safe to say that we shared a general disregard for science. The only two things that I remember from high school science classes are (1) coming across the phrase "the typical liverwort" in my biology text and deciding it would make a great name for a rock band, and (2) waking up on the floor in my chemistry class after falling asleep with my stool tilted against the wall. So I wasn't off to a very auspicious start.

In high school the art department was my refuge, and my first college degree, in 1977, was a BFA. Like many artists, out in the working world I completely dichotomized work done for personal satisfaction (etching and silk screening) and work done to support my printmaking habit. After moving to New York, I supported that habit by working as a freelance graphic artist for package design companies. This was a big step up from some of my previous jobs, because I could make a lot more money in a lot less time. Although I never cared for the graphics industry, freelancing provided me with the freedom to create my own artwork, to travel and broaden my horizons, and, ultimately, to develop new interests. I became involved with non-profit organizations including Public Image Gallery and Manhattan Graphics Center. Having moved to the East Village during the "Good Old Bad Old Days," when every vacant lot was an unsavory dump full of promise, I also joined a community garden. Sixth

Street and Avenue B Garden fostered my interest in plants. That interest was ramped up during several trips to the tropical rain forest and sustained with continuing education Botany courses at The New York Botanical Garden.

Okay, fine, so I'm developing new interests—but why subject myself to the downward financial mobility and all-round humbling experience of returning to school and making a mid-life career change? First, the package design business, my money-maker, was changing (becoming computerized), and because I was a freelance artist I was expected to retrain at my own expense. It was difficult to rationalize spending time and money to stay in a business I didn't like. Second, my non-profit administrative activities were cutting into my allotted printmaking time—so I was losing the obsession I'd used freelance graphics to support. Third, I encountered several individuals who were actually making a living doing work that they loved. I'm still not sure why this seemed such a novel concept, but it was certainly giving me some interesting ideas.

Two experiences catalyzed my resolve for change. While traveling through Peru in 1989, I visited Tambopata Lodge and had an opportunity to use rope-climbing techniques to climb a huge tropical tree. Looking out over the endless sea of green really made me think about how radically humans have transformed the planet, and about our relationship with the environment. The following year I participated in a volunteer-funded research expedition to French Guiana led by Scott Mori and Carol Gracie. Although I was increasingly intrigued by tropical rain forests, and amazed to see how the heat and humidity influence plants and insects, I personally found the climate debilitating. In retrospect, the French Guiana trip enabled me to gauge my potential to work in an environment that I found stimulating, but physically so uncomfortable.

The decision to return to graduate school went from fanciful impulse to in-your-face reality at record speed (and I give the City University of New York some credit, because few schools would have been willing to take a chance on such an unconventional candidate). I didn't have a specific research plan… just a notion that I'd like to work on a project with narrative potential, probably involving plant-animal interactions. The entire first year of classes (1992) passed in a haze as I shuttled among six campuses, at the very bottom of the learning curve, acquiring a scientific vocabulary, and constructing an ever-lengthening list of things I didn't want to do. At the end of the first year I embarked on a pilot study of seed dispersal by bats; I figured bats were cool, and thought a study of nocturnal organisms would be a good match for my nocturnal habits. By the end of the short study I realized that I didn't enjoy handling frightened animals who, with every justification, wanted to bite me. I'd also discovered that nocturnal projects entailed specimen preparation at 7 a.m., mid-day sessions cutting mist net lanes through the forest with a machete, and very little sleep at all!

The second year of graduate school found me at loose ends; if bat-plant interaction wasn't the ideal project, what was? When Scott Mori (specialist in the systematics and ecology of trees in the Brazil nut family, and my graduate co-advisor), suggested that I study the wood-boring cerambycid beetles that reproduce in these trees, I was distinctly underwhelmed. I didn't want to be relegated to a career spent chasing bulldozers, far from my idyllic vision of pristine rain forest. Nevertheless, a happy series of mishaps landed me in David Grimaldi's entomology class at the American Museum of Natural History. Obliged to spend plenty of time outdoors, stalking and seizing insects to identify, I was irrevocably smitten. Many of my collections were made at my community garden; that tiny oasis in the heart of Manhattan supported an amazing diversity of insects. I was intrigued to learn about the finely tuned associations that arise between plants and insects, and the idea that plants produce a huge variety of chemicals capable of manipulating insect behavior captured my imagination. The following semester I spent one day a week sketching my way through the cerambycid collection at AMNH. By the end of the semester, armed with a self-created field guide to cerambycids and reassured that I could conduct my study far from the bulldozers, I was devising a thesis project specifically designed to require a year in the rain forest of central French Guiana… and that is where I lived from September 1995 to August 1996!

Flirting with Domesticity

There are three ways to approach the vast roadless interior of French Guiana, one of three lesser known political entities on the northern coast of South America (a bit smaller than Indiana). One option is walking from the coast (approximately three weeks), a second is taking a canoe upriver (ten days) and then walking (two days), and the third option is flying in a small propeller plane from the coast (less than an hour). As a solitary traveler laden with 86 pounds of supplies, I opted for the plane. My journey had thus far encompassed every conceivable mishap. Gérard Tavakilian (Fig. 2-1), the entomologist in the sleepy capital of Cayenne who first discovered the association between my group of beetles and their host plants, immediately proclaimed that I was Not Lucky. On my 40th birthday I ar-

Fig. 2-1. My home and work in French Guiana. Top left: Tavakilian displays Titanus gi-ganteus (Linnaeus, 1771), one of the largest beetles in the world. Bottom left: I check out my very airy house. Top right: Scott Mori and I prepare plant vouchers and take notes on the study trees. Photos by Carol A. Gracie. Middle right: I sort beetles by headlamp. Photo by J. Bonavito. Bottom right: Limitless solitude in the middle of the rain forest. Photo by Carol A. Gracie.

rived in the village of Saül in a state of intense euphoria, convinced that my luck was changing, and made the final seven kilometer hike to the homestead where I would spend the next year.

My new house, tucked amidst secondary forest but only meters from pristine old-growth forest, was enchanting. The lower walls were made of interwoven strips of wood, while billowing lace curtains sufficed as upper walls (Fig. 2-1). One of the two rooms was just large enough for a built-in platform bed and a clothes rack; the second was furnished with a simple wooden table, a bench, and a set of wooden shelves. The latched door was a mere formality; nothing hindered frogs, lizards, bats, hummingbirds, rats, opossums, columns of ants, whip scorpions, and a tremendous variety of other invertebrates from coming over, under, or through the walls. The black plastic roof was constructed with an overhang, partially protecting a clothesline and the exterior kitchen table stocked with a gas tabletop stove and miscellaneous pots and pans. My hosts had outfitted the house with all of the essentials for life without electricity or running water: a cassette player lacking a functional power source, a cantankerous gas lantern, a full-length mirror, several plastic washtubs, and a chamberpot! The house very much suited my rather stripped-down aesthetic, and best of all, it was a waterfront property.

The crystal clear stream, accessible via a short staircase carved into the bank directly outside my door, would serve as drinking fountain, dishwasher, and washing machine. This confirmed Manhattanite, accustomed to instant culinary gratification and laundry service, was clearly embarking upon an unprecedented domestic adventure. Deliberations about the relative merits of Burmese, Mexican, or Polish takeout a thing of the past, I would now prepare my own rice and beans or go hungry. I would launder my clothes with a scrub brush on a stone table in the middle of the stream, and scour my pans with the assistance of the ubiquitous dish-washing fishes!

Sensory Deprivation in the Midst of Plenty

Fifteen years of life in the Big Apple had left me with a perpetual low-level craving for solitude, and I had intentionally sought out one of the lonelier spots of the world. French Guiana was the site of the famed French penal colonies, including Devil's Island. It still maintains a rather unsavory reputation and the French civil servants that brave this tropical paradise merit generous hardship pay. The majority of the Europeans and Creoles live in modern towns strung along a coastal highway, and the rivers that define French Guiana's borders serve as the transportation systems for villagers living along the banks. At the time of my thesis study, about one hundred souls were sufficiently hardy (and/or eccentric) to stake a claim in the less accessible environs of Saül.

The first week of my long-awaited year in the rain forest was an exhilarating but unnerving period of introduction to the unfamiliar. My body metabolism was undergoing its own revolution while I adjusted to a novel quotidian routine. I investigated the local trails, attempting to grasp the spatial logic of the treefalls that obscured them, and began to observe the remarkable profusion of plants and insects characteristic of tropical rain forests. This brief idyll came to an abrupt end with the arrival of Scott Mori (Fig. 2-1), accompanied by a coterie of assistants and colleagues. The following days were a blaze of high-energy activity as I worked with arborist Bob Weber, one of the world's top tree-climbers, to sever branches from 25 trees, ranging from slender understory specimens to massive canopy emergents. These branches, from five different tree species, were intended to lure wood-boring beetles so that I could determine how selective different beetle species were when laying their eggs. No sooner was the project established and the crew departed, than AMNH entomologists David Grimaldi and Lee Herman arrived, and I enjoyed a heady bit of vacation enlivened by a crash course in tropical entomology. All too soon, I was escorting them back to the airstrip—my leaden gut reminding me that until Bob Weber returned in three months to cut the rainy season branches, I was on my own.

When I planned my project back in New York, I had a vague notion that it would be pretty tough to spend a year more or less alone, but I assumed I was up to the task; there was nothing in my previous life that could have prepared me for the sense of isolation I would experience (Fig. 2-1). The first month was stressful and exciting, conditions very familiar to most urban dwellers. I'd been among other scientists, and my sense of the passage of time had remained firmly intact. Once everyone departed, time slammed to a halt. There was very little to distinguish one day from the next and each seemed a vast blank slate. We had allotted the beetles three months to lay their eggs in the first set of branches. Although I'd attempted to devise projects to keep myself occupied until it was time to collect and cage the cut branches, I found myself for the first time in my life rising and setting with the sun… constantly fatigued, ravenously hungry, and disoriented. I was surrounded by an incredible profusion of visual and auditory information, but it was all of unfamiliar form. I lacked the experience that might have enabled me to interpret the overwhelming natural stimulation and, as my life was devoid of artificial stimulation, the net effect was one of sensory deprivation.

Each morning I was encouraged out of bed by the arrival of vicious biting tabanids (Fig. 3-11 in Chapter 3). These pestiferous flies were first detectable as an ominous buzz at 6:30 a.m., and by 7 a.m. they would attack in a swarm (particularly attracted to my blue sheets). My shady waterfront house proved to be ideally situated for diurnal mosquitoes that arrived shortly after the tabanids and inspired days of perpetual, if unfocused, motion. They lazily circled me whenever I sat down, were particularly bloodthirsty whenever I was sorting or pinning insects, and eventually I would feel their bites even when my eyes assured me that there were none to be seen.

It was a revelation to learn exactly how many hours were required to provide for my basic needs in a pre-industrial setting. In addition to the predictable domestic chores, constant vigilance was required to prevent creatures of the forest from reclaiming my small piece of turf. Unable to concentrate, I wandered about my tiny house starting a project, then being distracted by another before finishing the first. Periodic forays into the forest to inspect the branch samples for signs of insect attack yielded dismal results. The beetles did not appear to be cooperating with my carefully-designed experiment, and I suffered plagues of uncharacteristic self-doubt. I was sure that, given the same opportunity, any one of my colleagues would be making amazing scientific discoveries, writing books, or creating portfolios of spectacular illustrations. I attempted to spend as much time as possible walking the trails, convinced that, if nothing else, I was at least assembling a respectable Guianan insect collection.

Throughout the remainder of the dry season, there was a seemingly endless succession of febrile days and languorous evenings. Although the family running the homestead had invited me to visit in the evenings, it was virtually impossible for me to remain awake until they finished dinner. I did occasionally meet for a chat with Andy, the elder son. This gentle, ethereal soul (a changeling amidst his robust and pragmatic pioneer family) seemed profoundly depressed, and the meetings did little to elevate my spirits. Day after day passed with minimal human contact; the only human sound I would hear was the distant wailing of the family's youngest member. When I was finally called upon to speak, my voice would emerge an inharmonious croak.

At the onset of the rainy season, my admittedly fragile state took a turn for the worse as I abruptly found myself with clothes and towel wet, smelly, and full of mildew. When it rained, my clear bathing stream swiftly transformed into a muddy torrent full of debris. The roof overhang protecting my outdoor kitchen table wasn't quite deep enough, and I often prepared dinner holding an umbrella aloft. The limited amount of natural light was drastically reduced and, anxiety-stricken about the prospect of an entire year without income, I had already instituted a strict energy-rationing program: one hour each evening with the gas lantern, and a second hour of dim illumination courtesy of my precious headlamp batteries (Fig. 2-1).

The insects in my prized collection began to mold, depriving me of my one consistent source of satisfaction (and the only evidence I had that my time in French Guiana had been in any way productive). Ultimately the papers lining the wooden collection boxes were completely covered with fungus, the glue holding the boxes together started to disintegrate, and the boxes literally began to fall apart. Whenever the sun reappeared, I was faced with a host of competing demands. I attempted to wash my muddy clothes, or to continue the prolonged drying process (knowing that a momentary lapse of attention—even a walk to the outhouse—might permit a passing shower to once again soak the laundry that I'd been drying for days). Alternately I would dash about with my insect collection, tracking the sun, trying to inhibit the fungal growth by giving my insects a dose of light (knowing that at any moment a sun shower might counteract the potential benefit of the treatment). All I really wanted to do during the sunny interludes was take advantage of the good weather and disappear into the forest.

My journal entries from this period hold a detailed account of daily activities, but assiduously avoid any mention of my mental state, and thus I no longer know when the most distressing phenomenon arose. Whether it started as soon as my visitors departed, or whether it accompanied the early rains, I began to experience auditory hallucinations: repetitive noises in my head that only ceased when I was actively engaged with other people, or during my two brief jaunts back to the coast. I suspect that this phenomenon is not uncommon in the rain forest, where many of the genuine sounds are repetitive and rather mournful. Marie-Claude, the family daughter-in-law, informed me that during her first rainy season she heard the call of the screaming piha echoing ceaselessly in her head. In my case, the soundtrack was at one point stuck on a Christmas carol, but more typically restricted to Brian Eno-esque phrases. Adventurers 'lost and maddened in the rain forest' have undoubtedly endured similar experiences.

Waiting for Things to Fall from the Sky

The one activity that provided some sense of structure to my life was a weekly jaunt to Saül, where I would enjoy a rather chaotic shopping spree. Each Friday I would pocket a few francs, latch the door of my open-air house behind me, and walk to the village on the rutted dirt road cut through pristine primary forest. The outgoing walk took, at a comfortable pace, about 2 1/2 hours, and I marked my progress by tracking landmarks associated with past visitors (knowing, for instance, that I had just passed Debbie's *Caryocar*, but had not yet reached Lee's Creek). Once in the village I scurried around collecting my mail, stocking up on tuna and chocolate, visiting the Hmong produce stand, making any essential telephone calls from the lone telephone booth, and purchasing fresh baguettes just flown in from the coast. On the return—weather permitting—I dawdled as I searched for unknown flowers or fruits, and collected insects, turning the equivalent of a dash to the corner store into a full-day outing.

Flights arrived from the coast three times a week (Monday, Wednesday, and Friday) and transported, in addition to the baguettes, immense sacks of mail and cartons of supplies. Most of the villagers put in an appearance on airplane days, lending Saül an air of relative festivity. The airplane was audible in the village as it made its descent to the airstrip, and this was the cue for the entire European population of the village to materialize at the combination Air Guyane Office / Postal Bureau, waiting for things to fall from the sky. I felt sure that many of the local residents had once, like me, nursed romantic fantasies about a flight from civilization, only to find themselves, again like me, dependent upon and desperately awaiting its products. When a flight landed, the mailbags were immediately loaded, along with any newly-arrived tourists, onto a truck that sped back to the village. Each precious letter or package was pulled from a sack by Yolanda, the combination Air Guyane employee / Post Mistress (who commuted to Saül on her bright blue tractor). She barked out the surname of the addressee, and handed it on to the waiting hordes, where it was passed from hand to hand until it reached the eager recipient.

A few other business concerns opened up for an hour or two on airplane days. A group of Hmong, relocated in French Guiana after the Viet Nam war, brought a pushcart full of delectable produce from their nearby plantation and manned a small outdoor market across from the Air Guyane Office / Postal Bureau. Two tiny grocery stores sometimes opened their doors for business. My initial reaction to these stores was one of dismay. Neither was large enough to accommodate more than one or two customers, they were both dimly lit by ambient light entering through the doors, and the floor of the larger store was always covered with great clumps of mud left behind by the ubiquitous rubber boots. The labels of most items were covered with mildew, and the selection of non-refrigerated products was depressingly sparse. In the entire year, there were two great moments at the local groceries: I once encountered a batch of freshly-laid eggs at Monsieur Agasso's, and Monsieur No-No once took pity on me as I cast a sad eye at his empty shelves, and presented me with a gift of two ripe avocados.

The first couple of months I spent in French Guiana I could never figure out why all the villagers leaving Madame Marie's on airplane days were bearing armloads of baguettes, while there never seemed to be more than, at most, a single baguette for me. I eventually learned that the villagers had prepaid for their orders, and thereafter I also paid in advance and picked up two baguettes each week. These lasted for three days, after which the bread became moldy; the remainder of the week I heaped my tuna or peanut butter onto crackers. Madame Marie was also purveyor of the *telecartes* required to make a telephone call from the local phone booth. The telephone operated like a radio phone: it was possible to trade snippets of information (leaving pauses between question and response), but if one party began to respond without waiting for the other to finish, the transmission would be interrupted. It was a viable method of communicating with anyone familiar with the system's peculiarities, but not especially reliable. All too often the telephone was out of order when the *telecartes* were in stock, and when the telephone was functioning, the *telecartes* were sold out!

Fridays in Saül were more than a long walk through the forest and a few new edible treats; they were also the main chance I had to try to chat with people and practice my halting French. I made friends with the Dumas, the one family in Saül that seemed to fullheartedly appreciate and enjoy their unconventional lifestyle and their motley assortment of neighbors. The Dumas had already constructed and lived in several houses since moving to French Guiana, but ultimately had settled right in the middle of the village. Their house was built using natural materials, but with an airy, loft-like design. Hugette was a petite, pipe-smoking woman who worked her own small gold-mining claim. She also did the cooking, painted scenes of village life on wooden panels, and created displays of forest flowers and fruits that rivaled anything you might encounter in a toney SoHo boutique. Her husband Gérald directed the construction of nearby trails, raised poultry, maintained an extremely productive garden, and made wonderful wood carvings. I would stop by each week after finishing my errands and hang out for a while as other villagers stopped by to visit, be fed some snacks or lunch, and check out the latest changing exhibition. They offered me a much-needed opportunity to make light of my various trials and tribulations, and were interested in my small accomplishments. I began to refer to the Dumas as my Emotional Rescue, and can only suppose that I might have become completely unhinged without their kindness and humor.

I never lingered too long in town, because I wanted to get home before the sun set, promptly at 7 p.m. Even during the rainy season mornings typically remained clear, but the afternoon rains often caught me just as I was leaving Saül, or when I was midway home. They could be quite forceful, and would have been blinding had I not always traveled with a small folding umbrella to protect my glasses. The dirt track sometimes turned into a veritable stream, and the sound of water rushing over rocks made a melodic accompaniment as I sloshed along. I ultimately became quite fond of those long walks in the pouring rain, even though I realized that they presaged a bath in a muddy and turbulent creek.

Thanks to the unflagging efforts of family and friends, I seldom returned from mail call empty-handed. When I finally arrived home, I'd take a quick dip to rinse off the sweat, and then bolt back to my house and settle down with my haul of cookies and mail. These communiqués I devoured were virtually the only contact I had with the outside world, but they supported my sensation of having come unstuck in time. Mail sometimes traveled to French Guiana by long and circuitous routes, and was often subject to lengthy delays (due to the propensity for either strikes or celebrations). Letters seldom arrived in consecutive order, and had to be accepted as a very non-linear narrative. Nevertheless, I was profoundly grateful for each page, and treasured each as material proof that I had not entirely ceased to exist!

The Emerging of the Scamps

My fears that the beetles were snubbing the cut branches proved to be totally unfounded. Unobserved, they had in fact mated and laid a multitude of eggs. At the end of December, my friend and visitor Christine Johnson helped me collect the dry season branches. We first sawed the branches into lengths to carry out of the forest, and pupae were literally dropping out from underneath the bark. No sooner were the branch sections placed into cages I'd fashioned from plastic screen, than adult beetles began to emerge. The little scamps were adept at gnawing through the screen as if it were no more than an additional layer of bark. During the next eight months much of my time was spent patrolling the cages, hoping to thwart potential escapees. Although the dreariest part of the rainy season and the year's most emotionally stressful experiences were still ahead, I now knew that I had data for my thesis, and I was never again quite so demoralized.

Over the course of the year, the 400-plus branch sections gave rise to 1,813 cerambycid beetles belonging to 38 different species (Berkov & Tavakilian, 1999)! Some of the beetles appeared to be tourists, while other species were represented by hundreds of individuals. Every single emergence, even of yet another individual belonging to one of the most ubiquitous species, was a thrill; I never opened a cage to extract an adult without feeling that I was unwrapping a gift. With so few external distractions, I felt uncannily attuned to my branches and beetles. Some of the feistier scamps would stridulate like mad when handled; others were comparatively passive. I had a pretty good sense of the time of day that each species emerged; some came out during the middle of the night and would be clinging to the tops of their cages in the morning, others emerged during the afternoon, and yet others tended to emerge shortly after dark. Even with the many cages surrounding my house, I would often get a peculiar feeling that a particular beetle was due to emerge from a particular branch, and these hunches almost inevitably paid off.

Hoping to figure something out about the chemical cues the beetles used to evaluate a potential host plant (Rovira et al. 1999; Berkov et al., 2000), I kept the adults alive for experiments with wood extracts. My table remained covered with a stock of cerambycids enclosed in individual plastic cups, leaving me only a small clearing for meals and specimen preparation. The experiments were frustrating and fruitless. I also maintained a selection of living insects collected in the forest, including a succession of pet assassin bugs. I took an evil satisfaction in capturing the biting tabanids and presenting them to my pets. The assassin bugs promptly subdued the oversized flies by trapping them with their strong forelegs. They inserted their sucking mouthparts between the sclerotized plates forming the exoskeleton, and moved from one position to another, sucking fluids until the fly was nothing but a lifeless husk.

These assassin bugs were quite abundant, both as nymphs and as adults, and I noticed that individuals often had their robust forelegs covered with some sort of sticky substance. I found an early instar inside a fallen *Clusia* fruit and wondered if these predators covered their legs with viscous *Clusia* resin to help trap their prey. Although I supplied my pets with a variety of blossoms and fruits, the bugs never showed any inclination to sample resin (the fruits may already have been too old). Eventually there was a fresh fruitfall from a hemiepiphytic *Clusia* near one of my study trees. Inside almost every mature fruit I found a large assassin bug, upraised forelegs clearly coated with the aromatic resin oozing from the fruit's central disk, awaiting insects attracted by the alluring smell. I subsequently learned that this bug, *Manicocoris rufipes* (Fabricius 1787), belongs to a tribe of "resin bugs" that prey on resin-collecting bees (Melo et al., 2005).

Vertebrate visitors were attracted to my house by the abundance of dead wood piled about, the scent of overripe papaya, the easy source of insect protein, or merely to escape the rain. This resulted in a number of unsavory nocturnal surprises. One night I was abruptly awakened at 4 a.m. when a hefty frog lost its grip on a support beam, and landed directly on my face. Another time, I woke up in the middle of a rainy night with the sense that *something* had just moved. I sleepily assured myself that a bat must have flown by, when I *definitely* felt something moving towards me on the bed. I was up in a flash, turned on both of my headlamps, and found myself face to face with a wet and indignant four-eyed opossum (*Philander opossum* [Linnaeus, 1758]). I promptly grabbed my insect net and chased it out of bed and out of the house, but it was to become quite the regular nocturnal guest.

I began to call the incorrigible beast *Mauvaise Habitudes*, because it shared most of my bad habits. Not content to raid my fruit, it would come in at night and unpack the bits of sausage that I occasionally added to my meals of rice and beans to create the illusion of meat. I once stumbled out of bed and into the other room to find the possum licking the lid it had removed from a jar of Nutella (chocolate-macadamia nut spread); so cocky that it refused to leave until it had finished. It would open up petri dishes to steal beetles and, after cleaning me out, return the next night, apparently expecting me to have provided a fresh supply. Desperate for company in any form, I actually began to leave *Mauvaise Habitudes* the occasional (non-Coleopteran) treat.

The Giant Wild Forest Pussy Cat

Not all of my encounters with wildlife took place indoors. As a veteran of excursions to the tropics in the company of avid birders, I assumed that I would see a great diversity of birds, but would rarely see mammals. I failed to anticipate the tremendous patience required to see and identify most arboreal birds. Although I periodically surprised large ground birds—including troupes of gracefully prancing trumpeters—as I prowled through the forest, I crossed paths more frequently with mammals. Oddly, I seldom saw the more abundant mammals (tapirs or peccaries), but I had the good fortune to stumble upon two of the more elusive denizens of the forest: a giant anteater (*Myrmecophaga tridactyla* [Linnaeus, 1758]) and a jaguar (*Panthera onca* [Linnaeus, 1758]).

It was more than an hour's walk to the trailhead of my favorite path, making it difficult for me to visit as often as I would have liked, particularly during the rainy season. On one of my eagerly anticipated out-ings, I noticed a side trail branching off to the right, and turned to explore this previously unnoticed path. After about 15 minutes, I decided to give up the exploration and return to the main path, which led to a small creek that I liked. I retraced my steps, and just as I reached the intersection with the main trail, I looked ahead and saw a jaguar about 15 meters away. It was smaller than I had imagined, but whereas an ocelot has spots fused almost into stripes, the jaguar is distinctly spotted. I froze, awestruck, as it strolled regally across the trail and into the forest without giving me as much as a glance. Although I had no inclination to aggravate this giant wild forest puss, the encounter was not particularly frightening to a person raised with cats. Like other cats, the jaguar seemed profoundly interested in its own business, and domestic relations aside, humans just aren't a noteworthy part of a cat's world.

My meeting with the giant anteater was actually more unnerving. In this case I suddenly found myself face-to-face with a large animal that was completely outside of my frame of reference; I could only describe it as looking like a cross between an Afghan hound and a vacuum cleaner! It had a long, tube-like snout, an amazingly long and bushy tail, and forelegs sporting three oversized claws. As it lumbered towards me, I intentionally made enough noise to make sure that it didn't get any sudden surprises. It continued its approach, but then decided to surrender the path. The sides of the trail rose up about five feet and the great beast looked as clumsy as I would have had I been the one heading off into the forest, climbing a bit then sliding partway back down, raking the bank with its claws. I subsequently learned that the giant anteater is generally fearless because it is avoided by predators, unappealing even to jaguars because of the formidable claws and its unsavory antlike flavor.

La Carte de Sejour Vrai

Many of my typical minor health complaints disappeared while I was in French Guiana, leading a healthy life with plenty of exercise and an incredibly low-fat diet. There were, of course, a number of distinctly tropical maladies that I hoped to avoid. My greatest fear was the flesh-eating disease leishmaniasis (Fig. 3-14 in Chapter 3), caused by protists transmitted by the bite of minute blood-sucking sand flies. If treated, leishmaniasis is not one of the most dangerous tropical diseases, but it is potentially one of the more disfiguring. My other great fears were of botflies (internally feeding fly larvae hatched from eggs that could be transmitted via mosquito bites; Fig. 3-13 in Chapter 3) and hemorrhagic dengue (breakbone) fever.

I didn't experience any significant or prolonged physical discomfort until I'd been away for five months. Then one day I noticed that one of my knees was sore, and assumed that I'd injured it while collecting branches. Within days, the pain had selectively spread to other joints, including my hips, wrists, and thumb joints. One of the Creoles in Saül diagnosed me with rheumatism, brought on either by bathing in the stream, or by overheating my feet in rubber boots. An alternate suggestion was a "*petite dengue*," and I was warned not to take aspirin, which could be fatal should a dengue become hemorrhagic. The pain, kept under control with a few soggy Advil, was never totally debilitating. It did make it difficult to perform many of my customary tasks, including swinging my insect net, flipping the lids off of the film canisters I used when collecting my beetles from the cages, and washing my dishes crouched at the stream. About a month after the onset, I made my second brief trip back to Cayenne. No sooner had I stocked up on enough pain-killers to see me through the next six months, than the pain disappeared as precipitously as it had arrived (perhaps supporting the 'petite dengue' hypothesis)!

As the rainy season progressed, I became less and less optimistic about my chances of avoiding leishmaniasis. Arborist Bob Weber had contracted such a severe case that he ended up hospitalized with 38 lesions (Fig. 3-14 in Chapter 3) and required a series of 28 injections. Several other visitors had developed less severe cases. Every week when I arrived in Saül, I saw yet another villager sporting the characteristic persistent ulcers (over the course of a single year, at least one in ten came down with the disease). About a month after I recuperated from the joint pain, I began to develop an unattractive facial inflammation.

One of the ongoing stressful experiences up to this point had been dealing with endless bureaucratic confusion about my "*carte de sejour*," an addendum to the special visa required for a lengthy stay in French Guiana. My occasional costly telephone calls and my two trips back to the coast had been, almost exclusively, part of a prolonged quest for this elusive document. During my recent trip to Cayenne I'd finally gotten the last paper affixed to my passport, but the official document in no way conferred residency as effectively as the facial inflammation… at last I felt truly integrated into tropical life! My lesions were not typical of those associated with leishmaniasis, so whenever I visited the Saül I was stopped by villagers offering opinions as to their cause and the appropriate treatment.

France may maintain a frustrating and labyrinthine bureaucracy, but it also supports a most efficient medical system. Even Saül had a small health center staffed by a doctor who flew in for two days each month. I missed the doctor's visit when the inflammation first appeared, and by the time he returned there were more than 23 lesions clustered on one side of my face. The doctor, an English-speaking dermatologist, acted quickly to minimize the chance of permanent scarring by immediately treating me for both leishmaniasis and a staph infection. The two leishmaniasis shots were anything but pleasant. Each needle was attached to a reservoir that took 15 minutes to drain (for a total of 30 minutes), and throughout the ordeal a nurse checked my blood pressure, because patients sometimes have adverse reactions to the medication. The walk back to the homestead was in fact quite surreal, as the drug kicked in. Pleasant or not, by the time the culture results came back from Cayenne positive for leishmaniasis, the lesions covering my face were already healing. It was, of course, a singularly liberating experience to withstand my worst fear!

Seasonal Change in the Land of Endless Summer

I originally wanted to spend a year in the middle of the rain forest because I was eager to experience a complete annual cycle. After living in a climate with seasons well-defined by changes in day-length and temperature, I was curious about a climate with seasons differentiated by the amount of rainfall. During the dry season, when many trees and lianas were in bloom, one day had seemed very like another. When I missed the opportunity to take a photograph because I wasn't carrying my camera, I assumed that my subject of interest would remain intact until my return. I quickly learned that although the tropics may seem timeless, biotic interactions progress at a rapid pace and one seldom gets a second chance. The orchid spotted one day would be consumed by a caterpillar the next; the fabulous spider camouflaged on its web disappeared overnight; and the tinamou eggs I'd barely restrained myself from preparing as an omelet were eagerly devoured by a less inhibited predator.

The rainy season, initially imagined as a monotonous sheet of rain lasting for nine months, proved to be quite variable in nature. I recorded the rainfall on a daily basis, but many patterns are not discernible because they failed to alter the total amount of precipitation. During the first few months, periods of fairly heavy rains were interspersed with days or even a week of sunny skies. There was a different complement of plants in bloom, and there was still an abundance of insects. The following month was kind of drizzly and overcast; then there was a return to alternating sun and showers. This was a good period to observe the striking assortment of fruits fallen from the forest canopy and was also the peak flowering period for many monocots, but insect populations seemed depressed. The short dry season, "*le petit été de mars*" arrived in early April, and provided a welcome vacation. The final two months of the rainy season most closely approximated the stereotypical image of endless rain and mud, but the ground was covered with newly-sprouted seeds taking advantage of the abundant moisture.

Signs of the impending dry season began to appear even before the rains started to let up. During the dry season I'd pulled minute ticks off of my body on a daily basis, and while it's difficult to imagine greeting a tick with joy, the first one I found on my leg near the end of the rainy season filled me with an incredible sense of buoyancy. In the following weeks, I encountered more and more flowers and insects I'd not seen during the past eight months, and greeted each as a long-lost friend. The ticks had returned but the tabanids and mosquitoes were on the wane, and I began to rediscover the almost forgotten pleasure of setting up a camp chair outside my house to read during the brief tropical dusk. Emergences from my branches were quickly tapering off, and I made a final climb into one of my study trees to saw off one last branch as a farewell gift for my beetles. I knew that the long, hot summer was coming to an end and I'd soon be back in New York, incredibly ready for autumn.

N'habite pas à L'adresse Indiquée

When I arrived in New York, I literally kissed the ground. As it turned out, I left French Guiana in the nick of time; a month later my house was flooded by a meter of water. My mail support committee had continued generating letters right up to the bitter end, and those that arrived in Saül after I'd left were returned to the senders stamped "*N'habite pas à l'adresse indiquée.*" Years later, back in New York where time once again passes in an orderly fashion and I once again firmly inhabit my person, those envelopes still grace my refrigerator, in remembrance.

Epilogue

I completed my Ph.D. thesis two years after I returned from French Guiana, moved almost seamlessly to a post-doc at the American Museum of Natural History (Berkov, 2002), and from there to a faculty position at the City College of New York (also 2002). Fourteen years have passed since I reinvented myself and realized my dream to spend a year in the middle of the rain forest. If I could turn back the hands of time, would I do it all over again? YOU BET... but I'm glad I don't have to! Well, most of the time. Back in 1995, when arborist Bob Weber developed his world-class case of leishmaniasis, he convinced his colleague Christopher Roddick to travel to French Guiana in his place to complete the rainy season cut. Chris's young protégé Alec Baxt recently wrapped up a year-long study of canopy microclimate using the very trees I once visited, climbed, and sampled, and I certainly suffered the occasional pang of jealousy. Central French Guiana has changed. Many residents departed Saül during a two year hiatus in flights from the coast, and there has been an influx of Brazilian goldminers. The miners purchase their necessities from Monsieur Agasso and Monsieur No-No with gold, and leave the waters at Les Eaux Claires less than clear. The footprint of my little house can be discerned only as a line of decorative Hawaiian ti plants hidden within dense secondary growth. On the up side, assassin bugs are still sitting inside fruits that have fallen from the very same *Clusia* plant, still awaiting prey with upraised forelegs coated with resin!

As a City College faculty member, I am now constrained to rain forest on an academic calendar. The original beetle project in French Guiana paved the way to additional rearing projects in Peru (Berkov et al., 2007) and Panama, and we have described our first new cerambycid species (Berkov & Monné, 2010). Side projects started as time-fillers during my longest year have blossomed into Masters projects and contributed host plant data to other studies (Feinstein et al., 2007; Feinstein et al. 2008; Robbins et al. 2010; Whigham et al. in prep.). Although much of my subsequent research has focused on the beetles, which are even more poorly known than the plants, I still conjure up mental images of the trees I studied in 1995. It is both gratifying and frustrating to have my home forest—the forest in which I have spent the most time and feel the most at ease—an expensive three-day trip from the city where I live. But there are compensations: each time I see a student look out the window of a bush plane onto that unbroken sea of green, it puts a smile on my face that feels as wide and deep as the forest below. Each time we attempt another audacious project, I leave the rain forest crowing "I can't believe we just DID that!" I want my students to understand that while they are living in a time of unprecedented environmental threats, it is also a time of boundless opportunity. They simply need the imagination to recognize that opportunities exist, the confidence to seize those opportunities, and the persistence to reap the rewards.

Chapter 3
Tips for Tropical Biologists

by Scott A. Mori

This chapter describes the dangers and annoyances that my colleagues and I have experienced during expeditions to the Neotropics. For a general introduction to the tropics, I recommend *Tropical Nature* by Forsyth and Myiata (1984), and for more specific discussions about tropical diseases and annoyances consult *A Neotropical Companion* by Kricher (1997). Although the tropics harbor many dangers, other than a number of bouts with intestinal parasites, the only serious disease I contracted over the course of 45 years of field research was leishmaniasis. Although when in the field I always keep the potential dangers of the tropics in mind, major problems are relatively infrequent.

Informative biological collections can be made practically anywhere in the Neotropics; for example there is still much to be learned from collections of common weeds and animals. Nonetheless, many biologists are drawn to explore remote and poorly collected areas, where the chances of finding plants and animals not yet known to science are greater. Consequently, in their efforts to document plant and animal diversity, biologists may be based in varied accommodations, ranging from comfortable hotels and biological research stations to rudimentary forest camps. One of the most important things to do before any expedition is to contact local biologists for information about diseases and potential dangers found in the areas you will be visiting.

On expeditions to remote areas, improvised camps might be reached by plane, helicopter, boat, hiking, or a combination of these methods of transport. In this chapter my goal is to make the work of other tropical biologists easier, more comfortable, and safer. Although I emphasize my expeditions to wilderness areas, much of what I write about also applies to biologists taking day excursions from established camps such as research stations.

Hiking to Camp

Botanical and zoological exploration in the tropics often requires rigorous hikes along poorly maintained trails. Most hikers average two to three kilometers per hour walking along rain forest trails while carrying heavy loads, but that rate varies considerably depending upon the condition and slope of the trail, the weight carried, and the physical abilities of the hiker. Expedition members should be able to carry all of their own gear, be in good condition, and wear boots that are well broken in. Loads above 15 kilograms (33 pounds) markedly increase the chance of blistered feet and accidents. On an expedition to Camp Inselberg in central French Guiana, one of the strongest participants was confined to camp for a week because of swollen and blistered feet, due to carrying a heavy load while wearing poorly broken in boots.

Each hiker should carry a single pack; thus empty day packs should be accommodated in or tied to the backpack until the destination is reached. Wearing a day pack on the chest becomes more burdensome with each step of the hike, and my former graduate student, Dr. Amy Berkov, learned this the hard way on a hike to Camp Inselberg from Saut Pararé in the Nouragues Nature Park. She started out optimistically with a full pack on her back and a day pack on her chest, but, after the first kilometer, it was apparent that she would not reach Camp Inselberg before nightfall, so we retreated to Saüt Pararé and started again early the following morning.

Each participant should carry at least one liter of drinking water for every five kilometers walked. In very sparsely inhabited neotropical forests, stream water may be safe to drink directly from the source, but it is always best to treat the water first. Even in the most remote areas one can get sick by drinking untreated water, especially if it is stagnant. For example, Nathan Smith and I contracted both girardia and amoebic dysentery after drinking contaminated water on an expedition to Pic Matécho to the northeast of the village of Saül in French Guiana—even though this is one of the most isolated areas in South America, and we had treated the water with hydroclonazone. Before drinking water of a dubious nature, it should be chemically treated, filtered, UV sterilized, or boiled for at least five minutes. A 15 liter Katadyn drip filter produces enough pure water per day for at least five people, but because of its size it is only practical for easily-reached base camps. Smaller, hand-pumped filters are less effective for purifying large quantities of water because they become clogged with sediment, and require frequent cleaning. UV water purifiers are now available and have the advantage of being small and not leaving an after taste to the water, which is a common adverse effect of some chemical treatments.

As a result of gold mining along many of the waterways of Central and South America the water is sometimes contaminated by mercury; thus, to play it safe, avoid taking drinking or cooking water from streams where gold is or was mined. On a trip to the Sinnamary River located in coastal French Guiana (in the early 1990s, to study the relationships between longhorn beetles and the wood they consume, see Chapter 2), we had to carry bottled water for consumption because the river was overrun with gold miners and contaminated with mercury and full of silt, thereby unsuitable to drink and unpleasant for bathing.

The combination of wet weather, perspiration, and lots of walking often causes a fungal rash (called jock-strap itch by athletes) on the inside of the upper thighs that can be very uncomfortable, and even painful. This type of rash can be kept under control by applying baby powder or Vaseline to the inside of the thighs. If a rash develops it can be treated with an anti-fungicidal cream such as Lotrimin.

Camp Sites

When setting up camps along waterways, always consider the possibility of rising water and flash floods. French botanist Christian Feuillet lost a potential new species of Gesneriaceae (the African violet family) due to a flood at his riverside camp on the Mataroni River in northeastern French Guiana. After collecting the prized plant he hung his camera bag and other valuables out of harms way. He did not, however, elevate the gesneriad collection high enough to clear the rising waters that occurred that night, and when he awoke the next morning, he was dismayed to find that the specimen had washed away. Unfortunately, Christian was not able to re-collect the species, so his photographs are the only documentation of this putative new species.

Neotropical rain forests are mosaics of forest at all stages of plant succession because gaps of varying sizes are opened by falling limbs and trees. Before setting up the sleeping shelters, camp sites should be carefully examined for dead trees and large limbs ("widow makers") that could fall and cause injury. Special caution is called for during the rainy season because forest gaps are most commonly formed at the beginning of the wet season when increased water weight causes branches and trees to fall.

For more permanent camps, frames can be constructed (Fig. 3-1), and tarps put over the top of the frames (Fig. 3-2). Because construction of these camps involves felling trees, they should not be established in protected areas.

On expeditions in which camps are moved frequently, each participant should carry everything needed to sleep under an individual shelter. The hammock (Figs. 3-3, and 3-4), an invention of Amazonian Indians, is the most comfortable

Fig. 3-1. Camp construction. Camps of this sort should not be constructed in conservation areas. Drawing by Michael Rothman.

way to sleep in the rain forest. The lightest are called *garimpeiro* hammocks, which refers to their use by Brazilian gold miners. They are made of synthetic fiber and weigh only 360 grams. In combination with a mosquito net and a 3–4 m (10–12') × 2–2.5 m (6–8') ultralight backpacking tarp, the *garimpeiro* hammock is ideal for trips requiring long hikes in the forest. One problem with this set-up is that during heavy rains water runs along the hammock ropes and into the hammock if the ropes are tied around tree trunks and extend beyond the tarp. If cotton strings are tied at right angles to the hammock ropes (Fig. 3-4A), they will intercept rain water as it runs down the ropes; when water reaches the cotton strings, it trickles down them and drops to the ground. There are also hammock/mosquito nets/rain fly combinations that can be purchased (e.g., Hennessy Ultralite back packing hammocks). Setup is simple and rapid, but I do not find them as comfortable as the time-tested hammock setup illustrated in Figures 3-3 and 3-4. If weight is not a concern, then a larger and more comfortable cotton hammock (1400 grams) can be used.

Some first-time visitors to rain forests imagine that it will be so hot that blankets will not be needed at night—but they are unpleasantly surprised when the temperature drops in the early morning hours and it becomes difficult to sleep because of the cold. A summer sleeping bag, that

Fig. 3-2. Rainforest camps. Top: Camp along the Sinnamary River, French Guiana. The person in the picture is the author. Photo by Carol A. Gracie. Bottom: A less permanent camp in the vicinity of Saül, French Guiana. The person in the photo is Brian Boom. Photo by Scott A. Mori.

can be unzipped and thus regulated for temperature changes, has served me well for sleeping in all kinds of tropical environments. A disadvantage of a sleeping bag is that it is bulky and takes up a great deal of room in a

Fig. 3-3. *Hammock enclosed in mosquito net. Note how the net is pulled upward caus-ing a trough into which the bottom of the net is dropped—this makes it difficult for arachnids and insects to enter the net from the bottom. Instead of a single string, a more efficient way is to tie a string at the bottom of each end of the hammock. Drawing by Michael Rothman.*

pack so, when pressed for space, I use a silk sleeping bag liner in combina-tion with a light jacket, a pair of running pants, and socks. This combina-tion is more efficient of space because the silk liner takes up about 10% of the space of the sleeping bag, and the jacket, pants, and socks have other uses. When I do use a sleeping bag I usually begin the evening with it open, but often zip up the bag as the night progresses. Clean, warm, dry, and protected from insects, one can enjoy the sights and sounds of the for-est—the approaching green and yellow lights of different kinds of fireflies, the hauntingly beautiful middle-of-the night song of the common *potoo*, the early morning roaring of howler monkeys, the dawn whooping of the blue-crowned *motmot*, and the raucous chatter of *chachalacas*. Some forest sounds, however, can make it almost impossible to sleep. For example, one of my expedition participants was so irritated by calling frogs that he got up in the middle of the night and stalked them, in an effort to silence their tormenting croaks. Another, who had chosen to sleep in the forest away from the main camp, was terrified by the calls of howler monkeys because

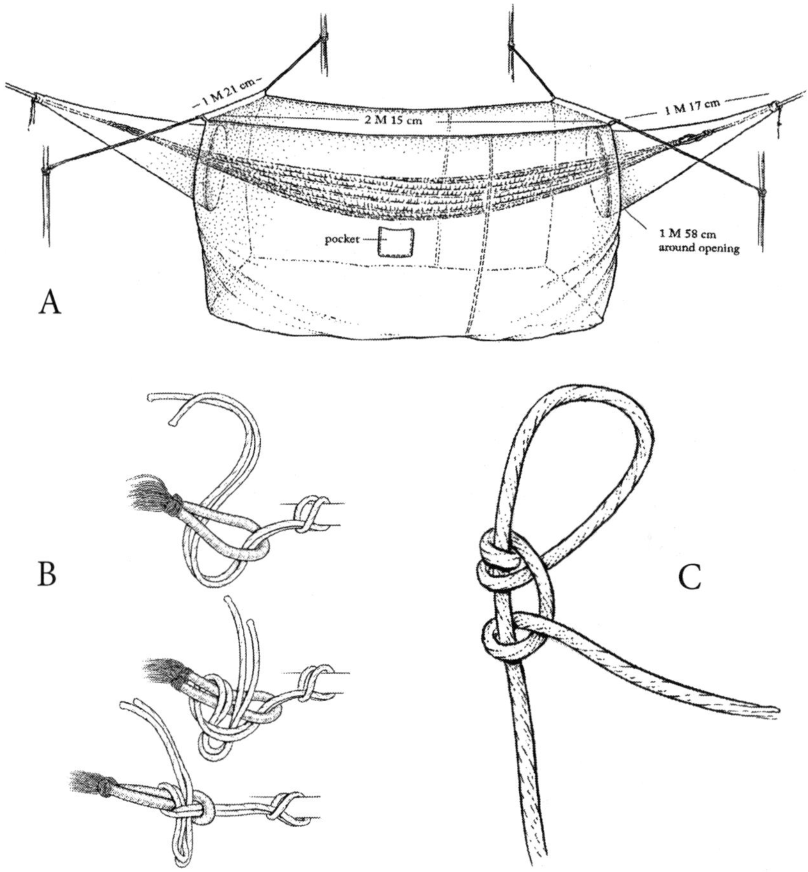

Fig. 3-4. Directions for making a hammock mosquito net and for tying knots associated with hammock sleeping. Top (A): Plans for a hammock mosquito net. Note the pocket on the inside of the hammock for placing items such as a small headlamp. Bottom left (B): Three stages in tying a common knot used to attach hammock end ropes to their supports. This knot is easy to release when the hammock is taken down. Normal knots become very tight and are extremely difficult to untie. Bottom right (C): A taut-line hitch used to tie the mosquito net to its supports. This knot allows for adjusting the tension of the line without untying the knot. Drawing by Bobbi Angell.

he believed their unnerving howls were roars of jaguars.

Inexperienced hammock-sleepers sometimes flip out while entering the hammock, or crash to the ground due to improperly tied knots. Figure 3-4B illustrates a knot that will not slip but can be easily released when the hammock is taken down. The knots should be carefully tested before putting one's entire weight into the hammock. The secret to sleeping comfortably in a hammock is to lie at an angle, rather than parallel to the long axis of the hammock, so that the body is more or less flat and not bowed. A small pillow also helps me get a good night's sleep.

Although areas with black water streams are relatively free from vampire bats and insects, it is always advisable to sleep inside a mosquito net to guarantee protection from these and other pests. Figure 3-4C illustrates a knot (called a taut-line hitch) that I use to tie the mosquito net to supports. This hitch is useful because it allows the tension of the line, and position of the net, to be adjusted without untying the knot. Strings attached to each of the two bottom corners of the mosquito net, and tied to supports above the hammock, will allow the bottom edge of the net to be dropped inside of the trough formed (Fig. 3-3). This prevents insects from entering the mosquito net from beneath the hammock. Glasses, reading material, and a head lamp can be placed inside the trough, or in a small pocket sewn onto the side of the mosquito net. To maximize the efficiency of the net, the trough can be further sealed with clothes pins or paper binder clips, or a zipper can be sewn in along the bottom for even more secure protection from uninvited visitors. The Epco Tropicscreen II is a very fine mesh netting that even excludes tiny biting no-see-um midges (*Culicoides* species) from entering the hammock. Fine meshed netting is highly recommended for areas where no-see-ums occur (which can be almost anywhere at certain times of the year).

In addition to excluding insects, mosquito nets keep vampire bats (*Desmodus rotundus* Geoffroy, 1810) from inflicting wounds on unwary sleepers. The vampires lick blood from these wounds and, on a number of occasions, I have seen blood clots hanging from the noses or ear lobes of people sleeping without nets. This occurs more frequently in populated areas where cattle, chickens, and other domestic animals provide vampires with a reliable source of food.

Camp Life

In forest camps the toilet is the forest, and the shower is the stream. Beetles and flies rapidly clean up animal excrement, and, as long as the toilet paper is burned, all signs of human waste will soon disappear. When a primitive outhouse is available, it is prudent to check first for creatures that might fly out of the pit. Once, when using the toilet at the ORSTOM field station in Saül, I was so startled when bats flew out of the hole that I jumped up, knocking down the box supporting the toilet seat as I bolted forward to get out of the way. Euglossine bees sometimes enter toilet holes as they search for scatole, a substance found in feces that is used by male bees to attract females as part of their courtship ritual. The buzzing sound that bees make as they attempt to exit the pit is an unwanted surprise for anyone sitting on the toilet!

Many disease-transmitting insects, such as the sandflies that carry leishmaniasis and the mosquitoes that bear malaria, are active at dusk or night. If biting or stinging insects are prevalent, it is advisable to wear long pants—with the legs tucked into socks—and a long-sleeved shirt, even in camp in the evening. Field clothing made of very lightweight, easy-to-dry fabric is now readily available—pants and shirts made of lightweight fabric take up less room in a pack and dry rapidly after they get wet in the rain or are washed. Less clothing can therefore be packed for a trip. Synthetic field cloths are, however, expensive, and they may not provide enough protection when one is climbing trees or walking through the underbrush in vegetation with plants armed with spines, prickles, or thorns. I organize my gear into different sized, zip-lock, thin-gauged compressible travel bags that can usually be found in the luggage section of department stores. In camp, I use two small gear hammocks to store my personal items.

A botanist's workday usually begins at daybreak with the removal of dried specimens from the plant presses. To take best advantage of daylight hours, I try to depart for the forest by 8:00 am. For day-long collecting trips, each expedition member should bring lunch in a plastic box and enough water to last the day. In some areas one can refill water bottles at streams, but along ridges water might not be easily available; thus, at least one liter of water should be carried. Tree climbers will need at least two liters of water per day, especially during the dry season. On rainy days, I carry a small tarp to make a temporary shelter; this allows me to press plants without getting the specimens wet. I try to return to camp at around 4:00 p.m. to prepare the collections and bathe before it gets dark.

Appendix C provides a checklist of the personal equipment that I take into the field; I refer to that list every time I pack for an expedition to make sure that I don't leave essentials behind. This list should be modified to accommodate one's personal needs. In addition, Appendix D provides a checklist of the equipment I bring for collecting plants.

Dangers and Annoyances

Accidents. While hiking through tropical forest, falls are common and can be dangerous, especially if one happens to land on a sharp, woody stem cut during trail construction. Crossing a stream on a slippery log is challenging, even for the most athletic expedition members. If the log is high above the water, or slippery, I either use a support pole, crawl along the log on all fours (Fig. 3-5), straddle the log and inch my way across, or wade through the stream. If you opt to wade, one smart but time-consuming strategy is to change into shoes that can become water soaked and dry rapidly (Crocs® are perfect for this); on the far side dry your feet and put the boots back on. Even if your feet get wet while crossing streams in hiking boots, the ensuing blisters are a small price to pay compared to the pain and difficulty of getting out of a remote area with a broken limb caused by falling from a slippery log bridge. Calf high rubber boots allow hikers to cross most streams with little problem, and offer some protection against snake bites.

Fig. 3-5. Nathan Smith crossing the rain swollen, turbulent Arataye River on a natural log bridge. Because of the danger of crossing a slippery log bridge, he is creeping along the log on his hands and knees. Drawing by Michael Rothman.

Unfortunately rubber boots, if they are not a perfect fit, can cause blisters; moreover, they are often hot and do not provide adequate support for walking on steep and rocky trails or climbing trees. Recently, I discovered that Muckboots® are much more comfortable than other rubber boots I have used, and even offer sufficient support for use with climbing spikes. I now use hiking boots in the dry season and Muckboots® in the wet season, and if I need to limit myself to only two pairs of shoes for the field, I now choose Muckboots® and a pair of Crocs®.

Walking on slippery surfaces, such as the algae-covered rocks of inselbergs or trails of wet, compacted clay, is especially dangerous because a fall can not only result in an arm or leg injury, but even in an uncontrolled slide over the edge of a cliff. German bryologist Ingo Holz came close to tumbling to his death on Pic Matécho. Engrossed in his search

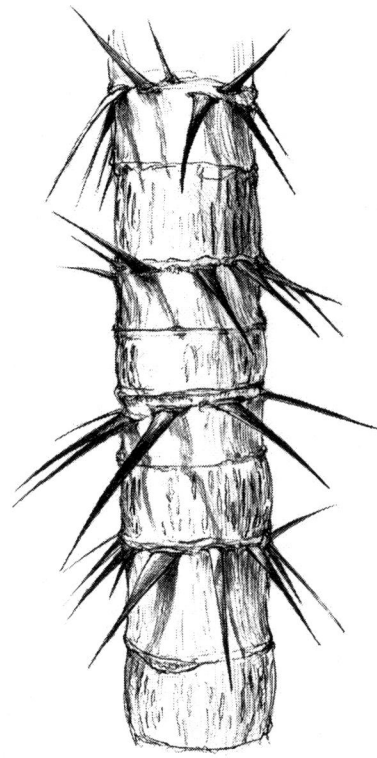

Fig. 3-6. The dangerous spines of an Astrocaryum palm. All species of this genus have long, flat, sharp spines that cause considerable discomfort when a stem is inadvertently grabbed while descending a slippery slope. Drawing by Michael Rothman.

for hepatics, he slipped on a patch of algal slime and slid toward a precipice with a drop of several hundred meters. We thought that Ingo was on a sure slide to death but, fortunately, he got caught up in a patch of vegetation that halted his slide toward the precipice. For the next week he hobbled around camp while the rest of us continued collecting. Luckily, at the end of the expedition Ingo was able to walk the 43 kilometers back to our base camp at Eaux Claires (others helped carry his gear so he was not burdened by a heavy load). He returned to Germany with a hematoma on his posterior that became so uncomfortable that it had to be drained of blood by his local doctor.

Machete cuts are perhaps the most common injuries among biologists. In 1971, while traveling in Costa Rica with Mike Nee in search of species of *Gustavia* (Brazil nut family), I made the mistake of using my machete as a cane while descending a steep slope in a torrential downpour. I slipped, my hand slid down the blade, and my right index finger was so badly cut that even today it is still partially numb. My long-time collaborator Jean-Jacques de Granville once came close to cutting off his finger when he slipped and his hand slid down the blade of his machete (Box 3-1). I know of several other biologists that have inflicted their own machete wounds. In Ecuador, a Danish botanist attempted to cut the iron-like wood of a palm trunk. His machete was knocked out of his hand and ricocheted into the air; as it dropped he tried to catch it and severely cut several of his fingers. On a trip to Cerro Pirre in Panama, one of our guides sliced off the tip of his finger while cutting stems for camp construction. As can be seen from these examples, machete accidents are so frequent that extreme caution should be used when working with one.

The forests of the Neotropics are rich in spine-bearing palms of

---

Box 3-1. Machete as a Cane
by Jean-Jacques de Granville

I was very seriously injured by my machete in 1984 in "La Trinité" mountains. My machete had just been sharpened and I was using it as a cane, walking on a steep bare granite slope of an inselberg just after a rain. As you can imagine the rock was very slippery, but I tried to walk along a slightly prominent quartzite vein. Unfortunately, my foot missed the vein and I slid about 10 meters down the slope. The blade of the machete slid into my hand and cut both flexor tendons of two fingers of my right hand! The fingers were cut up to the bone and were hanging! The following day I had to walk to the first camp where our drop zone was located and call for a helicopter. I was operated on in Cayenne, two days after the accident, but I cannot flex these fingers anymore. They remain slightly bent and stiff. The nerves were also cut but I progressively recovered the sensitivity several years later. Well....this was my story! Now, I always put my machete into my rucksack, but it is a bit too late!

---

the genera *Bactris* and *Astrocaryum* (Fig. 3-6). Special care must be taken not to grab hold of, back into, or fall onto these palm spines, which can easily reach 10 cm or more in length. They are especially wicked because after entering the flesh they can break off and fester for weeks, or even months, before they work their way out. Hikers sometimes receive an unpleasant surprise if their trailing leg catches on a palm petiole, and drives its spines into the back of the leading leg.

When machetes are used to clear trails and camps sharply pointed stems that can cause serious puncture wounds are created. On the 1988 Flora of the Guianas expedition to Mont Tortue, German botanist Paul Hiepko tripped and fell onto a spear-like stem that entered just below his left armpit causing a deep wound. Although it did not strike a major blood vessel, he was confined to camp for a number of days and it took several weeks for the wound to heal. A more serious accident befell Daniel Sabatier, who, while crossing a stream, fell from a log onto a cut stem that entered his leg and perforated the femoral vein. He initially feared that the stake might have cut his femoral artery, which is intertwined with the vein. Although the artery was not in fact cut, blood loss was considerable and it took months for him to fully recover.

Lost in the forest . Getting lost is a constant danger in tropical forests because trails may be difficult to follow, especially when tree falls obscure the way. In order to get around an impasse without losing the trail, one person should scout ahead to locate the trail on the other side of the obstructing tangle of vegetation, while others remain together on the trail. If it is necessary to leave the trail, mark the way with colored flagging so you can find your way back (but remove the flags as you return). Each person should carry a whistle and, if lost, blow three short blasts, followed by three long blasts, then three short blasts (the international code for SOS). Unfortunately, whistles only work for short distances because sound does not carry very far in the rugged terrain of most rain forests. If lost, the best strategy is to stay in place and wait for help to arrive. Each morning expedition members should inform the leaders where they intend to be on a given day so if they have not returned by 6:00 p.m., the leaders will know where to start a search. Searches will typically start the following morning to avoid the complication of losing additional expedition members in the dark.

In 2000, 36 years after I started collecting plants in the Neotropics, I got seriously lost for the first time. My research assistant, Nathan Smith, and I were collecting at the base of Pic Matécho and became so disoriented that we were forced to spend the night in the forest. Nathan describes our very long night in Box 3-2 (Lost in the Forest). This experience was an unpleasant reminder of the importance of carrying basic survival supplies in daypacks—a light tarp, rope, cigarette lighter, compass, GPS, maps, and extra food. Because visitors to Camp Inselberg at Nouragues have gotten lost and caused great concern, inconvenience, and expense, the station now provides all newcomers with walkie-talkies, transmitters that allow rescuers to pinpoint the location of lost researchers, maps, and first day orientations.

Dangerous waters. Biologists travel many rivers and streams that are interrupted by rapids, and therefore, special caution must be exercised to make sure that destinations are reached during daylight, when dangerous currents are most easily navigated. Sandals (such as Crocs®) or other aquatic shoes should be worn when traveling on turbulent rivers. These allow passengers to leave the boat and help the crew push or pull it through shallow spots or past rapids and waterfalls. If the boat should capsize, aquatic foot gear will protect the feet should a passenger be swept downstream (feet first in order to avoid hitting one's head against boulders). Boots, especially if they are rubber and fill with water, make it difficult to swim and increase the chances of drowning.

Box 3-2. Lost in the Forest
by Nathan Smith

Everyone has stories of being lost; such as the first time they wandered the streets of a major city, or the time when they became turned around while hiking in a forest near home. In a city there are people to help with directions and in many countries the forests are now quite small, divided by roads, major waterways, people, and dwellings. But there are some places in the world where you can walk for weeks in any direction and each day you will only encounter more of the same forest. Here, the towering trees block most of the light and the terrain is dissected by many small streams that look similar to one another. In these forests, even an "experienced" individual can easily become disoriented and, eventually, lost.

Such was the time when Scott Mori and I were working in central French Guiana. I was Scott's research assistant for an expedition creating an inventory of the biodiversity of Pic Matécho, a granitic outcrop surrounded by hundreds of miles of virgin lowland rain forest. The nearest settlement was a three day walk, if you knew the way.

One afternoon, after collecting along the edge of Point de Vue, an enormous cliff, Scott and I packed our equipment and began the 1.5 km trek back to our base camp. This should have taken us 30 minutes and probably took our guide, who left earlier that day, even less time. However during our hike back Scott and I took a wrong turn. By our calculations we had an hour before sunset and we knew we were fairly close to the camp. Unfortunately, we never found the camp that day, or that night.

When Scott and I realized we were lost we began slashing tree trunks, to mark our way, as we headed in "the right direction." We tried using our GPS but the canopy was too dense, we thought of yelling but the forest was too noisy, and we did not have our topographic map of the region. Thirty minutes before dark, we ran directly into a wasp nest that was concealed under a large leaf. Scott, who is allergic to wasps, was stung in the face. Usually one of us would carry a first aid kit with epinephrine, but we did not have a kit that day because we were collecting so close to our camp and decided to pack light. Instantly, Scott's face began to swell and there was nothing we could do. Ten minutes later, feeling anxious and helpless, we gave up and started to prepare for a night in the forest.

We found protection between the buttresses of a tree, made a mat of large fronds from a palm, and began to unpack the few use-

Fig. 3-7. Nathan Smith (left) and Scott Mori (right) spending a miserable, wet night lost in a forest in central French Guiana. They were rescued the next morning by their guide. Drawing by Michael Rothman.

ful things we were carrying. We had little food, water, or extra clothing, and no flashlight. Fortunately, as Scott and I conceded to the darkness, I found a lighter in my pack and we began to collect what dry wood we could find. Within a short time we started a fire and sat down for the night. Scott's face had stopped swelling and it did not appear to be a life-threatening reaction. We knew the fire would keep us warm and the light would keep animals from bothering us, but Scott was still worried about sandflies (the vector of leishmaniasis). Not long after the fire was started it began to rain, hard. Contrary to what many people believe, the lowland rain forests of tropical America can be cool at night, especially if it is raining and you're wet. We both had hats and Scott had a raincoat and nylon plants, but I only had a long-sleeved cotton shirt and cotton army pants. Within minutes, the down pour had soaked both of us and our fire was sputtering and fighting to stay alive. We cut open a plastic collecting bag and used four sticks to spread it over the fire just in time to protect the dying flames. The plastic sheet kept the water away from the fire and probably kept us from getting hypothermia.

As the night passed, I became very sick with dysentery. Appar-

text continues on next page

Box 3-2 continued

ently the thick pea soup I had eaten for lunch that day had turned bad while sitting around after our previous dinner. It was probably one of the most uncomfortable situations I have ever been in. Sick, wet, and cold, I would wander out into the darkness each hour to relieve myself, feeling my way so I did not trip, hoping that I would not surprise a snake or some other animal with my groping hands or stumbling feet. If anything, it was a humbling experience.

Because Scott had the proper clothing, he stayed dryer and warmer than I did and was able to sleep for a few hours that night. I spent most of my time tending the fire, thinking about the previous time I was lost, and remembering how I said if it ever happened again I would be prepared. For what seemed like eternity, I could not forgive myself for being so foolish—in remote places you should always, I thought, carry what you need to survive, and next time I will.

At the first sign of light we were able to find our way back to Point de Vue and shortly after that our guide found us. We were lucky. That morning I remember looking for the place where Scott and I made the wrong turn, but I could not find it. We both silently wondered how we could have gotten lost as we followed our guide on the clearly marked trail back to the camp.

On the outward leg of our hike to Pic Matécho in 2000, the Arataye River was swift but easily crossed by wading with the aid of a pole. On our return, however, rains had transformed the river into a raging torrent that could only be crossed on a log bridge. We elected to cross the log on all fours (Fig. 3-5), and were especially cautious because we realized that falling into a raging river with a heavy pack could easily lead to disaster.

Drowning is one of the most frequent causes of death of tropical explorers. The late João Murça Pires, one of the greatest Amazonian botanists, witnessed the drowning of his colleague George Alexander Black. After bathing in a natural canal leading from a lake to the Amazon River, George tried to scramble up the bank, but slipped back into the canal, which had a rapidly moving current. Murça jumped into the water to help but was not able to reach his friend, who was carried to his death.

On an expedition to Guyana in 1976, Mike, the driver of our Land Rover, violated a cardinal rule by diving head first into unknown waters: the dark and swift Mahaica River. We had been collecting that morning in the savanna near St. Cuthbert, and stopped at a beautiful place along the river where we took a refreshing swim before lunch. Mike asked me to photograph him standing along the bank, after which I turned and started to prepare the morning's collection. I had pressed only a few plants when a young villager informed me that Mike had resurfaced after the dive, then went back under, and did not resurface again. Several hours later we recovered Mike's body 25 meters downstream, wedged under logs at a bend in the river; he had hit his head on the bottom, crushed the front of his skull, and either died from the blow or drowned.

I have always been tense when traveling on rivers with rapids and falls because of the tragic fate of Walter Egler (Cunha, 1989). At the time he died, Dr. Egler—a geographer and botanist—was the Director of the Museum and the Chief of the Department of Botany at the Museu Goeldi, located in Belém, Brazil, at the mouth of the Amazon. In 1961, he and Howard Irwin of The New York Botanical Garden were the leaders of a botanical expedition along the Rio Jari. After nearly a month in the field, the group reached the Macacuara rapids and falls, one of the most dangerous portions of the river, so dangerous that they carried their gear around the falls. After that difficult job, they were elated after reloading their two large canoes above the turbulent waters. Shortly after they pushed off from the shore, the motor of one of the canoes suddenly stopped, and the canoe and its cargo were swept toward a falls with a large drop. Although most of the men were able to jump out of the canoe to safety, Egler remained aboard and was carried over the falls. Two days later his battered body was found five kilometers downstream. His death was especially tragic because, at the age of 34, he still had many years of Amazonian exploration ahead of him.

Once, returning to Cayenne from Nouragues, we narrowly averted disaster on the Approuague River. We had been stationed at Camp Inselberg, an arduous eight kilometer walk from Saut Pararé on the Arataye River, a tributary of the Approuague. We were returning by pirogue to Regina, where a van would be waiting to transport us back to Cayenne. Normally the trip between Saut Pararé and Regina takes five hours; thus, for the boat to make the round-trip journey in daylight, the pirogue must leave Regina early and travel directly to Saut Pararé. The Approuague has some dangerous rapids, and to avoid shooting them after sunset, the pirogue must arrive at Saut Pararé before 1:30 p.m. When we finally reached Saut Pararé at around 3:00 p.m. we knew that we would be traveling at night, but calculated that we would pass the last major rapids before it got too dark. When we started the trip sky was bright blue, but just after nightfall a sudden thunderstorm left us cold, bedraggled, and even farther behind schedule. Nevertheless, after the storm the beauty of traveling down a tropical river at night captivated us; the stars were brilliant and every turn in the river revealed a different picture drawn by the riverside tree canopy against the nighttime sky. This serenity was rudely interrupted, first by Saut Athanase. We marveled at how the boatmen skillfully navigated the rapids, a difficult job even in daylight. Shortly thereafter, we began to hear the roar of Saut Mapaou, the most treacherous rapids on this stretch of the river. Our boatmen guided us through to the point were there was only a single chute left. We entered it with a feeling of euphoria that was precipitously crushed when a wall of water crashed into the pirogue, soaking everyone; then we hit another, and yet another. The pirogue, difficult to manage when filled with so much water, was at the point of capsizing; but by some miracle our guides managed to get the boat into shallow water, and jumped out and pulled it to the shore where we bailed out the water. Exhilarated and thankful that we had avoided disaster, we continued on smooth waters the rest of the way to Regina and then on to Cayenne where we celebrated by eating hamburgers at McDonalds! Yes, you read it correctly, McDonalds in an overseas department of France—what a cultural tragedy! I can only defend our restaurant selection by adding that it was the only one open when we finally reached Cayenne at midnight.

Arachnids

Chiggers. Chiggers are the larvae of several mite species. They are also called red bugs because adults are frequently, and the larvae sometimes, bright red. Chiggers are not dangerous and, at least in the New World, are not known to carry disease, but they can be a source of major discomfort. They attach themselves to reptiles, birds, and mammals—including humans—as the animals move through the vegetation. Weedy pastures and grassy areas around human dwellings, and mossy logs used as a place to sit, are especially good places to pick up chiggers. Once on a host, chiggers penetrate the skin with their piercing mouthparts, but, contrary to popular belief, they do not burrow into the skin. They inject saliva that is rich in proteolytic enzymes; then suck up the resulting "soup" generated from the breakdown of host tissues. Once satiated, chiggers drop from the host and continue their life cycles in the soil, where they feed on insects and insect eggs. Unfortunately, they leave behind the small tubes formed by the host in response to the chigger enzymes. It is these proteinaceous tubes that cause incessant itching and lead to uncontrollable scratching. The scratches may in turn become infected. According to Hogue (1983), chiggers that normally infest reptiles cause a more severe reaction in humans than those that infest other mammals.

On my early trips to the tropics I was invariably plagued by chiggers, which prefer areas where clothing is tightly pressed to the skin; for example under socks at the ankles, waistbands of underwear and bras. My worst case was on Barro Colorado Island in Panama, which, due to high animal density, is infamous for its abundance of chiggers. It was impossible to sleep, and the only way that I found relief was to fill the bathtub with cold water and completely submerse myself. As I have grown older, I seldom get a bad case of chiggers; although I still have less severe reactions.

Chiggers are best avoided by staying away from their preferred habitats mentioned above. One can also tuck pant legs into socks, wrap wide masking tape around the tops of the socks, and dust boots, socks, pant legs, and waist with sulfur powder. A thorough scrubbing with a rough bathing glove and strong soap at the end of a day of field work also helps keep chiggers under control. Ascabiol® (a French product) alleviates the effects of chigger bites, especially if applied at the first signs of itching (but do a preliminary spot test, or risk finding out the hard way that an Ascabiol can induce an allergic reaction worse than the chiggers). If excessive scratching leads to open wounds, they should be treated with an antibiotic such as Neosporin® to prevent infection.

Scorpions. Scorpions can inflict painful and occasionally dangerous stings. Most areas of lowland rain forest have relatively few scorpions, but they are abundant in drier areas, such as the Palo Verde National Park in Costa Rica. The two stings that I received were the result of handling objects with concealed scorpions. The first occurred in Panama when I was preparing a collection of *Gustavia grandibracteata* Croat & S. A. Mori. This beautiful tree in the Brazil nut family has large, terminal rosettes of leaves and bracts subtending tightly congested flowers, fruits, and accumulated debris. The center of the rosette provides an ideal habitat for insects, arachnids, and other animals. The scorpion was hidden among the fruits, and defended its turf when I trimmed the leaves to make a herbarium specimen. I received another sting when I picked up a plant press in our camp along the Bartica/Potaro road in Guyana. Both stings were painful, but after 24 hours I no longer felt their effects. Scorpions also conceal themselves in clothing, shoes, and packs, so I recommend carefully inspecting and shaking these items before putting them on or reaching into them.

Spiders. Spiders are ubiquitous in the tropics, and all have fangs used to inject venom into their prey or enemies. The largest in the world, which may reach 25 centimeters (almost 10 inches) in diameter, is the goliath bird-eating spider, actually the tarantula *Theraphosa blondi* (Latreille, 1804). There are many tarantula species, and in addition to inflicting painful bites, tarantulas defend themselves by brushing clouds of urticating hairs from the dorsal sides of their bodies towards their enemies. If the hairs reach their target, they can cause a great deal of discomfort, especially if they get into the eyes. Hairs that penetrate clothing are impossible to remove, and the clothing should be burned. Although tarantulas elicit fear in most people, it is comforting to know that they attack only when provoked and, even if a person is bitten, the likelihood of a bite doing permanent harm is slight. The best way to avoid spider bites in the tropics is to look carefully before picking things up.

Ticks. Ticks, at all stages of their life cycles, feed on the blood of amphibians, reptiles, birds, and mammals. They wait on stems and leaves of plants for potential hosts, whose presence they detect by exhaled carbon dioxide, body temperature, and even shadows. Ticks attach themselves by a specialized holdfast with downwardly curved teeth called the hypostome, not, as is commonly believed, the head (Hogue, 1993). Daily bathing is essential to remove ticks, which not only cause discomfort, but can carry diseases such as spotted and relapsing fevers. The longer the ticks remain attached, the greater the chance of disease transmission. A rough bathing mitt and strong soap helps get rid of these unwelcome hitchhikers. When small ticks are picked up during the day, they can be removed with tape; but this only works if ticks are spotted before they embed their hypostomes. If that happens, I carefully pull larger ticks out with tweezers, and wipe the area with an alcohol swab. It is important to grasp the tick at the point of attachment and pull slowly, while slightly twisting, to ensure that the hypostome is not left behind to cause infection. My wife Carol Gracie, an expert at removing ticks and botflies, has helped me get rid of many ticks that I could not reach, or were attached in places that I would never ask others to even look at let along remove a tick from. The best way to avoid ticks is to spray field clothing with Permathone®. To avoid tick infestation during sleep, never crawl into a hammock while still dressed in field cloths, especially if that hammock does not belong to you!

Insects

Ants. According to Edward. O. Wilson (1991), there are about 9,000 described species of ants, and 9,000 to 18,000 species remaining to be described. He also estimates that at any given moment there are a million billion ants in the world—they, along with other social insects such as wasps and bees, account for only 2% of the species but 80% of the world's biomass. Many ants sting and/or bite, and can cause considerable discomfort and sometimes even death. The most feared is the lesser giant hunting ant (*Paraponera clavata* [Fabricius, 1775]), with a body over three centimeters long, it is one of the largest ants in the world. In Brazil paraponerine ants are known as the *vinte-quatro formiga* because the pain from their stings can last for 24 hours. They are often seen in small groups foraging for insects, but are much more abundant around their nests at tree bases. They also climb shrubs and trees to harvest nectar from extra-floral nectaries. The single sting I received from a giant hunting ant was inflicted on my leg just after I had started to climb a tree, and the pain lasted for less than an hour.

At the other end of the size spectrum are fire ants, whose bodies are never over five millimeters long. People from the southern United States are familiar with burning stings delivered by the introduced red (*Solenopsis invicta* [Buren, 1972]) and black (*S. richteri* [Forel, 1909]) fire ants. I have been stung countless times after treading upon their ground nests, which are often found in open areas around camps or in villages. Fire ant stings are less frequent when your feet are protected by boots. Although they are more of a discomfort than a danger, they are dangerous to people who are allergic to their venom.

A number of ants have evolved relationships with plants. For example, species of *Azteca* occupy the hollow trunks of *Cecropia*, and other ants inhabit the domatia of species of *Maieta* and *Tococa* and some species of *Piper*, but their stings cause little concern. On the other hand, the hollow stems of species of *Tachigali*, *Triplaris*, and *Pseudobombax munguba* (Mart. & Zucc.) Dugand are home to ants with more painful stings. Some people react to ant bites or stings by producing liquid-filled blisters at the site of the sting.

Bees and wasps. Of all the insects in tropical forests, I make a special effort to avoid the stings and bites of bees and wasps. On a collecting trip to the Mexican state of Oaxaca in 1965, I stepped onto a wasp nest and was stung so severely that my face swelled to the point where my eyes closed. Six years later, when Mike Nee and I were searching for species of the Bra-

Fig. 3-8. Left. A wasp nest of which there are many in tropical rainforests. One needs to be careful where one walks and cuts with a machete because many wasps have painful and even dangerous stings. Right. Scott Mori with a swollen face after being stung several times by wasps. Photos by Carol A. Gracie.

zil nut family in Amazonian Brazil near Manaus, my machete disturbed a nest and wasps swarmed around my head taking vengeance with their stings. The area around each of my five stings swelled up, I broke into a rash, and almost immediately developed a burning fever; in order to cool down, I stripped down to my underwear and jumped into a stream. On the walk back to the car, I felt nauseous, experienced fainting spells, and had blurred vision. Periodically I had to stop and place my head between my legs to clear my head and focus my vision. Mike, who guided me back to the road, said this was the slowest walk of his life. Back in Manaus, Anne Prance administered antihistamines and directed me to my cot, where I slept for six hours. These experiences were the first indications that I was allergic to the venom of at least some species of Hymenoptera.

Since then I have had numerous interactions with wasps and bees (Fig. 3-8). In 1976, while collecting with Brian Boom in French Guiana, I climbed 25 meters into a tree using a "Swiss tree bicycle" to make a collection. The bicycle has two adjustable, circular steel bands attached to a leg-like platform for each foot. On this climb, I first expanded the bands to pass a swollen part of the trunk and then tightened them. To descend the trunk it was necessary to reverse the process and expand the bands again. At the top of my climb, I pulled up my clipper poles and extended the end with the cutting head to a branch of an adjacent tree that I wanted to collect. As I was about to make the cut, a swarm of wasps surrounded my head and stung me several times; knowing that I had no escape, I remained motionless and the wasps retreated. Then, with extreme caution, I again moved the cutting head over to the adjacent tree to make another attempt at cutting the branch. This time I carefully looked around to make sure that I did not disturb the wasps, and turned just in time to see the non-cutting end of the pole poke a nest, causing a black cloud of wasps to leave the nest and fly along the pole towards me. Several stung me, but once again I remained still and they flew away. On my third attempt, I successfully avoided the nest and made a collection of *Buchenavia nitidissima* (Rich.) Alwan & Stace (Combretaceae).

My most serious encounter with wasps was on an Amazonian ecotour. On previous tours we had learned from our guide Moacir Fortes that *Azteca* ants will respond to loud noises by leaving the interior of their large, stalactite-like paper nests en masse. Mo would yell in a progressively louder voice and, with each increase in decibels, more and more ants would swarm out of the nest; a sight that never failed to astonish our travelers. On September 10, 2001, we took our group to a known ant nest and Junior, Mo's son, began yelling to "wake up" the ants... but this time more than just ants were disturbed. A swarm of wasps from an adjacent nest was aroused, and although the wasps attacked Junior, nobody else was stung because we remained motionless while they vented their fury on him. Just as the wasps finally calmed, another member of our party yelled over to ask if the wasps had settled down, and his voice was loud enough to provoke them to attack. Once again I tried to remain motionless, but when one of the wasps crawled under the lens of my glasses I could not resist trying to rid myself of it. As I ripped off my glasses and ran away, I received seven stings around my face and neck. The next morning, the infamous September 11th, I woke up with the left side of my face paralyzed by Bell's Palsy. The major effects lasted for nearly eight months and minor effects, such as a slightly asymmetrical look to my face, persist to this day—nine years after the stings. Although there is no medical evidence that wasp stings bring on Bell's Palsy, I feel that their stings triggered my case of this poorly understood disease.

The largest wasp, and one that packs a vicious sting, is the tarantula hawk (*Pepsis heros* [Fabricus, 1798]). In French Guiana, *Pepsis heros* reaches 4–6 centimeters in length. Females hunt tarantulas to provide for their offspring. They first tease a tarantula from its nest by tapping on the edge of its gossamer lined burrow. Believing that a prey animal is making the vibration, the tarantula comes to the mouth of its nest, is seized by the tarantula hawk, and a fierce battle ensues. If the wasp succeeds in stinging its prey, the venom semi-paralyzes the spider. The tarantula hawk then excavates a burial chamber, places the spider into the chamber, and lays a single egg on the living, but paralyzed, tarantula. After hatching, the larva feeds on the spider while avoiding its vital organs, thereby maintaining a fresh food supply until it pupates. I first learned of tarantula hawks in French Guiana, when Brian Boom and I were drawn to the noise of rattling leaves caused by a battle between a tarantula hawk and a tarantula. Because the tarantula was much larger, we first thought that the spider had captured the wasp; however after several minutes the spider gave up the fight and was dragged away by the wasp. On an expedition to the Rio Falsinho in Amapá, Brazil, tropical ecologist and natural history author David G. Campbell had an encounter with a tarantula hawk (Fig. 3-9) that he describes in Box 3-3.

In the Neotropics, there are two kinds of native social bees, those that sting (bumble bees belonging to the genus *Bombus*) and those with atrophied stingers (stingless bees belonging to the genera *Melipona*, *Trigona*, and *Lestrimellita*). The European honey bee (*Apis mellifera* [Linnaeus, 1758]) is another stinging bee that was introduced into North America by colonists in the 1600s, but did not reach the Neotropics until introduced into Brazil in 1839 (Hogue, 1993). In 1956, the African honey bee (*Apis mellifera scutelata* [Lepeletier, 1836)]), also known as the "killer bee" because allergic reactions caused by numerous stings can be lethal, was brought into Brazil to be hybridized with European honey bees in an attempt to breed a bee that was better adapted to tropical climates. By accident, a number of queens escaped from captivity in Manaus. Subsequently the African honey bee became naturalized and started to spread, reaching French Guiana in 1974 and the southwestern United States by 1990. The African and European honey bees differ physically only in the smaller size of the former; they do, however, differ markedly in their behavior. The African bee is more aggressive, it reacts more rapidly to disturbance at greater distances from the nest, it pursues and attacks victims for much greater distances, it remains agitated for a longer time after disturbance, and it stings in greater numbers. When honey bees sting, the stinger and the venom reservoir are ripped out of the abdomen and the bee dies; thus, the stingers should be scraped from the skin as rapidly as possible to stop further injection of venom. Do not pull directly on stingers because this squeezes the reservoir and forces out more venom. Fortunately, on our expeditions and ecotours I and others with me have never been attacked by African honey bees. To be prepared for bee and wasp stings, I carry analgesics, anti-histamines, and epinephrine (in an auto injector such as the Epipen®) as part of my first aid kit.

Box 3-3. Tarantula Hawk
David G. Campbell

At the Falsinho botanical expedition camp in Amapá, Brazil a disoriented tarantula hawk wasp landed on my chest (I was pressing plants, after dark, under a Coleman lantern) and I jumped up and flicked it off with my hand. It then flew right back at my face, perhaps in anger, but more likely just confused by the light. I brushed it off again and ran away because I understood the wasp's attraction to light and wanted to get away from the lantern to avoid a sting. I've never been stung by a "black jeff"

Fig. 3-9. A tarantula just after it was stung and paralyzed by a Pepsis wasp. The wasp is dragging the tarantula to a "grave" that it has dug for the tarantula. The wasp will lay eggs on the tarantula which will hatch into larvae that consume the still-living but immobile tarantula. Drawing by Naomi Pitcairn.

(that's what they are called in the Bahamas where I grew up), but my dad was stung over his right eye, "terribly painful," he said. The eye swelled shut for a few days.

In fact, I caught quite a few tarantula hawks and kept them in captivity during my childhood years in the Bahamas. The best way to capture them without killing them is with a net, which is then placed in the refrigerator. Once the wasp is chilled, one can remove it from the net and handle it for a few minutes before it wakes up. I used to set up fights between captive wasps and tarantulas in screen-topped aquaria. Sometimes the spider won.

Although not dangerous, stingless bees can be an extreme nuisance as they collect sweat and other body secretions from any available human. On one of my trips to central French Guiana, I climbed a tree to make a collection and, at about 20 meters above ground, I disturbed a nest of stingless trigonid bees in the hollow trunk. They came out of the nest by the thousands and swarmed around my head, getting into my eyes, ears, and nose. I descended the tree as fast as possible and yelled at my companion, Anita Pepper, to refrain from swatting the bees because they release alarm pheromones, which attract others, when killed. Nevertheless she squashed one of them, as I had done while in the tree, and their alarm pheromones attracted hundreds of their colony-mates. We hastily threw our equipment into our packs and ran down the trail to escape the onslaught of bees. After a hundred meters we had left our nemeses behind, so we slowed to a normal pace, but another wave of trigonids soon swarmed down upon us, forcing us to once again run out of their range. We were swarmed one final time before the pheromones dissipated and we no longer attracted bees. The pheromones were apparently so strong that they attracted stingless bees from two other nests after we left the first behind! I had a similar experience while eating lunch with another botanist in Bahia, Brazil. We were sitting at the base of a tree that, unknown to us when we sat down, harbored a trigonid nest. Several bees came out of the nest and my companion swatted one of them, apparently inducing the injured bee to release pheromones that attracted hundreds of bees that formed a cloud around her head, some of which crawled under her clothes. I sat in the midst of the confusion and calmly ate my lunch because I had not been contaminated by pheromones.

Tree climbers should learn to keep their mouths shut. The first time that John Janovec used rope-climbing technology, he ascended a huge specimen of *Couratari macrosperma* A. C. Sm. (a species of the Brazil nut family) to cut a bait branch for one of Amy Berkov's beetle-rearing studies in southeastern Peru. Before climbing, John had repeatedly practiced changing to the descent device (a procedure surprisingly easy to forget at 50 meters). He made the climb and severed a branch but still didn't descend, and Amy was afraid that he was having problems with the descender. When John finally returned to the ground he was making awful choking noises and claiming that a trigonid bee was biting him inside his mouth. Amy was convinced that he must have been stung by some other kind of bee and that the culprit had already departed. In addition, they were nervous about getting back to the station before dark (they had carelessly failed to bring headlamps) so she did not check his mouth to ascertain if a trigonid was there. After returning to camp, John shined a light inside his mouth and Amy confirmed that a trigonid had indeed clamped its mandibles down on his epiglottis, and was still holding tight. He reached right in and tore the bee loose, Amy treated the bite with alcohol and it healed with no problem.

Whiplash beetles. Some female staphylinid beetles have vesicant compounds in their hemolymph (equal to blood in mammals). If they land on people and are inadvertently squashed, the vesicants induce a chemical burn. *Paederus irritans* (Chapin, 1926) is the most common whiplash beetle (Fig. 3-10), and massive outbreaks occur periodically along the Amazon River. The beetles are easily recognized by their elongate bodies, short

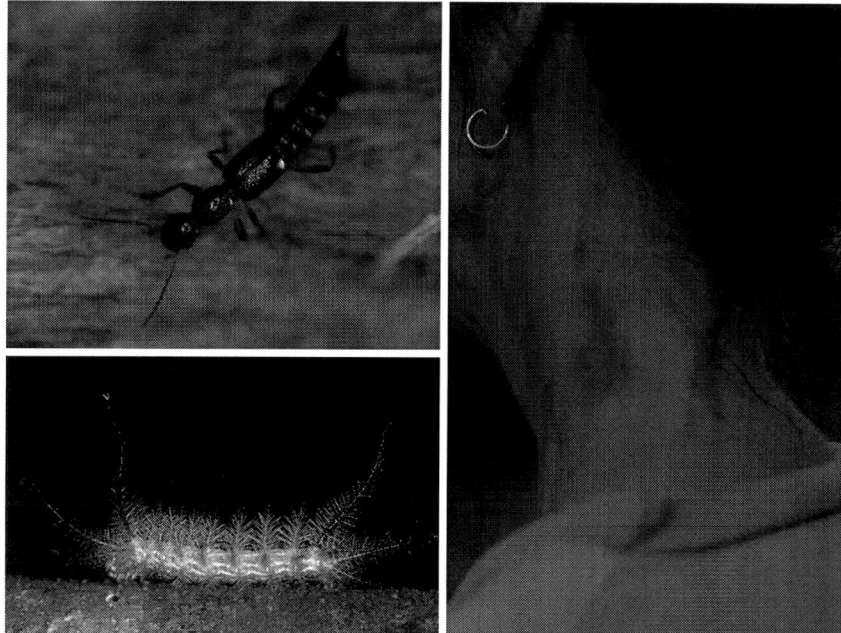

Fig. 3-10. Top left. A whiplash beetle. Bottom left. An example of a caterpillar, many of which have hairs that can be irritating if they are touched. Photos by Carol A. Gracie. Right: Burn on neck cause by a whiplash beetle in Amazonian Brazil. Photo by Scott A. Mori.

elytra that leave several abdominal segments exposed, and heads nearly as large as the prothorax (Fig. 3-10). I have seen many of these beetles on eco-tours to central Amazonia because they often visit lights. When they are common, they frequently land on humans and cause track-like pustules on the skin of unlucky individuals. The only way to avoid this discomfort is to flick the beetle off the skin with a forefinger instead of squashing it in place. Rashes (Fig. 3-10) caused by whiplash beetles should be treated with a topical antibiotic to avoid infection and, in time, will disappear, but if the hemolymph enters an eye, serious damage to sight can occur.

Butterflies and moths. The larvae of some butterflies and moths are covered by hairs that are irritating to the touch (Fig. 3-10). The greatest danger with these caterpillars is to touch one by mistake, or worse, stumble against a tree covered with an entire colony. Some adult moths also possess urticating, barbed scales that are released in self defense—the best-known and most toxic is *Hylesia metabus* (Cramer, 1775). Adult females of this species release the barbed hairs from their abdomens, and upon contact with human skin they cause an itchy rash. In French Guiana, this moth is most common in coastal areas because its larvae feed on the leaves of mangroves, but they can also appear in interior forests. On an expedition with me to central French Guiana, bat biologist Kirsten Bohn developed a severe rash over much of her body. Because we were a long distance from the coast we did not think that this could be *papillionite* (the French name for this infection). Fortunately this took place toward the end of the trip and, upon our return to Cayenne, Kirsten was admitted to the hospital, treated for several days, and released just in time to catch her flight home. Kirsten's case was particularly severe because she had already been covered with mosquito bites, and after exposure to the urticating scales, she developed a secondary staph infection caused from scratching the itches that were driving her crazy. The effects of a *papillionite* rash can be mitigated if a bath is taken as soon as the skin begins to itch.

Burrowing fleas (*Tunga penetrans*). Burrowing fleas are common in the sandy soil surrounding homesteads, especially if domestic animals are present. Female fleas burrow their way into skin on any part of the host, but in humans they are most frequently found under toenails and on the soles of feet, where they cause considerable discomfort. After a female flea is embedded in the skin of a host, a free-living male copulates with her and she swells up into an egg-laying factory as large as a pea. Fleas can be avoided by wearing shoes or boots, but if a burrowing flea does penetrate the skin, it can be easily removed with a sterile needle. Extraction must be done with care because if the flea body is ruptured, a severe immune reaction may occur (Hogue, 1993).

Fig. 3-11. *Flesh-piercing tabanid flies. Note that their beak is nearly as long as their body length. Drawing by Michael Rothman.*

Tabanid flies. Among the most annoying insects in the Neotropics are horseflies belonging to the Tabanidae, a family with more than 1,000 species (Hogue, 1993). Prior to laying eggs, females feed on blood obtained from the punctures they make in the flesh of their victims. Tabanids often occur in great numbers, and different species are active during different seasons, and at different times of day. They have not been shown to carry disease, but it is possible that they carry the eggs of botflies. Their painful bites and annoying swarming behavior are enough to make one take great pleasure in killing these pestiferous tormenters (Fig. 3-11). To avoid bites around the ankles, I usually wear two pairs of socks and tuck my pants into them. Although their beaks are as long as 7 mm, it is difficult for them to penetrate multiple layers of pants and socks. In addition, a hat keeps them out of my hair and a small towel can be used as a "horse's tail" to swat them away. I also avoid wearing blue clothes or using a blue-colored pack, because these pests seem to be attracted by blue. Their larvae are aquatic, so eliminating standing water around habitations reduces adult populations.

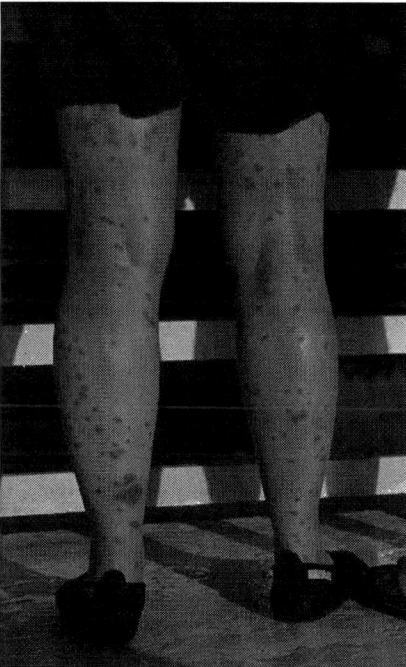

Fig 3-12. Top. Black fly. Bottom. Red welts on legs caused by black fly bites on the Rio Branco, Roraima, Brazil. Photos by Carol A. Gracie.

Black flies (Simuliidae). Black flies (Fig. 3-12) are not only so annoying that they drive people to the brink of insanity, they also transmit serious diseases such as river blindness (onchoseriasis) caused by filiarial nematodes (Katz et al., 1982) and dog heart worm, which also infects humans. Swarms of black flies occur along certain fast-flowing rivers, but not others. For example, on one of our Amazonian ecotours with Moacir Fortes black flies were absent along the Rio Negro, but when we ascended the Rio Branco they were present in such dense swarms that it was impossible to be in the open without long pants, a long-sleeved shirt, and a head net. They were, like other biting flies, less offensive when we were in moving boats. When black flies bite they leave small, red blood spots that cause minor discomfort (Fig 3-12), but when hundreds are present they can cause severe allergic reactions. Fortunately, black fly distributions are limited in time and place, the areas where they carry disease are known, and they are not present after sunset. Although the lower Rio Branco is rich in animal and plant life, we still remember the way black flies tormented us and have never returned with a group of ecotourists!

Botflies (*Dermatobia hominis* [Linnaeus Jr., 1781]). An extreme case of botfly infection is described by Grann (2010) as follows:

> "… Then he woke up to find what looked like worms in his knee and arm. He peered closer. They were maggots growing inside him. He counted fifty around his elbow alone. 'Very painful now and again when they move,' Murray wrote.
>
> Repulsed, he tried, despite Fawcett's warnings, to poison them. He put anything—nicotine, corrosive sublimate, permanganate of potash—inside the wounds and then attempted to pick the worms out with a needle or by squeezing the flesh around them. Some worms died from the poison and started to rot inside him. Others grew as long as an inch and occasionally poked out their heads from his body, like a periscope on a submarine. It was as if his body were being taken over by the kind of tiny creatures he studied [Murray was an entomologist]. His skin smelled putrid…"

Botfly infections are common in the New World tropics, and tree climbers seem to be particularly vulnerable to them but I have never seen them as bad as described in the above quote (Fig. 3-13). My wife, Carol Gracie, does not climb trees and has never been infected with a botfly, whereas she and others have extracted more than 30 botfly larvae from my flesh. Female botflies lay their eggs on the bodies of blood-sucking insects such as mosquitoes or tabanids. When those insects land on a mammal for a blood meal, the botfly eggs are brushed off and, in response to the warmth of the host, hatch into larvae that burrow under the skin. A larva develops for a short time before it opens a small hole in the host's skin to allow it to breathe. The botfly passes through several larval instars in a feeding pocket below the breathing aperture and, after reaching the final instar, it crawls out through the hole and drops to pupate on the ground. The interval from egg to adult fly is slightly longer than two months (Hogue, 1993); for example, a botfly raised to the pupal stage by a mammalogist friend started to emerge after 72 days of feeding on his flesh. He was jogging at the time of its "birth" and hastened his pace to get home in time to show his young son and wife the emerging botfly.

My botflies have initially announced their presence with a slightly raised and itchy spot on the skin, and have occasionally caused a slight but

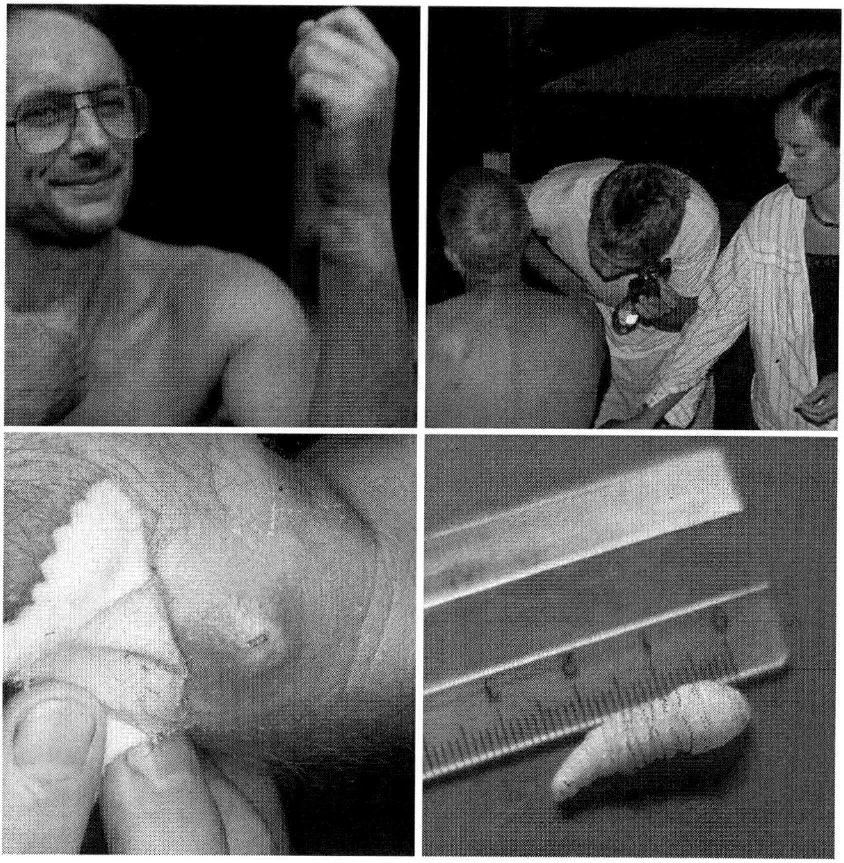

Fig. 3-13. Botflies. Top left. Scott Mori showing an infected bump on his wrist after a botfly had only been partially removed. Photo by Brian Boom. Top right. Students, Alexander MacFarlane and Heather Peckham, planning the strategy to remove a botfly from Scott Mori's upper right shoulder. Photographer unknown. Bottom left. A partially extracted botfly. Photo by Scott A. Mori. Bottom right. A mature botfly extracted from a dog. Note the circular black bands, which are made up of hooks that anchor the botfly in the flesh of the host. Photo by Scott A. Mori.

sharp pain similar to a needle prick. The presence of a botfly maggot can be confirmed by using a hand lens to watch the itchy spot; if there is a larva inside eventually its siphon-like breathing tube will protrude slightly from the hole. The maggots are impossible to extract while they are alive because they possess backwardly pointing spinules that anchor them into the flesh of the host (Fig. 3-13). If the maggots are dead, however, the hooks relax and the larva can be extracted by squeezing around the sides of the aperture, and then gently pulling it out with tweezers after it protrudes. An effective way to kill the larvae is to cover the breathing aperture with several layers of water proof tape or anything else that will stay in place and clog up the breathing hole (e.g., layers of fingernail polish and certain glues, etc.) for at least 24 hours; when the tape or other substance is removed the maggot is frequently stuck to it. The entire larva must be removed to reduce the chance of an infection and, after it is removed, the area should be cleansed and treated with a topical antibiotic.

I have never had a botfly surgically removed, but several of my colleagues have resorted to this method. Douglas Daly, the B.A. Krukoff Curator of Amazonian Botany at The New York Botanical Garden, wrote: "Here's one for the annals of medicine: yesterday I went to three doctors, ending up with the head of the tropical dermatology clinic at Bellevue. She took one look, said "botfly" and went to work. She couldn't drown it, so she shot the area up with an anesthetic, took out a plug of tissue, and started digging with a scalpel. I wasn't too happy as she kept on saying to her assistant "Wow, look how DEEP this is!" At last, she fished a big fat live larva from three cm into my left butt cheek and sewed me up with eight stitches." For an informative and entertaining description of a botfly infection see Forsyth and Miyata (1984).

Parasites

Biological field work in the Neotropics brings with it the risk of becoming infected by parasites such as nematodes (e.g., pinworms, hookworms, giant intestinal worms, and filarial worms), cestodes (e.g., the various kinds of tapeworms), trematodes (e.g., schistosomiasis), and protozoans (e.g., giardia, amoebic dystentry, and malaria). Most parasites are acquired by swallowing cysts or larval stages found in unclean water or contaminated food, by the larvae penetrating unbroken skin, or from insect vectors such as mosquitoes and sandflies. I have been fortunate to avoid most parasites because I chemically treat, filter, or boil dubious sources of water, use long-sleeved shirts and long pants when mosquitoes or sandflies are present, always sleep under a mosquito net, and never walk with bare feet. If I feel listless or have abnormal bowel movements for more than a week after returning from an expedition, I go to a doctor specializing in tropical medicine to be examined for parasites.

Dengue fever. Dengue fever is not caused by a parasite *per se*, but is a viral disease transmitted by the day-flying mosquito *Aedes aegypti* (Linnaeus, 1762). The initial symptom is a high fever that develops five to nine days after being bitten by an infected mosquito. As the disease progresses the fever persists and the following symptoms develop: headaches, pain behind the eyes, joint pain, muscle aches, swollen lymph nodes, weakness, nausea, vomiting, and sometimes a rash begins three to four days after the fever. Not much can be done to treat the disease other than taking medicine to reduce the fever (aspirin is contraindicated because of its anti-coagulant properties), getting plenty of rest, and drinking lots of fluids. The best way to avoid dengue is to avoid mosquito bites by wearing long-sleeved shirts, long pants, and applying repellant to exposed areas of the skin.

Dengue is usually not a deadly disease, but it can incapacitate even the most enthusiastic expedition members. Over the years Michael Kennedy and Ted Gill, university students that accompanied me on separate expeditions to central French Guiana, contracted dengue fever. Ted was confined to his hammock but recuperated in about 10 days. Michael's case was more severe; he was not able to walk the mile from the village to the landing strip, and had to be transported there by car. Once he reached Cayenne, he did not have enough strength to take his scheduled flight back to New York. In most cases, those afflicted completely recover after two weeks. A variant, dengue hemorrhagic fever, is a much more serious disease that can result in fatalities.

Cutaneous larva migrans. Cutaneous larva migrans (caused by several species of *Ancylostoma*) is a minor nematode infection that I have contracted frequently, especially in areas with high monkey densities. Other potential hosts of these nematodes are dogs, cats, and other domestic animals. In humans the larvae penetrate unbroken skin, but are incapable of developing into adult parasites (Katz et al., 1982). They do, however, migrate in subcutaneous tissues and secrete proteolytic enzymes that cause a burning and itching sensation. An infection is easily recognized because in the course of their wanderings the larvae "draw" patterns on the skin, and for this reason the infestation is called "creeping eruption." On a number of occasions I have contracted cutaneous larva migrans, and on one expedition I had numerous infections that occurred in separate places on my chest and arms. None of the non-tree climbing participants on my expeditions have been affected, suggesting that this nematode is frequently carried by arboreal animals, probably monkeys. Topical applications of thiabenzadole readily eliminate the symptoms.

Leishmaniasis. This disease can have serious consequences, as described by David Grann in *The Lost City of Z*, an engaging book describing the perilous Brazilian expeditions of the infamous English explorer Percy

Harrison Fawcett (Grann, 2010 page 134).

> "Next, Costin contracted espundia [= leishmaniasis], an illness with even more frightening symptoms. Caused by a parasite transmitted by sand flies, it destroys the flesh around the mouth, nose, and limbs, as if the person were slowly dissolving. 'It develops into … a mass of leprous corruption,' Fawcett said. In rare instances, it leads to fatal secondary infections. In Costin's case, the disease eventually became so bad, as Nina Fawcett later informed the Royal Geographical Society, that he had 'gone off his rocker.'"

This is the most serious parasite I have contracted during my career as a tropical botanist, and is second only to malaria among protozoan-caused diseases (Claborn, 2010). The vectors of Leismaniasis are flagellate protozoans transmitted in the New World by phlebotamine sandflies of the genus *Lutzomyia* (in the Old World other genera are the vectors). The tiny sandflies, which are one-third the size of a mosquito, are delicate with hairy whitish wings. They are active in the evening and night; during the day they rest, often among the buttresses of trees. Alternate hosts of *Leishmania* include domestic animals, rodents, marsupials, and probably many other mammals. Signs of infection start to appear between two weeks and several months after being bitten by an infested sand fly, and are often mistaken for a bacterial infection (especially if the symptoms appear several months after returning from the tropics). The first manifestation is often a small pimple-like pustule that itches or burns. Most commonly, the pimple increases in size and soon looks like a crater with raised edges (Fig. 3-14); the wound can be moist or dry and encrusted. Most of the lesions in cutaneous leishmaniasis appear on exposed parts of the body such as the head, arms, and legs (Herwaldt et al., 1993). Other manifestations of the disease are widespread pustules over much of the body and disfiguring lesions, especially around the nose, mouth, and ears. Leishmaniasis has three forms: cutaneous, muco-cutaneous, and systemic; each is a progressively worse form. Fortunately, all of the cases that I have seen have been the cutaneous form. Even though cutaneous leishmaniasis may spontaneously disappear it is important to get treatment to eliminate the possibility of more serious problems in the future.

My case of leishmaniasis first appeared as a small, itchy pimple on the lower part of my back, several weeks after returning from an expedition to Amapá, Brazil in 1985. The pimple expanded into a circular, encrusted lesion about four centimeters in diameter, surrounded by a large red area with smaller, satellite, encrusted lesions. The lesions were first diagnosed as a bacterial infection, but when they did not clear up after taking antibiotics, I suspected leishmaniasis and consulted a specialist in

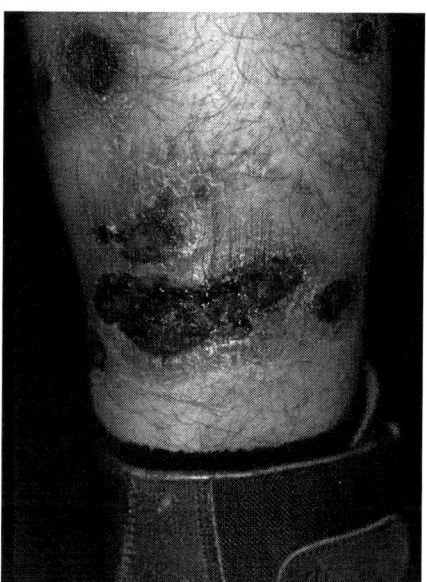

Fig. 3-14. Leishmaniasis lesions on the ankles of a person who slept several nights without netting in a tree canopy in central French Guiana. Photo by Carol A. Gracie.

tropical parasites. The tests based on smears and serology yielded negative results, so he took a tissue plug from the lesion and cultured it, and the culture yielded a colony of flagellated protozoans. When I looked through his microscope, I saw an entire swarm of wiggling protozoans, and was dismayed to realize that I was host to many, many more of them. The positive diagnosis left me with a choice of aggressive treatment, or letting the disease take its course in hopes that it would spontaneously disappear. After being advised about the possible complications caused by the parasites, or of lesions reappearing years after their disappearance, I choose the aggressive treatment.

The parasitologist turned me over to medical doctors, who prescribed twenty consecutive days of intravenous injections of Stilbogluconate Sodium (Pentostam). Other countries, for example Brazil and French Guiana, have shorter and less expensive treatments with drugs not approved for use in the United States; thus there are advantages to being treated in areas where leishmaniasis is endemic. Although Pentosam has been used in other countries for many years, it is of minimal commercial interest in the U.S. and pharmaceutical companies have not obtained Food and Drug Administration approval for its distribution. The Center for Disease Control provides Pentostam free-of-charge, with the provision that there is close monitoring of its administration, and, therefore I was required to spend a week in the hospital to make sure that I did not have an adverse reaction to the drug. I received the remaining injections as an outpatient. With each injection, I developed more severe side effects, including nausea, fatigue, and aching joints, and by the end of the treatment, I could barely lift my arms to put on a shirt. But the leishmaniasis lesions started to disappear shortly after the end of the injections.

One of my expedition members suffered a much worse case. In exchange for a trip to the tropics, a professional tree climber agreed to climb trees and set up Amy Berkov's study of cerambycid beetles and their host plants (see Chapter 2). Because of his skill in climbing even the most difficult trees with ropes, the climber soon finished his work with Amy and dedicated himself to helping with general collecting and learning more about tropical rain forests. One of his goals was to sleep in the canopy, so he secured a sleeping platform in a tall *Caryocar* (Caryocaraceae) tree and spent two consecutive nights 30 meters from the ground. He planned to observe birds and other animals coming to the fruits of a hemiepiphytic *Clusia* (Clusiaceae) shrub growing in a tree next to the *Caryocar*. Amy's climber was bitten by insects while on this precarious perch, but thought little of it when he left French Guiana at the end of September to return to his tree maintenance business. By the middle of October, he was concerned about several sores on his hands and ankles that did not heal; by the end of October the lesions were so severe that his socks stuck to his ankles and had to be removed by soaking them in the shower. By the time I finally saw him again in December, he had developed 35 lesions on his face, hands, lower legs, and ankles; I was so shocked by their severity that I immediately took him to see a tropical parasitologist. The climber underwent the Pentostam treatment and the lesions eventually cleared. When Amy herself developed leishmaniasis several months later, she had the good fortune to be treated at the Saül dispensaire with two consecutive shots on a single day.

Malaria. This mosquito-borne disease is still a major health problem in the tropics of the world; almost 250 million cases were reported in 2008, including fatal infections of 1,000,000 children (Hurtley et al., 2010). Some vectors have become resistant to anti-malarials (e.g., chloroquine, fansadar, and artemesisinin-based therapies) (Enserink, 2010), and vaccine treatments have only been partially effective (Vogel, 2010). The most effective control is still the use of insecticide-impregnated bed nets (Roberts, 2010). On the positive side, malaria has been eliminated from some areas and there is some hope that, with aggressive programs of control, it can be eliminated in others and eventually be completely eradicated throughout the world (Roberts, 2010).

Malaria, caused by protozoans in the genus *Plasmodium* and transmitted by night-flying *Anopheles* mosquitoes, is the most deadly and debilitating disease of the tropics. This disease can appear as early as seven to nine days, and as late as a year after being bitten by an infected mosquito. A classic case of malaria goes through a cold stage characterized by fever, shaking, and chills; a hot stage caused by a high fever, accompanied by a headache, nausea, vomiting, dizziness, pain, and delirium; and a sweating stage characterized by a fall in body temperature.

Before visiting new areas, information about the presence of malaria should be obtained from someone who knows the region and, if recent cases have been reported, anti-malarials recommended by the Center for Disease Control in Atlanta, Georgia (http://www.cdc.gov/) should be taken before, during, and after the trip. Special care is needed when sleeping in the vicinity of strangers, for example around villages and gold mining camps, because the inhabitants may be carriers. In regions dominated by acidic black water, mosquitoes are less abundant and malaria is not as much of a problem as it is in places with alkaline waters. In eastern Amazonia, just north of the Amazon River near Macapá, hordes of mosquitoes appear for about one-half hour at dusk. This cloud of mosquitoes is so predictable that local people call this time of day the *reforço*. When that time approached, I was always safely tucked under a mosquito net until the *reforço* stopped. Our camp in this area was at a farmer's home and at least one family member had malaria when we were there, so it was especially important for us to avoid mosquito bites.

I never took anti-malarials when exploring for plants in central French Guiana because relatively few cases of malaria had been reported there. Nobody on my expeditions or ecotours has contracted malaria, most likely because we all sleep under mosquito nets, even in hotels if mosquitoes are present. Those malaria-free days are now over in Saül, where most villagers now experience recurring malaria, presumably introduced by the influx of Brazilian gold miners.

Other biologists have been less lucky. The renowned botanical explorer Ghillean T. Prance, and all but two members of his expedition, contracted malaria while exploring the headwaters of the Purus River in Amazonian Brazil. Everyone on the trip was taking anti-malarials, but, unknown to Prance, mosquitoes in this region harbored a resistant strain. The presence of resistant malaria was probably why he couldn't find guides to travel on the upper Purus, yet no one told him about the problem— so even when they got sick they did not realize what was causing their symptoms. Malaria first struck a Ph. D. student who was sent downriver to Manaus where he could get the best possible medical attention. Next, several days before embarking on a five day trek across the divide between the Purus and Madeira Rivers, Iain started to feel sick. Because the party was short on boats and it seemed easier to cross the divide to the Madeira watershed than to retrace their path by descending the Purus, they decided to continue as planned and portage their equipment across the divide, then seek help in Pôrto Velho on the Madeira River. After a grueling day-long hike, difficult even for a healthy person, they spent the first night in the home of a *caboclo*. The following day they obtained a mule and, because of Iain's rapidly deteriorating health, he went ahead with a Brazilian field assistant. The rest of the party planned to follow at a slower pace because they were burdened with the expedition's equipment and plant collections. Shortly after Iain and his field assistant left the main party, the assistant also came down with malaria and was so sick that Iain gave him his place on the mule. By some miracle, they found their way to Pôrto Velho and flew directly to Manaus. Iain was soon diagnosed and treated for malaria, and it looked like the field assistant, who had returned to his home in Belém for treatment, also survived (Langmead, 1995).

Snakes

Poisonous snakes are always in the back of the mind of even the most experienced biologists. The rattlesnake (*Crotalus durissus* [Linnaeus, 1758]) normally does not frequent rain forest, but is found in savannahs. Coral snakes can be deadly, but most bites come from handling them; therefore biologists with no reason to touch snakes are not likely to be bitten. The true coral snake (*Micrurus surinamensis* [Cuvier, 1817]) is brightly patterned with red, black, and white bands, but other coral snakes with different banding patterns are also dangerous. Although the false coral snake (*Oxyrhopus petola* [Linnaeus, 1758]) can be handled without risk, even herpetologists may have difficulty distinguishing it from the true coral snake; thus no snakes, especially brightly colored ones with alternating bands of color, should be handled.

The *fer-de-lance* (*Bothrops atrox* [Linnaeus, 1758]), bushmaster (*Lachesis muta* [Linnaeus, 1766]), and other vipers are the most dangerous snakes found in neotropical rain forests, and they account for most of the bites reported each year. These vipers are easily identified by their triangular heads and heat-sensing pits located between their eyes and nostrils. They hunt for rodents, frogs, lizards, birds, and other snakes by waiting for prey to pass in front of them. Viper venom immobilizes or kills prey by breaking down tissues and causing bleeding from the blood vessels; thus symptoms of a severe viper bite include swelling and internal bleeding. Although the actual strike may be relatively pain free, with time the puncture site becomes very painful. The area surrounding the wound begins to swell and, in 25% of the cases (Chippaux et al., 1988), swelling is followed by local hemorrhaging. If the bite is severe, blood begins to appear in the victim's urine and saliva.

The *fer-de-lance* and the bushmaster are ever present dangers, but they are seldom seen because they occur in low densities, are mostly

Fig. 3-15. A species of fer-de-lance, probably Bothrops brasiliensis. The camouflage of most poisonous snakes in the Neotropics makes them very difficult to see. Drawing by Michael Rothman.

nocturnal, and blend in with the background because of their camouflage (Fig. 3-15). The *fer-de-lance* is more common around human habitations, where rodent populations tend to be high. Poisonous snake densities also increase along streams during the dry season, because the vipers wait for prey to come to drink. The bushmaster is particularly dangerous to humans because of its larger size and greater quantity of venom. In addition, a viper that has not recently taken prey accumulates a maximum amount of venom, thereby making its bite more potent. Although these snakes are not usually aggressive toward humans, they are more likely to strike during their mating season, and when they are more densely aggregated along water courses during the dry season. Snake bites occur most frequently when the victim steps on a snake, or disturbs it while picking something up from the ground. The chances of getting bitten by vipers can be reduced by not stepping blindly into patches of vegetation, or reaching down to pick up fruits, flowers, or other objects without inspecting the ground carefully. One should also look on the other side of a log before stepping over it.

Some poisonous snakes, such as the two-striped forest pit viper (*Bothrops* [*Bothriopsis*] *bilineata* [Amaral, 1930]) and Schlegel's viper (*Bothriechis* [*Bothrops*] *schlegeli* [Berthold, 1846]), are arboreal and sometimes occur on low branches. Fortunately, these snakes are not aggressive and usually strike only when threatened, for instance if someone brushes against the branch.

Guy Bourdot, a French tour guide working out of the village of Saül, was searching for a lost tourist when he stepped on a *fer-de-lance* that struck at the back of his foot. One fang entered his heel and the other harmlessly penetrated his shoe. Guy initially thought that he had kicked a palm spine into his heel, and did not feel any significant pain for about 15 minutes. After discovering the bite, he walked slowly for two hours to reach the village, only to find out that there was no anti-venom on hand. He decided to tough it out in Saül rather than fly to Cayenne for treatment. After three days Guy began to feel better, but for the first week he was confined to bed with his swollen leg colored mostly black, but also displaying shades of green, blue, and red. During the second week he could stand for short periods, by the third week he was able to walk, and after two months he no longer felt any ill effects from the bite.

During 45 years of working in the Neotropics, I have had few close encounters with poisonous snakes, but feel that I have been lucky. One of the most alarming features of vipers is that they are extremely difficult to see. On several occasions a colleague has pointed out a nearby poisonous snake that I was able to see only after careful scrutiny. One near-miss occurred on an expedition to the *tepui* area of Venezuela when I was collecting with Mario, a Venezuelan mycologist. I spotted a polypore mushroom in the genus *Amauroderma* and, instead of collecting it, called for Mario to see it *in situ*. As he bent down to pick up the mushroom, he perceived the slight movement of a *fer-de-lance* coiled around its base and carefully pulled his hand back before the snake could strike.

None of the several hundred people that have accompanied me on expeditions or ecotours to the Neotropics have had a close encounter with a snake. Nevertheless, biologists are sometimes bitten. Robert K. Colwell (1985) wrote a spell-binding account of his life-threatening bite by a *fer-de-lance*. Colwell's snake bite was the first among faculty and students of the Organization for Tropical Studies, after 450,000 man hours of field work in Costa Rica.

Because of this low but ever-present danger, biologists need to be prepared for a snake bite and information about their bites should be discussed at the beginning of each expedition. Anti-serum can be purchased from the Instituto Clodomiro Picado in San José, Costa Rica and the Instituto Butantan in São Paulo, Brazil, as well as in pharmacies in many neotropical countries. The directions for testing for hypersensitivity to the anti-serum and the protocol for injecting it into the victim should be reviewed at the beginning of each expedition, especially on trips into remote areas. If a bite does occur, Chippaux et al. (1988) recommends that the following protocol be followed:

1. Try to kill the snake so that it can be identified and the correct anti-serum administered if it becomes necessary. This is very important because the venoms of snakes are different, and the anti-serum that works for one snake may not work for another. For example, viper venom is a hemotoxin while that of the coral snake is a neurotoxin, and each is treated with a different type of anti-serum.
2. The victim should remain as quiet as possible and the limb that has received the bite should remain immobile. If the victim must be moved back to the camp, do this by calmly walking. One person should remain with the victim, and another should go ahead to the camp to start making plans in case evacuation to a hospital is needed.
3. Clean the wound with a topical antibiotic because snake fangs are usually contaminated by bacteria.
4. Do not apply tourniquets or cut into the wounds. The only first aid that should be applied at the scene of the snake bite is aspiration of the wound with a commercially available rubber suction tube. If this is done immediately, some of the venom might be removed, thereby lessening the severity of the bite.
5. If signs of venom poisoning occur, and the victim can be transported to a hospital within three hours, further treatment should be directed only by a physician.
6. If professional help can not be reached within three hours, then anti-serum should be injected, but only after the patient has been tested for allergy. The directions that come with the anti-serum should be followed closely, and the injection should be made into a large muscle such as the *gluteus maximus*, but not near the bite, because this can decrease circulation around the wound.

Although the folk remedy of using electric shock treatment to treat poisonous snake bites has been touted as effective, its efficacy has not been scientifically demonstrated. On our earlier expeditions and ecotours, we carried an electric stun gun and our demonstrations of its use never failed to terrorize some of our travelers.

Fortunately there are relatively few snake bites among biologists, and many of the bites that do occur are venom-free because the snake had either recently killed prey and its venom had not yet been replenished, or because the snake did not happen to inject venom at the time of the bite. Permanent damage caused by a snake bite is rare and deaths occur infrequently—but both do occur!

Mammals

Among mammals, jaguars, pumas, peccaries, monkeys—and other people—can either be a nuisance or even dangerous to biologists working in rain forests.

Jaguars. Because of its ability to kill prey with a single crushing bite to the skull, the jaguar (*Panthera onca* [Linnaeus, 1758]) is the most feared of neotropical mammals. It is the largest New World cat and, among the world's cats, is second in size only to the lion and tiger. The closest I have come to observing a jaguar in the wild was in French Guiana when Brian Boom and I were sitting on a log eating lunch. We were looking in different directions and Brian spotted a jaguar passing in front of him, but in the instant it took to turn my head it had disappeared. While spending the night in a French Guiana *carbet*, my wife and I once heard sounds of a struggle in the adjacent stream that we surmised was probably a jaguar killing a caiman. A few days later we found a dead caiman with its belly slashed open upstream from our water source. I probably haven't seen a jaguar in the wild because I make such a commotion collecting plants; however, others on my expeditions have come face-to-face with jaguars.

In 1982 Jef Boeke, now a prominent molecular biologist, spent time with me in French Guiana. Jef, a fearless tree climber with an intense curiosity about natural history, had joined our expedition to help document pollination of Lecythidaceae (Mori & Boeke, 1987). On several occasions, Jef slept in a small *carbet* in the forest on Antenne Nord de la Fumée (five kilometers from Saül) so he could observe pollinators early in the morning. It was easier for him to sleep near the trees than to spend 90 minutes walking to them while it was still dark. Alone in the camp, Jef climbed into his hammock early. Later that night he heard hoarse cough-like sounds coming from the darkness surrounding the camp. He looked from his hammock and spotted two reddish eyes looking in his direction; curious, he turned on his flashlight and the light illuminated a jaguar, which moved quickly into the darkness. With only a mosquito net separating him from the jaguar's powerful jaws, Jef barricaded his hammock with the clipper poles, his climbing equipment, and anything else he could lay his hands on, and then spent the longest night of his life, waiting for daybreak.

Pierre-Frank Dubernat, a gold prospector and author of *Les Solitaires Crèvent Seuls* (Dubernat, 1982), has lived for years at an isolated homestead called *Certitude*, 35 kilometers northeast of Saül. Pierre-Frank spent countless hours walking between the village, where he had a house, and his forest homestead. He told me that a jaguar once attacked him from behind and injured his leg. On another occasion he was fording a stream when a female jaguar circled and then jumped towards him; according to Dubernat, he raised his gun and killed the cat in mid-air. On several of our ecotours to the Amazon, a *caboclo* related the tale of how he killed a jaguar with his machete after it had attacked his dogs. Although he told the same story and displayed the jaguar skin to many visitors, each time he recounted his adventure with great enthusiasm and detail, and that made him a local celebrity and an attraction for the few tourists that passed his way.

Jaguars pose much less of a threat to humans than humans pose for them. On my first trip up the Amazon from Belém to Manaus in 1971, I saw two black jaguars in a cage waiting to be shipped to a zoo. At that time, there was also an active and legal trade in jaguar pelts. Tio Romeu, the most experienced *mateiro* (woodsman) in our study of Lecythidaceae for the Biological Dynamics of Forest Fragments Project, had previously supplemented his income by hunting jaguars, otters, and caimans for their skins. One of the techniques that Romeu used to capture jaguars was called *tronqueira*. In this method, he used treble fish hooks attached to a chain; one hook baited with monkey meat was placed lower than the other. After shooting a monkey for bait, Romeu smoked it, and than dragged the carcass over the ground to the *tronqueira*, which was suspended several meters from the ground. When a jaguar came into the vicinity, it would follow the scent to the meat, leap up to tear it lose, and become hooked in the mouth. In an attempt to dislodge the hook, the cat would slap at the chain and get its paws caught on the upper hook. When Romeu came to inspect his trap, he would sometimes find a jaguar hooked in its mouth and front legs and spinning around on its hind legs.

Pumas. The puma (*Felis concolor* [Linnaeus, 1771]) is the second largest cat in the Neotropics and, other than man, has the most-wide spread distribution of any mammal in the western hemisphere, ranging from Canada to the southern tip of South America. It occupies habitats ranging from lowland rain forest to deserts and high elevation coniferous forests in western North America. Once in Suriname, I was sitting and measuring fruits from a tree of *Couratari stellata* A. C. Sm., a member of the Brazil nut family, when I heard an agouti make an abrupt bark. I turned in time to see a puma disappearing into the undergrowth, and assumed that it had been stalking the agouti. There are numerous accounts of pumas attacking and killing humans in the western United States, as well as stories of pumas threatening man in the tropics. On one of botanist Jean-Jacques de Granville's expeditions in French Guiana, a puma jumped from a tree, knocked his field assistant to the ground, and immediately ran away. They did not know if this was an attack, or an attempt by the puma to escape the presence of humans. Pierre-Frank Dubernat also told me of a 16 year old boy in French Guiana who was killed by a puma that, having been previously wounded by hunters, could no longer hunt its normal prey.

Peccaries. Throughout most of lowland Central and South America, there are two species of peccary (Fig. 3-16), the collared (*Pecari tajacu* [Linnaeus, 1758]) and the white-lipped (*Tayassu pecari* [Link, 1795] [Link, 1795]) (Haemig, 2002). Although collared and white-lipped peccaries resemble pigs, they are placed in the family Tayassuidae along with the Chacoan peccary (*Catagonus wagneri* [Rusconi, 1930]), while the 16 species of true pigs are in the family Suidae. Peccaries are found only in the New World while pigs were formerly restricted to the Old World (Sowals, 1984).

The collared peccary is smaller (17–30 kilograms) than the white-lipped (25–40 kilograms), and can also be recognized by the collar of pale yellow hairs around its neck, while the white-lipped peccary has white patches on its cheeks. (Emmons, 1990; Box 3-4). These peccaries also have very different behaviors. The collared peccary inhabits a restricted home range while the white-lipped is wide-ranging; the former travels in small bands of several to fewer than 10 individuals whereas the latter travels in bands of up to 300. The collared peccary feeds by digging up rhizomes and tubers, and has a relatively weak bite force; in contrast, the white-lipped peccary uses its snout and tusks as a kind of plow to push through the upper few centimeters of soil and litter to search for food, and has a strong bite that enables it to crush a wider variety of hard fruits and seeds. All peccaries have scent glands, one below each eye, and another on the back a few centimeters above the tail. They rub their scent glands on plants, as well as against the dorsal scent glands of other individuals. Secretions

Box 3-4. *Tayassu pecari*
    by Naomi Pitcairn

On the day that my story takes place, Tatyana Lobova, Scott Mori, and I were out trudging through Amazonian mud. We were on an expedition based at the Los Amigos research station in Madre de Dios, Peru where we were collecting plants dispersed by bats and studying the natural history of the Brazil nut family. We had been walking for hours already without seeing much besides tree trunks, lianas, and mud.

We, or at least I, were a little bored. That is, until we began to hear a noise far off in the distance. When it first started up, we barely even heard it. But it kept getting louder until nobody could help but notice. It was a thunderous, reverberating, bellowing kind of sound. After a few seconds, we figured out that it couldn't be what it sounded like: an army of tanks or a fleet of bulldozers. We were a long way from roads. Earthquake? I, who had never experienced an earthquake, considered the possibility. But it wasn't.

What was moving towards us was loud, roaring and ground-shaking, but not any kind of vehicle, this was pure animal. Lots and lots of animals. "Pigs," I said (note: actually peccaries, which are not really pigs exactly but I am not a scientist and at this particular time nobody attempted to correct me.)

Now Scott, in a noble effort to protect us, said that he would go alone towards this hidden, but apparently massive threat. Tatyana and I were strongly in favor of everyone running together in the other direction. It was two against one, so we ran.

"Get up high, get up on something high," advised Scott. We were now running full speed in the direction we had just come from, but where were the seemingly endless fallen trees we climbed over and under along our way? They were now nowhere in sight. We finally saw one flimsy fallen tree but that would barely hold one person. The others offered it to me, and selfishly, I took it. They were forced to take refuge on a lower tree trunk. The noise raged toward us.

Some pale-winged trumpeter birds charged out from the underbrush, squawking in panic. We looked at each other and started to laugh. Had we been so silly as to be frightened by birds even more scared than we were? But wait, what were they so damned scared of? The noise got louder, and was now accompanied by a powerful odor. What does a herd of white-lipped peccary smell like? A herd of white-lipped peccary

text continues on next page

Box 3-4 continued

Fig. 3-16. White-lipped (left) and collared (right) peccaries drawn to scale. The author and his colleagues N. Pitcairn and T. Lobova were forced to run and climb onto the fallen tree trunks of a forest gap to escape a large band of white-lipped peccaries. Drawing by Naomi Pitcairn.

smells like shit or like old garbage left to rot. It is a nasal assault.

Soon we began to catch glimpses of ghostlike black forms through the trees. The rumble split into a cacophony of individual squeaks and grunts, breaking sticks and hoof-falls. White lipped-peccary are known to travel in herds of up to 100 or more individuals. We couldn't count but it seemed to be about that many. Maybe I exaggerate, but they just keep coming and coming and coming.

Finally, after what seemed like a long time but really wasn't, one broke through the vegetation. He (I think of him as a "he") was dark brown, looked about two feet high, and was covered with bristly hair. He poked his snout out from the dense vegetation, looked up at us and immediately began squealing like, well, like a stuck pig. Wheee! Wheee! Wheee!

The whole herd joined the panic and started running the other way. Wheee! Wheee! Wheee! Snort. Grunt. Tooth-clack. Thunder. Rumble. Wheeeeeeeeeeee! After about 15 minutes, maybe a lot less, they all finally seemed to be gone.

Now that we were safe, I immediately wished they would come back so I could get a better picture. We cautiously crept down off of our perches and started back along the trail in our original direction. We walked over peccary carnage. The earth was pocked with divots and gouges scooped out by peccary snouts. Everything, including the rare plants we hope to find, had been uprooted. But, I was no longer the least bit bored.

from the glands probably mark territories and facilitate group recognition, making it easier for the band to stay together. A band of peccaries moving through an area leaves a characteristic pungent odor. Collared peccaries are less dangerous than white-lipped peccaries, and almost always scurry away when they perceive humans. In contrast, the large bands of white-lipped peccaries can be quite aggressive. When alarmed, they raise their hackles and clack their tusks; thus appearing even larger and more dangerous than they really are. Though there are many stories of field biologists climbing trees to escape white-lipped peccaries, I have been "treed" only once (Box 3-4).

Monkeys. Although usually harmless, neotropical primates, especially spider [*Ateles* species (Linnaeus, 1758)], capuchin (*Cebus* species), and howler (*Aloutta* species) monkeys, sometimes attempt to intimidate humans by throwing sticks, or defecating and urinating on them. In most unprotected areas, the over-hunted spider monkeys are seldom seen. In reserves such as the Nouragues Nature Park in French Guiana, monkeys in general, and spider monkeys in particular, are common.

On one occasion a troop of spider monkeys (*Ateles paniscus* [Linnaeus, 1758]) became irritated when we watched them with binoculars, and pursued us for at least 20 minutes as we ran along the trail in an attempt to escape the debris they were throwing down at us. We finally decided to stop running and remain still to see if the monkeys would lose interest in tormenting us. To my surprise, two came to the tree directly above me and started to shake branches and toss down debris. At Nouragues, during the dry season when fruit is scarce, I had another encounter with spider monkeys in their arboreal habitat after I had climbed up a tree of *Pouteria venosa* (Mart.) Baehni (Sapotaceae) to take specimens of immature fruit. Hearing the commotion of my climb, a nearby troop approached the tree, and two came into the crown four or five meters above my perch and started to threaten me by shaking branches. When I poked my aluminum clipper pole at them they lost their courage and went away. I believe that the monkeys realized that I was competing for fruit that they were planning to eat, and wanted to drive me away.

My most contentious encounter with monkeys was with white-faced capuchins [*Cebus capuchinus* (Linnaeus, 1758)] on the Osa Peninsula in Costa Rica (Fig. 3-17). I had just returned with members of an ecotour group from an excursion to the Isla del Caño, and as we walked from the beach to the lodge we spotted a troop of about 50 capuchins. We followed them for a half hour as they foraged in the trees along the path,

Fig. 3-17. White-faced capuchin monkeys (Cebus capuchinus (Linnaeus, 1758) on the Osa Peninusula, Costa Rica. This monkey becomes a nasty pest when agitated. Drawing by Naomi Pitcairn.

and were so fascinated observing the monkeys at such close range that we did not notice that they were becoming less patient with our presence. Some of them started to jump up and down on branches, and others approached within several meters, baring their teeth and chattering at us in an agitated manner. It was clear that the monkeys were completely fed up with our invasion of their territory. One suddenly lunged at eco-tourist Maggie Carson, and grabbed her by the shirt; she turned and started to run with the monkey in hot pursuit. I grabbed another eco-tourist's walking stick and wielded it like a sword, which caused the monkeys to retreat just out of the reach of my furious swings enabling us to pass them and return to the lodge for our late afternoon drinks.

Humans. The most dangerous mammals in tropical forests are not peccaries or large cats, but humans. In an article about how botanists died while conducting field work, Stewart (1984) reports that 16 of 56 cases were due to encounters with belligerent people.

In spite of spending 45 years in the tropics, including a visit to the home of some of the last indigenous peoples still living in isolation in the Amazon (now the Yasuni National Park in Ecuador), I have never had a close encounter with Amerindians. Fortunately, I have never seen a shrunken head hanging in an Amerindian hut or witnessed a tribe consuming human flesh as described by Grann (2010):

> "Human meat was typically prepared in two ways: roasted or boiled. The Guayaki who practiced ritualistic cannibalism when members of the tribe died, cut bodies into quarters with a bamboo knife, severing the head and the limbs from the trunk. 'The head and the intestines are not treated according to the same "recipe" as the muscular parts or the internal organs,' explained the anthropologist Pierre Clastres, who spent time studying the tribe in the early 1960s. 'The head is first carefully shaved...then boiled, as are the intestines, in ceramic cooking pots. Regarding the meat proper and the internal organs, they are placed on a large wooden grill under which a fire is lit...The meat is roasted slowly and the fat released by the heat is absorbed gradually with the koto [brush]. When the meat is considered "done" it is divided among those present. Whatever is not eaten on the spot is set aside in the women's baskets and used as nutriment the next day. As far as the bones are concerned, they are broken and their marrow, of which the women are particularly fond, is sucked.' The Guayaki's preference for human skin is the reason that they call themselves Aché Kyravwa—'Guayaki Eaters of Human Fat.'"

See Clastres (1998) for an English translation of the original French version of this book, and Brown (2000) for a critique of the book. The latter concludes: "Published in 1998, three decades after the original fieldwork was done and twenty-six years after its initial French publication, this book is divorced from the context in which it was originally conceived and written. Because of its controversial narrative style, the absence of an editorial introduction or preface to [sic] this volume will likely inhibit its acceptance in academic circles."

Today the threat of cannibalism and head shrinking is essentially non-existent. By the time I started my career, Amerindians were confined to reservations or had taken up the ways of European colonists. My negative encounters with other humans have been limited considering the many years I have travelled in remote areas in the Neotropics. I have, however, been robbed on several occasions in larger cities and, as a result, am extremely careful when walking there at night or when I am taking public transportation.

My most harrowing experience happened in 1971 when Michael Nee and I traveled throughout Central America and northern South America studying trees of the Brazil nut family (see Chapter 1), especially the genus *Gustavia*, the topic of my Ph.D. dissertation. Mike had agreed to travel with me in exchange for being able to record our collections in his number series because he was, and still is, an inveterate botanical collector. In the history of The New York Botanical Garden, Nee's 57,173 (as of 17 July 2010) collections are second only to those made by the late Bassett Maguire and his associates (65,172 fide his field books). Close behind is William R. Buck with 56,435 collection numbers (as of 17 July 2010). Nee and Maguire collected mostly flowering plants and Buck gathers mostly mosses, their relatives, and lichens.)

My study of *Gustavia* herbarium specimens had drawn my attention to a series of specimens from the Magdalena Valley that I recognized as a species new to science; thus, Mike and I set out to visit Campo Capote, a joint Colombian/German lumbering operation located southeast of the Magdalena River between Puerto Berrio and Barrancabermeja (see Chapter 1 and Box 3-5). Because of *guerrillero* activity in the vicinity, Mike and I were forced to make two trips to the area, one in March and the other in July, 1971. Our efforts were rewarded as we collected flowering material of the new species, described in 1975 as *Gustavia romeroi* S.A. Mori & García-Barr, in memory of the great Colombian botanist, Rafael Romero-Castañeda. Excerpts from Mike's diary describe how we avoided contact with guerrillos in our search for specimens (Box 3-5). To this day,

---

Box 3-5. *Guerilleros,* Campo Capote, and *Gustavia*

by Michael Nee

March 25, 1971. We (Michael Nee and Scott Mori) stayed in Puerto Berrio overnight after a terrible train ride from Medellín. The next morning we waited for the train going toward Barrancabermeja and it finally came around noon. This train soon got us to "Las Montoyas," a small group of houses where the train made a stop. At Las Montoyas we waited a while and finally got a ride in a rickety old truck that made more-or-less regular runs back and forth to Campo Capote. The driver and a couple of passengers were talking excitedly, but we didn't find out till later what it was all about. The camp itself (the big sawmill, worker's houses, etc.) is at 6° 38' N 73° 55' W.

March 26. We had spent a long hot day (botanizing in the forest) with nothing to eat or drink until we finally got back about 3:00 p.m. It was then we found that all the excitement the day before was because four Germans and five Colombians had been kidnapped by a left-wing guerilla band. They had been rescued by the army, which had a big post at the edge of the camp. We were ordered by the army not to leave the immediate camp area. It was exciting, but we didn't worry us too much since very few people even knew we were there, and we weren't very famous or important.

March 28. We were anxious to get back to Medellín because we had not yet tried the formaldehyde method (to preserve the plants), and we were worried that our plants would get moldy. We eventually ended up getting an army helicopter ride from Campo Capote to Las Montoyas and eventually all the way to the military hospital near Cimitarra; we went with seven pretty miserable-looking soldiers who had caught malaria in the Campo Capote area. We had a long wait for a bus to Puerto Berrío, so they invited us to dinner and we ate with the commanding general of the region.

July 6, 1971. Scott and I took the train again from Medellín to Puerto Berrío and luckily this time it was not so bad—no derailing and it took a few hours less than before, but it was still a long and boring ride.

July 7. We stayed the night in "our" hotel in Puerto Berrío where we stayed before. My candle came in handy because the lights on our side of the street were out when we got there and didn't come back on until 10:00 p.m., which was nice because the fan in the room could then turn on and it was hot, like usual, in Puerto Berrío. We caught the train about noon going from Puerto Berrío to Barrancabermeja and took it to Las Montoyas where we went into Campo Capote again. The clearing was more recent and not as extensive along the road from Las Montoyas to Campo Capote (than between Puerto Berrío and Las Montoyas). As we saw the first time and were to see again from the helicopter, the rain forest is really not continuous away from the road, but rather a checkerboard of clearings soon to be entirely cleared.

July 8. We saw all the people (except the Germans) who were there before, bought some more *vales* (meal tickets) and settled down in the smaller of two houses, which we had all to ourselves. Again, all in innocence we had come to Campo Capote. It seemed that the situation was even more fraught with "danger" than before. The *guerilleros* had attacked the police station in Carare a few days before and killed two policemen and made off with ammunition and uniforms, etc. The Germans never returned to Campo Capote, and now even the work crews were restricted to a five kilometer or so radius around the camp because supposedly the army and the guerilleros were fighting it out periodically.

text continues on next page

Box 3-5 continued

We went out on a logging truck with Don Eduardo ("Carrielito") who had been our guide before. Don Eduardo cut down a big liana, cut off one meter segments with his machete, and showed us how good the water was to drink.

July 9. We hadn't yet found any *Gustavia* in flower. Scott offered don Eduardo the equivalent of $7.00 for each species of *Gustavia* we found in flower, so we took off farther than the previous day. We saw hillsides cleared the year before from the rain forest and planted to upland rice. They were harvesting it now; the farmers wearing yellow hard hats to show who they were in this "war zone." The forestry venture contracts the land out the first year to the rice farmers who may be regular employees of the company. The idea is to plant rice the first year and then replant selected timber trees so as to have a lumber plantation. I would not be surprised if the whole thing fell through and the *colonos* cleared the whole area for cattle pasture. ...We got back to camp about noon just as a helicopter let down in front of the dining hall. It was piloted by the same guy who had piloted us out before and he offered us a ride to Cimittarra. We packed all our stuff, ate a hurried dinner and took off. We went first to the military hospital near Cimitarra, but then decided they could take us to Barrancabermeja as they were going there anyway. The ride back over the Campo Capote area from Citmitarra to Barrancabermeja was interesting—it was almost all flat rain forest land along the Magdalena. From the helicopter we could see plumes of smoke rising around the whole horizon—funeral pyres of the rain forest. It was one of the most graphic sights of rain forest destruction that I had seen. At Barrancabermeja we went to the train station and purchased tickets back to Medellín on the "*tren de lujo*," a real misnomer. We had a lot of time so we left our packs at a little store, put our plants in formaldehyde, went to town for supper and then waited. At about 11:00 p.m. the train arrived and we were lucky enough to get seats. It had been a long hot tiring day and I tried to sleep on the straight stretch from Barrancabermeja to Puerto Berrío. This was the most uncomfortable bit of traveling I had yet experienced. The train went fairly fast—20 mph or something, but the track was so uneven that it was like riding a bucking horse. I was afraid if I were to sleep my neck would snap off, but I was so tired that I had to. I was thankful for the mountains between Puerto Berrío and Medellín for that meant the train had to slow down and the ride was much smoother.

July 10. We got back to Medellín very tired but happy to have gotten two *Gustavia* species in flower from Campo Capote.

Colombia, one of the most biodiverse areas of the world with many new and poorly known species of Lecythidaceae, remains a dangerous place to collect plants.

Conclusions

Tropical forests present numerous dangers, each of which occurs infrequently. The most efficient way to avoid problems is to ask local scientists or other residents before you depart if malaria, leishmaniasis, other diseases, or guerilla activities are common in the area you plan to visit. The chance of contracting disease is considerably reduced by always wearing boots, long-sleeved shirts and long pants, using insect repellent when biting insects are present, drinking properly treated water if there is any doubt about its purity, and sleeping under a mosquito net. If guerilla activity has been reported do not venture into the area, because no plant or animal collection is justified knowingly putting your life on the line.

Once in the forest, there is always the possibility of an insect, snake, or arachnid bite; thus your first-aid kit should include analgesics to relieve pain, antihistamines to combat minor allergic reactions, and epinephrine (adrenaline) in an auto-injector (e.g., the Epipen®) to treat severe allergic reactions that affect the ability to breath. Snake bites and scorpion stings are avoided by carefully watching where you place your feet and hands—never step or reach into places you cannot see! Gather things from the ground carefully, and use caution when swinging your machete to avoid disturbing wasp nests, which are often constructed on the undersides of banana-like *Heliconia* leaves. Most important, snakes should not be handled under any circumstances!

Self-inflicted machete wounds can be avoided by always knowing where your limbs are in relation to the blade, and by never using a machete as a cane. Don't follow directly behind another person, especially someone cutting a trail, because if a wasp nest is disturbed, he or she will turn and run without thinking about where the machete is pointed. When traveling by river, make sure that your footwear is light enough so you can swim in case the boat capsizes (rubber boots should never be worn while traveling in boats because they fill with water if the boat capsizes). In addition, never travel at night on rivers with rapids.

You can avoid getting lost by making sure that you are well oriented to the area in which you are working, and carrying a map if one is available. Each expedition member should have a compass and a whistle. A GPS is also essential, especially new ones that enable readings to be taken under the canopy. Before departing for a day in the field, tell the leader or (if at a larger research station) leave a message indicating where you will be working that day. If you leave the trail, mark your way with flagging, and if you do get lost, stay in place and wait for help to come to you.

Chapter 4
From the Field

by Scott A. Mori

Species studied in systematic botany, plant anatomy and morphology, cytology, molecular biology, economic botany, conservation biology, and ecology ideally should be documented with herbarium specimens so that it will be possible for the work to be verified by others. Repeatability, one of the most important tenets of scientific research, requires that the results of research be open to acceptance or rejection (i.e., testing) by other scientists. Without specimens it is difficult for other researchers to confirm what species of plants were studied and, thus, a study not documented with voucher specimens violates the law of repeatability.

One of the most important, but tedious, tasks in writing monographs and Floras (see Box 1-1 in Chapter 1) is including lists of collections that were studied; these lists are called *exsiccatae*, and the specimens are known as vouchers. In monographs and Floras, lists of *exsiccatae* (= vouchers) assist others in understanding the species concepts of the authors. Vouchers allow others to confirm or reject the identifications of the plants studied. Boom (1996) defined vouchers as follows:

> "A voucher specimen is one that serves as proof or supporting evidence for the identity of a particular species or infraspecific entity, and for the existence of that taxon at a certain location at a certain time. It is the 'anchor' for all reference to that taxon's occurrence, and it provides means for checking the identity in the future. In essence, the voucher specimen is the most basic element in plant science research. No matter what the particular field of inquiry—ecology, genetics, systematics, physiology, ethnobotany, pathology, or biodiversity prospecting—if one does not collect a plant specimen and deposit it permanently in an herbarium, then there is no way for any work to be verified as to the identity of the organism involved."

As an example of the importance of vouchers, early studies of chromosome numbers were often based on samples (either buds or root tips) that were not vouchered by herbarium specimens, and those studies are therefore essentially worthless. Goldblatt et al. (1992) state:

> "...because of the absence of voucher material, literally thousands of early chromosome counts, published over the past 50 years, have had to be discounted, and we suspect that fully half of the counts published before 1965 are based on plants the identity of which is questionable and cannot be verified."

Imagine that a molecular systematist removed a sample of leaves from a plant growing in a botanical garden; the name of the plant was obtained from the plant's label but no voucher was collected, genes were sequenced and the sequences served as the basis for a phylogenetic analysis, and a cladogram including the "species" was published in a scientific journal. Sometime later, a monographer of the genus recognizes that the relationship of the species shown in the molecular cladogram is not congruent with a tree based on morphology. Because no collection was cited to voucher the DNA sequence, the only way for the monographer to confirm the identification is by studying the plant itself. The monographer returns to the botanical garden to search for the plant, but instead finds a new visitor center occupying the site where it used to grow. If a voucher had been prepared and deposited in a herbarium, the monographer could have examined it to confirm or reject the determination upon which the molecular sequence was based. Without such a voucher, the true identity of the species can never be known. Because molecular studies play such an important role in plant systematics, many peer-reviewed journals require voucher citations as a condition for the publication of papers including phylogenetic analysis based on DNA sequences.

Vouchers should also be deposited for ecological studies. For example, Lopes and Ferrari (1994) reported that a white *uacari* monkey (*Cacajao calvus* [I. Geoffroy, 1847]) ate a plant identified as *Eschweilera albiflora* (DC.) Miers (Lecythidaceae). Because no voucher was cited, it is not possible to confirm that *E. albiflora* was actually the species that the monkey ate. If this information had been added to the *E. albiflora* species page on the Lecythidaceae Pages (Mori et al., 2010), and the plant was in fact *E. ovalifolia* (DC.) Nied. (also eaten by the white *uacari* in the same area), the *E. albiflora* species page would be based partially on information about *E. ovalifolia*; in short a species page including information for two different species! The only way to verify which plant had actually been consumed would be to examine a voucher specimen deposited in an accessible herbarium.

Authors of papers that cite plant species should send duplicate vouchers to specialists to make sure that they have the most accurate identifications for the plants they study. Failure to do this sometimes results in errors that could have been corrected by the specialist. For example, a paper published in a well-respected journal of tropical biology reported that fish dispersed the seeds of *Gustavia speciosa* (Kunth) DC (Lecythidaceae). Because this species is distributed in Andean valleys and the study was done on the lower Amazon River in Brazil, I suspect that seeds were probably from *G. augusta* L., which is common in the riverine habitat where the study was made, and has seeds with hard coats (adapted for passage through the digestive tracts of fish). It is, however, impossible to check the plant identification because the authors failed to cite a voucher specimen. The authors and the journal are jointly responsible for the publication of this apparent misinformation; the authors because they failed to collect vouchers, and the journal because its editorial policy did not require citation of voucher specimens as a condition for publication.

When information based on incorrect identifications is published, the erroneous information may become the basis for subsequent studies, and sometimes even becomes accepted as fact. In this age of electronic communication, the propagation of errors is increasingly common, because the review process is often less rigorous than it is for hard copy publication. This is especially problematic in the production of species pages, which are sometimes prepared by persons who lack the expertise needed to evaluate the information they add. When information based on incorrect determinations is added to electronic species pages, the species pages may include elements from two or more species and, thus, become "mixed" pages. Species pages created using Wiki technology need to be viewed with special caution, because anyone can modify them with additional text or images, increasing the probability that the pages will end up including information for more than one species.

The cascading effect of incorrect species determinations has been recognized for a long time, but Bortolus (2008) is one of the first authors to provide data on the prevalence of errors in the ecological literature due to misidentifications. This author found that 62.5% of the modern ecological studies he analyzed lacked supporting information that allowed others to confirm or reject the species identifications. More importantly, only 2.5% of the papers backed up the identifications with specimens deposited in herbaria, which the author notes "terminates all prospects for conducting further taxonomic confirmations in the other 97.5%" of the studies. Bor-

Fig. 4-1. A bryologist collecting mosses. Bryologists use a hand lens to make sure that they do not include more than one species in their collection, a problem seldom associated with collecting flowering plants. The collection of bryophytes is not covered in this book. See Buck and Thiers (1996) for more information about bryophyte collection. Drawing by Michael Rothman.

tolus (2008) provides examples of how misidentifications cause a variety of negative biological, ecological, and sociological consequences, and recommends the collection and citation of vouchers as a way to circumvent these problems.

Because voucher specimens provide such important documentation for plant studies, this chapter describes methods of collecting specimens in the field. See Chapter 5 for details about entering those collections into herbaria, and Chapter 6 for information on making the information available online. This and the following chapters are not, however, a primer for plant collection and herbarium management because these topics are adequately covered elsewhere (e.g., Alexiades, 1996; Bridson & Forman, 1992; Fidalgo & Ramos Bononi, 1989; Fosberg & Sachet, 1965; Harvard University Herbaria, 2009; Lawrence & Hawthorne, 2006; Lot & Chiang, 1986; Metsger & Beyers, 1999; Mori, 1987a). In addition, several articles have been published in the *Annals of the Missouri Botanical Garden* on collecting for specific groups of plants (e.g., Dransfield, 1986 [palms], Jørgensen et al., 1984 [passionflowers], and Mori & Prance, 1987c [Lecythidaceae]). Nor do I discuss collection techniques for special groups

Fig. 4-2. A mycologist making sure that her collections of ascomycete fungi gathered earlier in the day are suitable as specimens. Mycologists need to confirm that the microscope features they use to classify fungi are present before they make a collection. The collection of fungi is not covered in this book. See Halling (1996) and Cannon and Sutton (2004) for more information about the collection of fungi. Drawing by Michael Rothman.

of plants, such as mosses (Fig. 4-1) and liverworts (Buck & Thiers, 1996), mushrooms (Halling 1996, also see http://www.nybg.org/bsci/res/hall/boletes/collecting%illustrated.pdf) or microfungi (Fig. 4-2) (Cannon & Sutton, 2004).

Collection Numbers and Duplicates

A collection is identified by the name of the collector and a number assigned to it by the collector. In this chapter, "collection number" refers to the combination of the collector's name and the associated number.

The series of collection numbers for a given collector ideally starts at one and continues consecutively throughout the career of the collector. If the collector's name changes, for example because of marriage, the collector might choose to start a new number series associated with the new name, or continue to collect under the original name. Exactly the same name should be used on the label for all collections made by a collector. For example, S. A. Mori is the name assigned by Brummit and Powell (1992) in their *Authors of Plant Names* (an electronic version of this important publication is available at the International Plant Names Index [IPNI] Internet site of the Royal Botanic Gardens Kew). For collectors appearing in Brummit and Powell, I recommend that the standardized name be used on labels, with the surname(s) given in full if abbreviated by Brummit and Powell. To reduce confusion among collectors with similar surnames, it is best to include initials or, in some cases, spell out the given names.

Spanish and Portuguese collectors need to be especially diligent in not using different permutations of their names. Take, for example the fictitious Spanish name José Oscar Fernandez Calderón in which the first two names are given names, the next is the family name of the father, and the last is the family name of the mother. Unless there is another botanical collector with the exact same name, the name on the label would normally be abbreviated as J. O. Fernandez Calderón. However, sometimes one of the family names is dropped (e.g., J. O. Fernandez when the mother's is dropped or J. O. Calderón when the father's is dropped). In addition, sometimes the two family names are hyphenated (e.g., J. O. Fernandez-Calderón) or the mother's name is abbreviated (J. O. Fernandez C.). The situation is similar with Portuguese names but the mother's family name usually comes before the father's. When a collector's name is added to a database or cited it should be added exactly as it is on the label (see Chapter 5) and this includes the accents, without which the name is misspelled. Names in other languages have similar problems.

The collector's name on the label should be that of a person and not an organization. On a trip to Ecuador in 1982, the collections were numbered in a series called *Studies in Ecuadorian Forests*, with the collector given as SEF. This acronym has meaning only for those who participated in the expedition, and is less informative than listing the actual collectors of the specimens. Collections for the Suriname Forest Service clearly demonstrate the perils of using an acronym in lieu of the collector's name; during different periods of its history the organization has been known as the Boswezen (BW), Bosbeheer (BBS), and Lands Bosbeheer (LBB). Collectors working at different times have gathered collections under these different acronyms. Some collectors had their names on the labels (others didn't); some had their own collection number series, and yet others produced labels with their names and collection numbers as well as the numbers of the Suriname Forest Service. Today, the herbarium of the Suriname Forest Service is part of the Anton de Kom University of Suriname, but it retains the acronym BBS. This convoluted history, which has been discussed by Lindeman and Mori (1989) and Ek (1991), makes additional work for those who study early collections made by the Suriname Forest Service collectors. People make the collections, not institutions, and in the interest of clarity specimens should be linked directly to their collectors.

Another confusing situation arises when a collector sends others into the field to collect under his/her collection numbers. In this case, the owner of the number series will often have the surrogate collectors add a prefix to the number to indicate that the collector whose numbers where used did not personally make the collections. This has been done in two ways: (1) the initial of the absentee collector is placed in front of the number, e.g., *Berg et al. P20000* indicates that C. C. Berg and his collecting team were collecting under the series of G. T. Prance, or (2) the name of the absentee collector may appear on the label and the initial of the actual collector is placed in front of the collection number, e.g., *Granville B1505* indicates that a French Guianan collector, Biogog, collected the plant under the series of Jean-Jacques de Granville. In both examples, without further information it is impossible to know what the abbreviation stands for— but at least the same number in the collection series of the collector will not be used for two different collections. To eliminate confusion, collectors making the collections should use their own names and collection number series on labels.

Sometimes two different collections are given the same number due to mistakes in the field. When such an error is recognized, it is necessary to distinguish between the two collections. One of the duplicate numbers can be suffixed by a letter of the alphabet; for example Mori et al. 25822 (a collection of *Lecythis rorida* O. Berg) and Mori et al. 25822A (a collection of *Gustavia hexapetala* (Aubl.) Sm.). This kind of numbering mistake can usually be corrected when plants come out of the drying press by checking the number written on the newspaper against the number in the field notebook.

Collections of different individual plants belonging to the same species should be given different numbers. If the same number is given to multiple conspecific specimens, subsequent researchers using molecular techniques might overestimate genetic variability within individual plants. In addition, if the same number were given to specimens collected from separate male and female plants, the species could be interpreted as being monoecious when it is, in fact, dioecious. These are several examples of how duplicates gathered from different plants and given the same collection number can lead to confusion.

Exceptions are when entire plants are small enough to be attached to a single herbarium sheet. But when that is done, attention must be given to making sure that the small plants actually represent the same species.

Another problem arises when flowers of one plant and fruits of another plant are given the same collection number because, if the two plants happen to be different species, it might lead a botanist studying the collection to describe a species based on a mixed collection in which the description of the flowers are from one species and those of the fruits are from another. Both examples given in this paragraph represent mixed collections.

Some plants are too large to be represented on a single herbarium sheet; take, for example a collection of a palm for which duplicates are collected for two different herbaria—one for the herbarium of the collector and another for the herbarium of a specialist. If the leaves are too large they will have to be cut into segments and the pieces should be labeled in the following manner: base of leaf one, middle of leaf one, and apex of leaf one for one duplicate and base of leaf two, middle of leaf two, and apex of leaf two for the other duplicate.

Specimens gathered from the same plant at different times of the year, e.g., flowers in April and fruits in October, should also be given different collection numbers. This seems obvious but there is a temptation to use the same collection number to document images of flowers and fruits from the same plant, even when the images were not taken at the same time. If flower and fruit images taken at different times are attached to a single collection number, it gives the false impression that flowers and fruits were both present when the original collection was made. The only way to document the phenological stages of a plant is to make separate collections and assign different numbers to collections made on different days. If a collection is knowingly made from a plant that has been previously collected, it is useful to provide the previous collector's name and collection number on the label of the new collection. In this case every effort should be made to ensure that the locality information, including the coordinates, is the same for the two collections.

How Many Collections?

If the material available is only sufficient for a single herbarium sheet, the resulting collection is called a *unicate*. For the most part, however, multiple duplicates are gathered with the intention of making multiple herbarium sheets; thus, each sheet of the same collection number is called a *duplicate*. When duplicates are collected from the same plant they are given the same collection number. Duplicates of collections are gathered because:

1.   duplicates are sent to other herbaria for identification by spe-

cialists;

2. botanists collecting in foreign countries are required to deposit one or more duplicates in the host country's herbaria;

3. duplicates housed in several herbaria are insurance against loss or damage of duplicates in other herbaria; and

4. duplicates are used for exchange with other herbaria as a way of increasing the collections of the collector's home herbarium.

When professional botanists are working in foreign countries, the minimum number of duplicates is typically three—one for the local herbarium of the host country, one for the consulting specialist, and one for their home institution. When I make a Lecythidaceae collection, for which I am a specialist, and it bears little information other than serving as a distribution record, I collect only two duplicates—one for the local herbarium and one for my home institution, The New York Botanical Garden (NYBG, note that NYBG is the abbreviation for the Botanical Garden and NY is the abbreviation for its herbarium). On the other hand, if a collection is of special interest due to new information it provides, I collect the original plus five duplicates, and distribute the extras to appropriate herbaria. For example, when I collected an interesting plant in Bahia, Brazil, I left a duplicate at the local herbarium (i.e., CEPEC), sent one to NY, and the others on exchange to MG, A, RB, and K (the latter is a herbarium with a history of curators who have studied the Brazil nut family as well as the flora of eastern Brazil). Finally, if I suspect that I'm making a collection of a species new-to-science, I collect as many as 12 duplicates to distribute to herbaria with frequently consulted tropical collections—especially to herbaria located within and adjacent to the country where the specimen was collected, and to herbaria with a history of studying the genus or family to which the plant belongs.

It is important to know and adhere to the laws of the country in which collections are made. These laws regulate the distribution of duplicates and may also state, for instance, that unicates must be archived in an in-country herbarium. Even if local law does not require it, it is good practice to leave unicates and types in the host herbarium. If a unicate represents a species of particular interest, the collector can make notes and take photos of the specimen, and include the information and image in a database record for future reference.

Although a single collection of a taxon may be sufficient to document its existence, support its description as new to science, or record its presence in a flora, it does not provide the information needed for the preparation of Floras and monographs (Anderson, 1996). For those products, multiple collections are needed to show: (1) geographic distribution, (2) ecological preferences, (3) abundance, (4) phenology, (5) all parts and stages of a species' life cycle (e.g., stems and leaves, flowers, fruits, seeds, seedlings, etc.), (6) morphological variation, and (7) economic uses.

What To Collect?

Most of the time, it is not economically sound for herbaria to archive multiple specimens of a well-known species collected from the same locality without some justification. Prance and Campbell (1988) demonstrated that many recent collections are of common species, while rare species are still poorly represented in herbaria. An example of an over-collected species is *Gustavia hexapetala*, an under story tree with showy flowers and conspicuous fruits, represented by 53 collections from French Guiana in the NYBG herbarium. This duplication is unnecessary, especially since most of these collections are also archived at the Herbier de Guyane (CAY) in Cayenne. No additional collections of this species from this area are needed unless the collection documents new information such as a new locality, vouchers for images, ecological studies or pollination or dispersal observations, or provides specially preserved tissues (such as pickled collections for anatomical or morphological study, leaves dried in silica gel for molecular study, etc).

One way to make sure that only information-bearing collections are added to herbaria is to have collectors focus on the floras of given areas (floristicians) or specific groups of plants (monographers). Experienced specialists can recognize those species for which additional information is needed, and refrain from collecting specimens that do not provide it. It is now possible, even while collecting in the field, to obtain information from Internet sites about which species are still in need of additional collections.

The addition of sterile voucher specimens to herbaria (Mori, 1998) is problematic because these specimens are usually difficult to identify. In floristic inventories, sterile specimens of species previously unknown from a locality are sometimes collected, and worth preserving. On the other hand, ecological inventories often generate large numbers of sterile specimens that place a financial and logistical burden on herbaria (see discussion below).

Because plant collecting is expensive and difficult, especially in the tropics, I recommend including collections for botanists who are studying other groups as well as for your own studies when that is possible. The importance of images in helping those who study the collections is discussed below. In addition, the preservation of flowers, fruits, bark, stems, and leaves in a liquid preservative (e.g., 70% ethanol with the addition of 1% glycerol to keep the material less brittle) for anatomical and morphological study and leaf fragments dried in silica for molecular analysis increases the value of a collection. In the past, FAA (= 10 ethanol: 7 distilled water: 2 formalin: 1 glacial acetic acid, v/v) was used for pickled collections, but this is no longer recommended because the formalin is too caustic to handle. In both types of collections, the specimens should have labels with the collector, the collection number, and the name of the taxon on both the outside and the inside of the jar, and the labels should be on high quality paper written in lead pencil. It is especially useful and rewarding to gather additional material if there is a specialist carrying out monographs of the group collected or if there is a floristic study underway in the region where the collection is gathered.

Vouchers For Ecological Studies

We established a 100-hectare plot in central Amazonian Brazil to study the systematics and ecology of Lecythidaceae; it includes 7791 individuals of Lecythidaceae ≥ 10 cm diameter at breast height (DBH) (Mori & Lepsch-Cunha 1995). Of these individuals, 1539 represent the common and wide-spread species *Eschweilera coriacea* (DC.) S. A. Mori, which was almost always sterile during the course of the study. If all sterile individuals of *E. coriacea* had been collected and archived in herbaria, very little taxonomic information would have been added to those herbaria. At a conservative estimate of $5.00 per specimen (Anderson, 1996), the cost of including these sterile vouchers in the herbarium of NYBG would have amounted to $7695. Moreover, by law, a duplicate of each collection must be archived in the herbarium of the Instituto Nacional de Pesquisas da Amazonia (INPA). If all 7791 individuals of Lecythidaceae had been collected, the seemingly simple process of mounting and inserting them into the herbarium of NYBG would have cost $38,955! They would occupy 2.5 double herbarium cabinets, and take up 1.7 square meters of herbarium floor space indefinitely. Funding for adding this number of specimens to the herbaria of both NY and INPA was not available. Hence, specimens were collected from trees only if there was some doubt about their identity; the sheets of sterile collections were stored in boxes until the study was completed; and a limited number of representative collections (mostly fertile) of each species were mounted and archived at INPA and NY.

Researchers planning ecological inventories seldom consider the financial burden that their collections impose on taxonomic specialists and their institutions. Specimens are often sent to specialists for determination as "gifts," without previous consultation. Therefore, the specialist is not given the chance to consider the scientific merit of the project for which the vouchers have been collected, or to decide if the time invested in making the identifications will contribute to the knowledge of the group under study. Once the specimens arrive, even if the specialist returns the specimens without studying them, there has already been a significant cost in specimen handling. If the specialist decides to determine the specimens, additional time is invested in making and reporting the determinations.

Specialists who work on large plant families and make rapid and accurate determinations may end up spending considerable time identifying collections made by other researchers. An extreme example is that of the late Rupert C. Barneby, specialist in legumes and Menispermaceae at NYBG. He estimated (Barneby, pers. comm., 1999) that in a given year he might receive 4000 collections for identification (gifts, staff collections, unsolicited loans, etc.), and that it took him an average of 10 minutes to identify a collection. Consequently, he sometimes spent four to five months of normal work time (seven hours a day, five days a week) identifying collections, many of which were not directly related to his research. Rupert's diligence and productivity made the legume collection at NYBG one of the most complete in the world, and that was the reward that he and NYBG received in exchange for his work. Generally, specialists are pleased to accept well-prepared, fertile collections as gifts because they increase the information content of herbaria—in the same way that an art museum curator's acquisition of a new work of art enhances the content of that museum.

Not only is it excessively time-consuming for specialists to identify sterile herbarium specimens, but these identifications are prone to error if the diagnostic characters separating species are not reflected in leaf morphology. For instance, in central Amazonian Brazil, there are two species of *Corythophora* (Lecythidaceae) that are easily distinguished by striking differences in their bark (Fig. 4-3). These differences, in the field, allow them to be identified with 100% accuracy (Mori & Lepsch-Cunha, 1995). One species, *C. rimosa* W. A. Rodrigues, has very deeply fissured, thick outer bark and the other, *C. alta* R. Knuth, has scalloped, thin outer bark. In contrast, the leaves are so similar that identifying the two species on the basis of sterile specimens—without notes describing the bark—is difficult, if not impossible. Unfortunately, collectors for ecological studies

Fig. 4-3. Notes on bark differences should be recorded in the field because similar species sometimes have very different barks. Left: The scalloped bark of Corythophora alta from an unvouchered tree in central Amazonian Brazil. Right: The fissured bark of Corythophora rimosa subsp. rimosa from an unvouchered tree in same area. These species are difficult to tell apart based on leaves but they are easily recognized by their bark, flower, and fruit characters. Photos by Carol. A. Gracie.

do not typically describe the bark as part of their protocol; thus, errors are commonly made in distinguishing sterile collections of species with similar leaves.

While sterile specimens are often difficult to identify, they can still provide valuable information. Our 100-hectare Lecythidaceae plot included 364 individuals that were originally identified as *Eschweilera collina* Eyma, a species characterized by medium-sized leaves and smooth bark with trunk-encircling bands (called hoop marks). As our work progressed, we began to notice that specimens from trees identified as *E. collina* fell into two distinct groups, one with salient veins on both leaf blade surfaces (*E. collina*) and the other with salient veins on the abaxial surface only. When the trees flowered, we were able to correlate the difference in leaf venation with flower and fruit differences. As a result, our previous determinations had to be changed to accommodate a new species, *E. romeu-*

cardosoi S. A. Mori, named after one of the woodsmen participating in the study. Out of the 364 trees, only 55 were actually *E. collina*, and the other 309 were *E. romeu-cardosoi*; if vouchers had not been collected, 309 trees of *E. romeu-cardoosoi* would have been misidentified as *E. collina*.

Another problem with ecological inventories occurs when non-specialists are responsible for species identification, because a single common name is often used by them for several different species. For example, Procópio and Secco (2008) found that Brazilian tree spotters gave the same common name to three different, unrelated species of *Couratari* growing in the same area, and I have found as many as five different species archived as a single species in various herbaria. With such inaccurate plant identifications, ecological studies in tropical areas are sometimes so flawed that reliable and reproducible results are impossible to achieve.

It is easy to understand why a taxonomic specialist, without a vested interest in a particular ecological project, would decline to spend time identifying sterile specimens collected for another researcher's study (especially without salary support, and without inclusion as a co-author in the publication of the results). For example, 8000 sterile specimens of Amazonian Lecythidaceae once arrived—unsolicited—at NY for me to identify, but it would not have been a wise use of my time to identify them because of the months of uncompensated labor for me to make and report the identifications. Moreover, because these specimens would have made little additional contribution in understanding the classification of Lecythidaceae, there would have been little intellectual return for the effort. Although it is very important to have correct determinations for all plants in ecological studies, a single sterile specimen provides nearly the same amount of taxonomic information as 1,000 sterile specimens of the same species, from the same area.

Ecologists can entice systematists to make determinations for them by providing collections that contribute to the taxonomic knowledge of the groups that the systematists study. This means making a special effort to collect specimens with flowers or fruits, and relating the fertile collections to their sterile collections. Optimally, ecologists should sort the sterile collections into "morphospecies," gather fertile collections representing those entities, and then send only fertile collections to the specialists for identification. Ecologists can also assist specialists by making careful field notes, photographing features of the plant that will not be apparent in the dried herbarium specimens, contracting in advance for identification services, and including the names of those who identify significant numbers of specimens as co-authors in resulting papers.

In summary, sterile specimens should be gathered only when they add new information to the flora, or when they serve to document photographs, collections for molecular and phytochemical studies, or ecological studies with clearly defined goals. Ideally, leaves collected and silica-dried for molecular studies should be collected from fertile plants to maximize the probability that they will be identified correctly. The best way to obtain determinations from specialists is to send them duplicates of fertile specimens for deposit in their collections. In ecological studies, the best case scenario is a mutually rewarding collaboration between systematists and ecologists. The systematist adds considerable value to an ecological study by making sure that organisms are correctly identified, and the ecologist can add to the knowledge of collaborating systematist by collecting data that contribute to a better understanding of ecological relationships within their groups of study.

Equipment

In Appendix D, I provide a list of essential collecting equipment. Many other items could be taken into the field; however, computers and other electronic devices are practical only if there is a reliable source of electricity. Nevertheless, I bring my laptop on all trips and, if electricity is not available on a given excursion, I leave it in the local herbarium. A computer has become part of my collecting equipment because I depend on it for managing images, adding label data to the database, and for recording data from specimens in local herbaria. Because electrical systems vary from country-to-country it is important to research the systems utilized in the countries to be visited and to bring the required converters and plug adapters on the trip. Electrical systems used in other countries can be determined by consulting the World Electric Guide at http://kropla. com/electric2.htm. It is also important to carry a three- to two-prong plug adaptor to insure connectivity between three-prong plugs and two-prong outlets.

Clothing. Clothing color is important in tropical areas because insects are preferentially attracted to certain colors. For example, biting tabanid flies (Fig. 3-11) are attracted to blue, which should therefore be avoided in both clothing (such as blue jeans) and backpacks. A blue backpack is an open invitation to tabanids, which will proceed to swarm around the back of your head. I wear khaki field clothes because this color does not seem to attract insects. Neutral colors also allow you to be less conspicuous to birds and other animals, which is important if you are making observations of pollinators or seed dispersers. My standard field outfit consists of a double layer of brown, calf-high socks, long pants, a t-shirt, a long-sleeve shirt, and a cap with a wide visor. The cap protects me from both rain and insects; for example, if I am swarmed by trigonid bees, the cap keeps them out of my hair. Because I climb trees, I usually wear a pair of solid hiking shoes that protect my feet and ankles from the straps of the climbing spikes; however, I prefer rubber gardening boots called Muck Boots® for working in the rainy season. I tuck my pants into the two pairs of light socks because three layers of clothing prevent the tabanid flies (with their seven millimeter long beaks) from attacking my ankles—one of their favorite places to take a blood meal. This is also why I wear both a t-shirt and a long sleeve shirt (when I am walking long distances and want to stay cool, I put the long-sleeve shirt in my pack because biting insects are less bothersome when one is moving). I almost always put the long-sleeve shirt back on when I climb to protect against insects (attacking while I'm in no position to swat them!), but also to protect my arms from abrasions if I slip while my arms encircle the tree trunk.

Collecting equipment *(Appendix D).* For information about the equipment used to climb trees see "Climbing trees," and for instructions about how to make drying frames consult Boxes 4-3 and 4-4. The sections "Preparing specimens," "Specimen images," and "Drying specimens" individually provide information about the equipment needed for these activities.

On my belt, I have a Leatherman® tool, a 17-inch machete, and a pair of Felco 2® hand pruners. The former allows me to repair almost anything that goes wrong with the collecting equipment. I prefer a short machete, which does not get in my way as often as a longer machete while I walk or climb. When climbing down a tree with a long machete, the machete sheath can snag on a branch and be upended. The machete slips out of the sheath and plummets downward, posing a hazard to anyone working under the tree! A good quality hand pruner is essential for cutting away small branches during a climb and for preparing specimens. Of special importance is to buy pruners with a lock mechanism that keeps the knife blades closed when the pruner is not in use; in other words, a cheap hand pruner ends up causing so many problems that paying the price for more expensive ones is a good investment.

When walking from one tree to the next, I carry my climbing belt around my waist with the two lanyards draped around my shoulders and my clipper pole balanced on my shoulder. My spikes and plant press are attached to the outside of my pack. Fortunately, the collecting gear is usually parceled out among the various participants in the day's excursion, so

I seldom end up carrying all of the gear myself.

In my pack, I carry a field notebook in a sealable plastic bag, a mechanical pencil, a spare pencil, and a red felt-tipped, indelible pen for numbering the specimen sheets. I always have a small water tight box with a compass, toilet paper, a ruler for measuring the size of flowers and fruits, paper rulers used to provide scale in photographs, spare pole buttons, band-aids, an antibiotic ointment, and two epi-pens in case of an adverse reaction to wasp stings. I also carry a GPS, a rain poncho, a 3 × 3 meter (10 × 10 foot) tarp with attached cords for making a temporary rain shelter, a water tight box with lunch, and, on days that I will be climbing trees, at least two liters of water. The amount of water I carry depends on whether I will be working on ridges, where it is difficult to obtain water, or in lower areas where there is an ample and often pure supply. In most areas, however, the water needs to be treated.

In the past, I used record books with 300 pages for writing up my collections while in the field. These books are large (10.5 × 8.2 inches) and, thus, take up considerable room in a day pack. When a book was filled, I had it rebound and my name, the number of the book, and the collection number range in each book printed on the spine. At the beginning of my career, labels were typed directly from the field books with a manual typewriter, but now they are typed directly into the NYBG database. The field notebooks, which can be purchased in most office supply stores, serve as a permanent record of my field collections and will be deposited in the library of NYBG when I retire.

With databases to keep track of specimens in herbaria and from which labels are printed, it is my opinion that large note books are not as important as they were in the past. As a result of this new philosophy, I now use small notebooks (7.5 × 4.75 inches with 160 pages of water resistant paper) thereby reducing the weight and space taken up by the larger record books. I use the Rite in the Rain® all-weather journal No. 390F which accommodates about 800 collection numbers. The same company makes a small waterproof carrying pouch that can be attached to a belt or carried in the day pack. The smaller field books will also be deposited at NYBG, but they will be less frequently consulted because information about my collections can be obtained from the Virtual Herbarium of NYBG. The notebooks, however, will serve as a backup if there is a failure with the database system. All of my notes are written with a mechanical lead pencil because the writing is more permanent than most pens which have ink that smears in the rain.

Specimen Collecting

Finding collections. The goal of plant collecting for monographic and floristic projects is to document all species of plants in a particular group (monographic collecting) or in an area of interest (floristic collecting), at all stages of their life cycles. All strata (e.g., ground, shrub, and canopy layers), in all habitats, at all times of the year, must be explored. After the specimens have been databased, graphs with specimen collection dates can be used to target times of the year that have not been adequately sampled. Because collecting is so much easier and pleasanter during the dry season, there is usually a collection bias favoring this time of year, so a special effort has to be made to collect during the rainy periods. Wet season collections are more likely to document fruits, or seedling morphology. In floristic studies, a species accumulation curve can be constructed as collections are gathered, and this will give an idea if sampling has been adequate to publish the Flora. If the curve is steeply inclined, more exploration is needed, whereas when the curve begins to level off, it indicates that the species in the Flora are relatively well represented by collections. Keep in mind, however, that tropical floras are so rich that additional exploration, especially in different places and at different times of the year, will almost always add additional species to an inventory.

When planning expeditions to collect specific groups for monographic research, researchers have to consider the phenology of the group targeted for collection. For example, Lecythidaceae seldom flower or have mature fruits during the peak rainy season (Mori & Prance, 1987), and I have undertaken some of my least productive and most uncomfortable expeditions searching for specimens during the height of the wet season. My least productive trip was in Costa Rica in late May and early June 2004; this was the onset of the rainy season and it rained constantly for two weeks. Although I was accompanied by highly skilled botanists, we mostly found sterile trees, a few individuals with a few flowers, and trees with a few rotten fruits on the ground. We were forced to walk long distances in the rain, wade in and ford flooded rivers, and spend an inordinate amount of time confined to shelters—watching sheets of water pour from the sky. As a result, I subsequently resolved to confine most of my Lecythidaceae collecting trips to times of year that herbarium sheets suggest that they will be in flower or fruit. It is still necessary to make some collections during the wet season, because some species either flower or fruit only at that time.

In forested habitats, one of the most efficient ways to find trees, lianas, and epiphytes in flower or fruit is to locate roads recently cut through

Fig. 4-4. After flowers and fruits are found on the ground, scanning the crowns of trees is an efficient way of locating the plants for collecting. Drawing by Michael Rothman.

the forest. New roads provide transects that can be easily walked while scanning the tree crowns for fertile material. As a road becomes older, dense secondary vegetation develops between the road and the trees, and it becomes more difficult to get to the trees to climb them. Another effective strategy is walking slowly along a forest trail scanning the ground for fallen flowers and fruits. We often walk for about 100 meters, take our packs off, put down the climbing equipment, and use binoculars to carefully look into the crowns of trees to the side of the trail for flowers and/or fruits (Fig. 4-4). We do this in both directions because we see different views of the plants, depending on the direction we're walking. We typically collect only fertile specimens, unless, as previously mentioned, there is a compelling reason to gather sterile vouchers.

Morphological variations that occur within a plant, or at different stages of its life cycle, should be documented with herbarium specimens. For example, in species such as *Quiina obovata* Tul. (Quiinaceae) and *Syagrus inajai* (Spruce) Becc. (Arecaceae), the leaves of saplings are remarkably different from those of adults. The former has pinnately lobed leaves as a juvenile, and simple leaves as an adult plant, whereas the latter has entire leaves as a juvenile, and pinnate leaves as an adult plant. In *Pourouma* (Moraceae), leaf blade shape may vary between juvenile and adult trees, as well as within the crown of adult trees—where some leaves are lobed and others entire. In some species of *Pourouma* the leaves associated with inflorescences are not lobed, while those not associated with them are (Kincaid et al., 1998).

Specimen notes. Collectors are responsible for providing the data needed for making labels that provide essential information in an easy-to-record manner. The information on plant labels should provide only details that cannot be seen on the collections themselves. For example, including "leaves simple, alternate" gives no information that cannot be determined by examining the specimen and should not be included on labels. Long ecological descriptions should also be avoided; for example, extensive lists of plants collected in the same habitat are seldom reliable or useful, because there is no way to verify the names. Nevertheless, listing a few species known as indicators of a particular habitat is helpful in understanding the ecology of species, especially if the cited species are well-known and easily identified. Collectors should not include superfluous information on labels, which creates unnecessary labor for those who database and study the collections.

For any plant collection, the most important features to describe are those that are lost in the collection process. Bark features and flower color are often inadequately represented in specimens, so they should be described on the label and photographed. If a color chart is used, the name of the color classification system employed and the actual name of color should be mentioned, along with the color code reference. Adding a color code to a label, such as GPF9, to describe the color of petals, has little meaning to those studying the specimen. Colors should be specific to the part of the plant described; for example, "flowers yellow" should not be used if "petals yellow" is meant. For some plant families, it is important to record the time of day when the flower color is observed, because color may fade later in the day. For example, the petals and androecial hood of *Eschweilera parviflora* Aubl. are both light yellow in the morning when they are visited by pollinators, but fade to white in the late afternoon (especially after they have fallen to the ground). If flowers are collected from the ground, this should be mentioned on the label because their color may not be the same as that of flowers still in the tree.

Nearly all collectors now use GPS devices to record the geographic coordinates of their collections. Ideally, the coordinates should be added to the label in degrees, minutes, and seconds because that is the most widely understood method. GPS readings should be rounded to the nearest second because most GPS devices used by plant collections to not have greater precision than that. If coordinates are given in a less well-known system, such as UTM (Universal Transverse Mercator), it will cause additional work for people entering coordinates in databases due to the fact that those data basing specimens are less familiar with this system and many users do not have the topographic maps needed to determine the UTM localities.

The best practice is to record coordinates at the locality of each plant collected. If a GPS is not available on the trip, coordinates can be added later from hard copy or electronic maps or gazetteers (a list of geographic localities, often with coordinates). Locality descriptions such as "three miles downstream from camp A" are of little use to those mapping specimens, so labels should reference at least one permanent geographic feature. Another problem arises when the collection locality is given as a distance and direction from a particular city. First, it is seldom clear if the reading was made from the center or the edge of the city, and moreover, cities grow in size. For example, a note made (in 1975) that the Reserva Ducke was located at km. 26 on the Manaus/Itacoatiara Road was difficult to interpret 34 years later, when I revisited the area and the limits of Manaus had nearly reached the reserve.

Online Gazetteers and Google Earth have made it is much easier to geo-reference collections retroactively. On a trip to Puerto Rico, *Mori & Gracie 27178* was collected on the property of Mimi's Guest House (ca. 4.25 km SE of Guanica). Although we did not have a GPS with us at the time of the visit, I was subsequently able to locate Mimi's Guest House using Google Earth, move the cursor to the spot where the plant was collected, and record the coordinates directly from the screen!

If herbarium labels are prepared with a program that automatically maps collections once the specimen data has been added to the database, incorrect coordinates will not be a problem, because the map symbol showing the locality can be checked for accuracy before the labels are printed. For example, we originally recorded coordinates for a series of collections (Mori et al. 26082–26108) made along the south shore of Saba that placed our locality in the sea instead of along the coast. Unfortunately, this mistake was not discovered until after the labels were printed and the specimens distributed. I then had to change the collection site coordinates in the database, and indicate why the coordinates on the labels conflicted with the current positions of the symbols on the Google map. This error could have been avoided by checking the position of the coordinates on a map before the labels were printed.

Serious problems also arise when collectors record the common names for plants, especially if collectors are not fluent in the local language. Many years ago, I was pressing plants with a group of botanists in Venezuela, surrounded by an audience of curious Amerindians from the village in which we were working. The lead collector, a native English speaker, would periodically yell out "*¿Como se llama esta planta?*" Someone from the audience would respond "it is a green plant," and that phrase was transliterated, recorded, and appeared on the label as a common name of the species. Collectors should not add a common name to a label unless they are literate in the language in which the name is given, and the name has been confirmed by several different and reliable informants native to the area where the collection is made! It is also useful to provide information about the source of the common names because that provides some indication of their reliability; e.g., "common name provided by Pablito Varón who provides medicinal plants to local market" or "common name added from the *Guide to the Vascular Plants of Central French Guiana.*"

Climbing trees. It may seem obvious to state that trees play an important role in neotropical forests—in central French Guiana they, along with the epiphytes and lianas they support, make up 68% of the species found in those forests (Mori et al., 2002b). For this reason, tropical plant collectors need to climb trees to adequately sample plant diversity. As an example, 14 of the 39 species of the Brazil nut family found in our 100-hectare plot in Amazonian Brazil were published as new to science in the last 30 years, directly as the result of canopy exploration (Mori & Lepsch-Cunha, 1995). Climbing trees is one

Fig. 4-5. *French tree spikes are an effective way to climb trees because they allow easy and fast access into the forest canopy where there are the greatest number of plant species. The drawing shows spikes for the right (left) and left (right) feet, a safety belt, and two lanyards, one of which should always be attached around the tree trunk. Drawing by Michael Rothman.*

of the most dangerous parts of tropical plant exploration, and those interested in collecting specimens from trees should take a course in tree climbing from a professional arborist, such courses are periodically offered by botanical gardens. It is especially important to master the art of tying knots, and learn to recognize and replace ropes that are old or frayed.

After finding flowers or fruits on the ground below a tree, I use binoculars to search for the fertile plant. When I locate it, I determine which tree should be climbed to give easiest access, which, depending on plant diameter, could also be a neighboring tree or liana. I carry two pairs of French tree-climbing spikes (Fig. 4-5), which enables me to climb as many different-sized trees as possible. French tree climbing spikes, called *griffes*, or *grimpettes*, are made by Lacoste et fils (12 av. Pasteur, 24160 Excideuil, France) and can be purchased through their web site at www.lacoste-outillage.com. They have become very expensive – for example a pair of 35 cm spikes cost $342.00 when I ordered my last pair (September 2008), a marked contrast to the first pair that I purchased in 1980, at about

Fig. 4-6. This drawing shows the author using French tree spikes and a clipper pole to collect specimens from the tree climbed as well as from adjacent trees. Drawing by Michael Rothman.

one-sixth that cost. With the smaller size (24 centimeters in diameter) I climb trees from 10 to 24 centimeters in diameter, and with the larger size (35 centimeters in diameter) trees from 26 to 35 centimeters in diameter (Fig. 4-6). Even larger trees can be climbed if they have irregular trunks, because the irregularities themselves can be used to anchor the spikes. For example, I can often climb the sulcate-trunked species of *Aspidosperma* (Apocynaceae) and *Chimarris* (Rubiaceae) or the fenestrate trunks of *Geissospermum* (Apocynaceae) that are larger than 50 centimeters in diameter. The most difficult trees to climb are those with a combination of hard bark, cylindrical trunks, and diameters near the limit of the spikes.

Earlier in my career, I used a Swiss tree climbing "bicycle" designed for collecting seeds from pine trees; but it worked well only for trees with clean, straight boles (Mori, 1984, 1987b). The biggest disadvantages of the bicycle for tropical tree collection are that the trunks have to be cleaned (causing unnecessary damage to epiphytes and lianas), and it is necessary to remove the rings surrounding the trunk one at a time when passing over a branch; a process that is dangerous and fatiguing. Spikes with a single spur, used to climb telephone poles in United States, have the advantage of providing access into all sizes of trees, including those with very large diameters. However, climbing with single-spurred spikes is more fatiguing, less safe, and they are much more difficult to use in combination with a clipper pole than the French "griffes."

For trees smaller than ten centimeters in diameter, a climber can use an adjustable loop called a *peconha* in Brazilian Portuguese (Fig. 4-7). A *peconha* was traditionally made of bark fiber, but is now made of canvas.

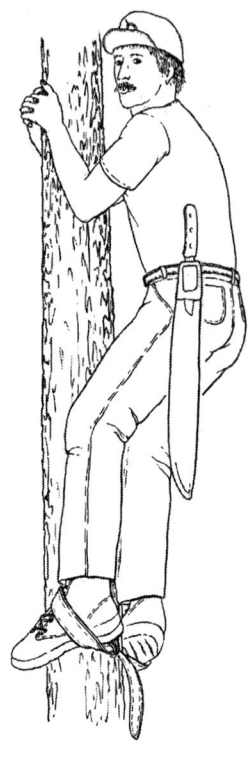

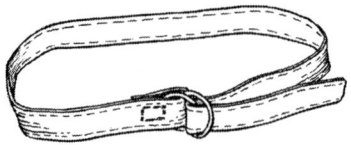

Fig. 4-7. The use of a Brazilian peconha for climbing relatively small trees is an effective, but physically demanding, method for climbing trees from 10 cm to about 20 cm DBH. The climber clasps his hands behind the trunk, keeps his feet attached to the tree by turning the ankles outward, and then "hops" up the tree. Wearing a long-sleeve shirt protects the forearms from abrasion. Drawing by B. Angell.

The climber places it around his feet, grips the tree with both hands and his looped-together feet, and essentially hops up the tree (Mori, 1987b). The *peconha* can be employed to climb trees up to 35 centimeters diameter, but the method is so fatiguing that it is unrealistic to attempt to climb very many trees of that size in a day. However, experienced peconha climbers in good physical condition can climb as many as 10 trees per day. The best *peconha* climbers come from eastern Amazonian Brazil, where they are accustomed to climbing *açaí* (*Euterpe oleracea* Mart.) and *bacaba* (*Oenocarpus bacaba* Mart.) palms to harvest edible fruits.

Mori and Boom (1987) summarized data on trees with diameters ≥ 10 centimeters DBH for six lowland, neotropical forests. In these forests, 89.8% to 97.9% of the trees measured less than 50 centimeters in diameter. However, even in central French Guiana, where tree size is larger, only 10.2% of the trees could not be climbed with the French spikes. In the case of the largest trees, it is often possible to make a collection by climbing an adjacent, smaller tree from which specimens of the larger tree can be reached with a clipper pole. At peak flowering season, a number of collections can often be made from a single climb. Generally, when I am climbing with spikes, I collect from fewer than 10 trees per day; searching for fertile plants, making the collections, taking photos, writing notes, and pressing the specimens make a full day's work.

After a fertile tree has been located and the tree to be climbed selected, the climber and ground support team prepare the pole for the climb. Aluminum pole segments can be purchased from TW Metals (27 Engle Road Dr., Monroe Township, New Jersey 08831 for $100.00 to $180.00 for the six segments, depending on the number of sets ordered (a set = three pairs of nesting pole segments) and cutting done by the supplier. The segments come from the manufacturer in 3.6 meter (=12 foot) lengths, so they have to be cut into the desired lengths, either by the dealer or the researcher.

A clipper pole consists of six separate pole segments, each 1.8 meters in length (Fig. 4-8). Each nesting pair of pole segments consists of an outer, larger-diameter socket section (29 mm diameter) and an inner, smaller-diameter nesting section (25 mm diameter), both of 2 mm gauge aluminum. In the socket section, I drill a 10 mm hole 5.2 centimeters from the end; and in the nesting section, a 10 mm hole 7.6 cm from the end. I insert a spring button into the inner pole section (Fig. 4-9) and the poles are connected by depressing the spring button and sliding the end of the inner section into the outer section until the two holes are aligned and the spring button pops into place (Fig. 4-9). It is important to use springs with the longest button available, because shorter buttons can fail to hold and the sections then separate and fall, potentially striking someone below. The best buttons are the "B" Style Snap Buttons (catalog number B-173), which are available from Valco Valley Tool & Die, Inc. (10020 York Theta Dr.; North Royalton, OH 44133; Phone: 440-237-0160). Prior to an expedition, the buttons should be tested to make sure that they hold the segments together securely.

If the pole or a series of connected segments fall from a tree, there may well be severe consequences (such as impaling someone below the tree). On a recent trip to the Biological Dynamics of Forest Fragments Project north of Manaus, I was nearly skewered when several of the climber's pole segments separated and fell to the ground while I was working under the tree. No matter how many times I remind myself that one should approach a tree only if the climber is aware of your presence and stops all activity, there is always the temptation to walk underneath a climber to take a photograph of the bark, get a closer look at an interesting insect, or pick up a specimen dropped to the ground.

In addition, if parts of the pole separate and fall, it can take considerable effort to retrieve the segments of the pole above the separation. This occurs if the cutting knife of the clipper gets stuck in a branch, the pole segments separate, and the upper segments (attached to the clipper head stuck in the branch) are out of the reach of the climber. In a worst-case scenario, the clipper head and the remaining attached pole segment(s) would have to be abandoned in the tree.

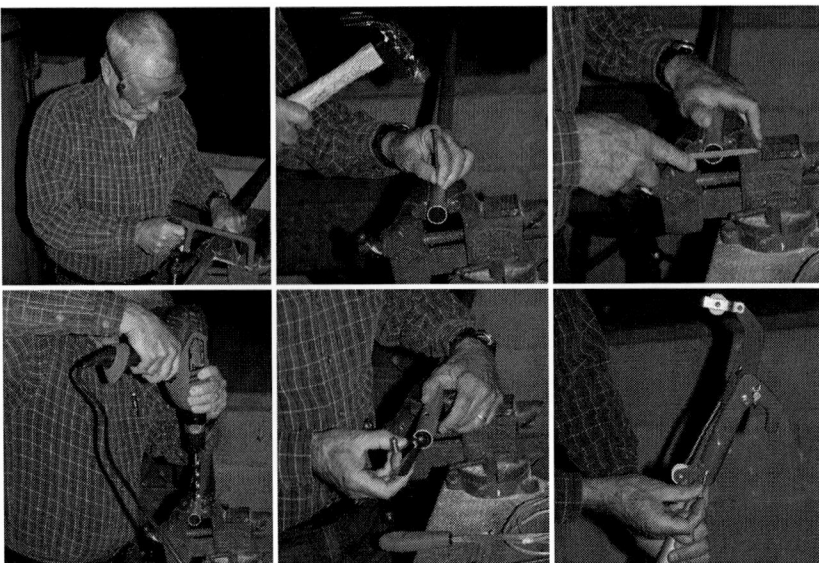

Fig. 4-8. *The author preparing a set of clipper poles. Top left: Cutting poles into the desired lengths. The poles are purchased in either six or 12 foot lengths and are used in six foot or shorter segments depending on the preference of the collector. The author prefers three foot lengths because they can be more easily transported to and from expeditions in a duffel bag. Top middle: A hole is punched in the smaller diameter pole so the drill bit will not slip. Top right: Sharp edges are smoothed to reduce the chances of the pole handler being cut. Bottom left: A hole is drilled which is slightly larger than the spring button, the button will pass through it and a similar hole drilled in the larger diameter pole. Bottom middle: A spring is inserted in the hole of the smaller diameter hole. See Figure 4-9 for instructions on inserting the spring button. Bottom right; a smaller hole is drilled entirely through the end of the smaller diameter pole. The size of this hole is slightly larger than the holes in the clipper pole head through which a bolt passes to attach the head to the pole. Photos by Carol A. Gracie.*

Clipper poles are heavy; therefore, to more easily maneuver the pole while cutting specimens, the cutting knife should be the lightest available. I have found the SnapCut® head: Model 33 to be the best, but it is becoming increasingly difficult to find. There is a tradeoff in using a more maneuverable head, the upper limit of branch diameter it can cut (about three centimeters) restricts its use to smaller branches. A heavier head, with a knife that can cut thicker branches, is extremely difficult to maneuver, especially if the assembled pole has to be moved off the vertical (which is the case when cutting specimens from the outer part of the tree crown, or when cutting from an adjacent tree). To attach the cutting knife, I cut holes completely through one of the inner diameter segments so that they line up with the holes used to secure the knife. (Note that the holes drilled to attach the knife will have to be drilled in different positions for each type of knife.)

When a light clothesline rope (Fig. 4-9) is used to operate the cutting blade, the spring mechanism of the clipper head allows the knife to open more easily. The spring mechanism should also be oiled periodically to allow the cutting edge to open and close more readily—there is nothing more irritating than having to bring the head down to eye level to open the blade by hand. One disadvantage to clothesline is that it frays more easily than heavier rope, and climbers have to be especially careful not to step on it with spikes because it is easily damaged. To insure that the rope will not break in the middle of a cut, I inspect it frequently and replace it as soon as I detect damage. To prevent fraying, both ends of the rope—especially the free end—should be whipped (Fig. 4-10), not knotted. If knots are used to control fraying, they will inevitably get caught in branch forks when the rope is retrieved following a climb.

To facilitate air transport, the 1.8 meter clipper pole segments can each be cut into two sections (this requires drilling twice as many holes, and inserting twice as many buttons). The advantage is that the shorter segments fit into a duffle bag and can be handled as normal baggage. When the segments are 1.8 m long they are considered oversize baggage—increasing the chances that they will be lost in transport! If transport is not an issue, it is better not to cut the poles because it is easier to work in the field with fewer but longer segments.

A fundamental rule for aluminum poles is never ever use them near power lines. Even if a pole does not touch the line, electricity can arc from the line to the pole and electrocute the person holding it. I know of at least one botanist who was killed, and another severely injured, while attempting to collect using metal poles near power lines. For the same reason, metal poles should never be used during lightning storms. The remainder of this discussion assumes that 1.8 meter pole lengths are being used.

When the pole segments are assembled on the ground before a climb, initially only three segments should be put together and lifted from the ground. The last three pole segments should be added by someone on the ground to the end of the pole, one by one, as the climber ascends. When the climber is high enough in the tree, he/she holds the assembled segments vertically, and the person on the ground inserts the next segment into the end of the pole. As each segment is added, the climber pulls the pole up another segment-length from the ground, and the subsequent segment is attached. As the button snaps into the hole of the socket section, the ground support person should make sure that the button pro-

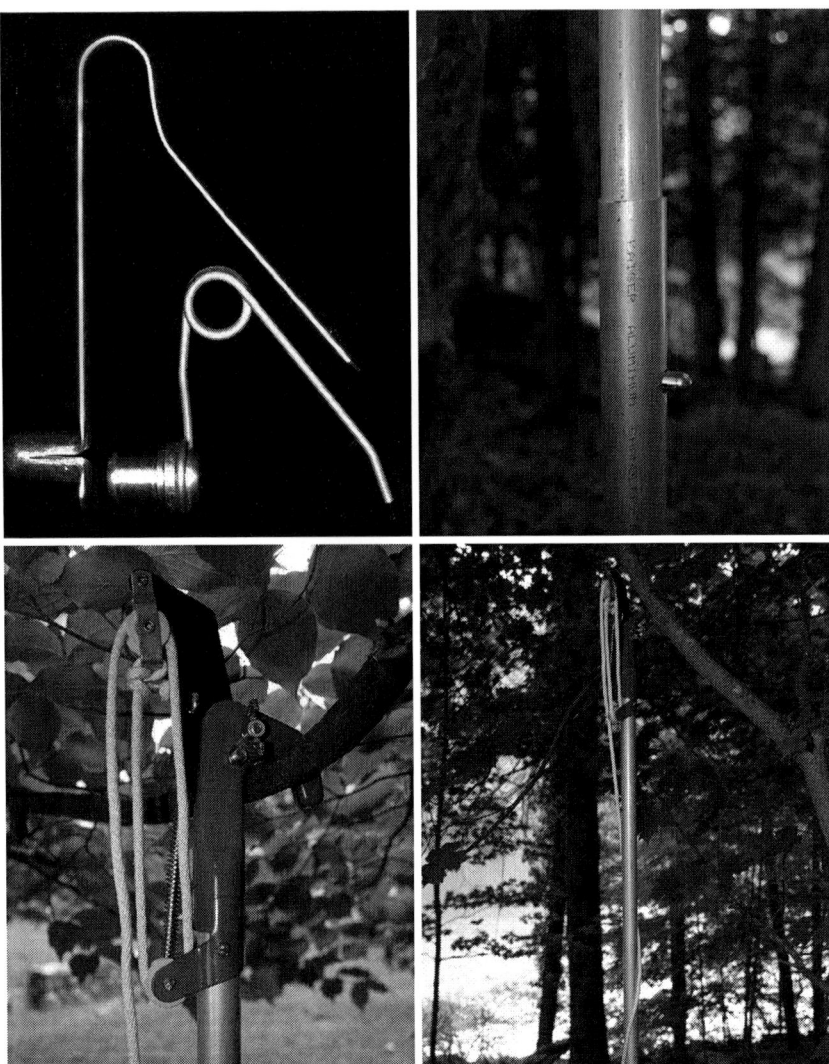

Fig. 4-9. Clipper pole tips. Top left: Two types of button springs for holding the smaller and larger diameter poles together. The spring on the top can be placed in the smaller pole in either direction but the spring on the bottom needs to be placed into the pole button end first because it is difficult to remove if it is placed spring end first. Top right: the smaller and larger diameter nesting poles are held together by the button passing from the smaller diameter pole through the hole in the larger diameter pole. Bottom left: To make maximum use of the pulley the rope is attached to the clipper head in this fashion. The rope is attached to the pulley using two or more half hitches. Bottom right: in order to lessen the chance of bending the poles the rope can be wrapped several times around the shaft of the pole when cutting branches. This is especially true when more than three pole segments are used and when a cut is made with the clipper pole at an angle away from the collector (i.e., the pole is not vertical). Photos by Scott A. Mori.

trudes from the hole, indicating that each new section is firmly attached to those already in place. If a climber first assembles all six segments on the ground, and then lifts the pole from one end, it will flex and this can bend the segments (especially if the pole is being raised from the uphill side). Bent segments will not nest properly, making them more difficult to carry, assemble, and disassemble. Worse still, it is difficult to manipulate a bent pole while cutting branches and this greatly increases the effort required to make a collection.

Before starting the climb, the head of the pole is placed at the base of the tree and the rope coiled so that it will easily unravel. The free end of the rope is attached to the climber's safety belt, and the rope on the ground unravels as the climber ascends the tree. When the last segment of the pole leaves the ground, the climber is at about 22 meters (the length of the poles plus an equal length of rope), providing a reference for estimating tree height (a pole consisting of six connected 1.8 meter segments is the longest that can be handled by most climbers). An ascending climber must take care that the pole doesn't get caught on a liana or projecting branch. Should that occur, ground support can guide the pole past the obstacle as the climber continues to pull it upward.

Once the pole is out of the hands of ground support, everyone should remain clear of the tree. On one of my expeditions, a student climbed into a tree alongside a giant *sapucaia* tree (*Lecythis zabucajo* Aubl.) that he wanted to collect. A sudden storm rushed across the valley bottom up onto the ridge where he was climbing. He looked about with terror as he witnessed trees whipping back and forth in the wind, and when the squall struck his tree he dropped his pole, hugged the tree, and prayed that the trunk would not snap. The pole fell like a spear and drove 20 cm into the ground. Fortunately, no one was underneath the tree. I have dropped pole segments, machetes, hand clippers, and branches while climbing, but luckily these never hit anyone below. One of the greatest modern tropical plant explorers, the late Al Gentry, was struck on the head with hand clippers dropped by a tree climber... but Al ignored the bloody gash and continued to collect plants for the rest of the day.

Before climbing a tree, it is important to make sure that it is alive and without cracks or major deformities in the trunk; that there are no large branches that could fall onto the climber or people below; and that there are no ant, bee, or wasp nests in the tree (the climber should also make periodic checks during the ascent). Safety lanyards must be attached at all times, and the climber must make sure that the spikes are securely planted, at both the apex of the curve and at the heel. Once a climber is accustomed to the French spikes, planting the spurs becomes second nature.

If bees or wasps attack, the best strategy is to stop moving and remain as still as possible. This does not, however, deter trigonid bees, which do not sting or inject venom, but bite; they are an incredible nuisance when thousands of them swarm around your head and—attracted to secretions—get into your eyes and ears (see Chapter 4).

If a tree is small in diameter, the safety lanyard will not break a fall; it will simply slide down the tree if a fall occurs. The lanyard should therefore be double wrapped around trunks that are less than 20 cm. As a safety measure the climber should employ two lanyards, especially when passing branches in the tree's crown. Upon reaching a branch, one lanyard remains secured around the main trunk while the second is released at one end, passed around the trunk above the branch, and reattached at the free end. The lower lanyard can then be removed and reattached above the branch. When climbing into and within the crown, it is sometimes necessary to cut away branches with a machete (or pruning saw). The most dangerous branches to remove are those that are vertically oriented and directly above the climber, because it is difficult to control their fall, which most of the time is straight down in the direction of the climber. My first student, Brian Boom, cut such a branch on an expedition to French Guiana in 1983; the branch broke his watch-band, which fell from his wrist and was never seen again!

When the climber is in position to collect, the next step is to select an appropriate branch to cut (Fig. 4-9). The branch should not be so large that the clipper blade cannot make a complete cut, otherwise the blade may get stuck and be very difficult to dislodge. If the branch is severed but the cutting head remains stuck in the falling branch, the weight of the branch can bend the pole if the climber continues to hold it securely. In this case, the climber might have to alert ground support, and let the pole fall to the ground along with the branch.

Cuts made while the pole is not oriented vertically are more difficult, and the chances of bending the pole are greater. The rope should therefore be wrapped around the shaft enough times to direct the force of the pull along the length of the pole (Fig. 4-9) rather than at an angle to it. An experienced collector can go for many years without bending a pole, but in just one climb a novice can bend it so severely that all subsequent use of the pole will be difficult.

Specimens often get caught in branches as they fall to the ground. Most can be retrieved by reaching out with the pole, grasping the specimen with the clipper head, tightening the rope just enough to secure it in the jaws of the clipper, and bringing it back towards the climber. A light clipper head is almost mandatory for this maneuver, and also to cut specimens

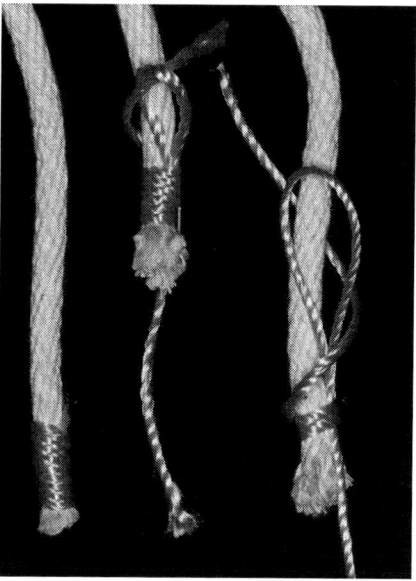

Fig. 4-10. Additional clipper pole tips. Left: The poles are more efficiently carried from one tree to another by wrapping them and the rope together with a velco strap at the top and the poles together with another velco strap at the bottom of the pole. Right: To stop the rope from fraying at both ends it is whipped as shown here (steps shown in order from right to left). Do not use a knot for stopping the rope from fraying because it gets stuck in the forks of plant branches when the rope is retrieved. Photos by Scott A. Mori.

from adjacent trees. The ground support person should watch carefully to see where the cut specimens land, because the ground may be obscured to the climber.

Before moving to the next tree the pole segments should be separated and held together with Velcro® straps (Fig. 4-10) or Bungee® cords. If the climber drags the assembled pole through the forest, it will be difficult to maneuver through the understory, and the segments may get bent as the pole is pulled around trees.

One downside to climbing with spikes is the potential damage to the tree. Trees with thick outer bark can be climbed with relatively little damage because the outer bark is dead, and therefore spikes do not bite into the tree's living tissues. When trees produce latex or resin, the exudates immediately fill the wounds and apparently inhibit the entry of insects, bacteria, and fungi into the tree. Castilho et al. (2006) evaluated the impact of wounding due to the use of French climbing spikes on the growth rate and survival of trees. In their study no trees died as a result of spike injury, and the growth rates of spike-climbed trees did not differ from those of control trees. Nevertheless, when trees are climbed repeatedly, which may be required for ecological research including pollination studies, spike use can lead to the death of the tree, even in species defended by exudates (Mori, 1995a). The gaps created when trees die and fall are a natural part of tropical forests, and a small increase in gap formation as a result of climbing with spikes probably does not have a negative impact on plant diversity (Mori, 1995a) or forest structure. But for ecological projects, in which trees will be climbed repeatedly, rope-climbing techniques or canopy cranes facilitate canopy access without harming the trees.

Climbing trees has rewards above and beyond the plant collections acquired. The climber passes out of the dark forest understory into an airy realm with breathtaking vistas over the forest canopy, and has the opportunity to see animals in their arboreal habitats. For example, I have had close encounters with howler and spider monkeys, and have also seen a harpy eagle land in a tree next to me to escape pursuit from smaller birds, a violaceous trogon take the fruit of a Sapindaceae while still in flight, mixed flocks of birds visiting *Moronobea coccinea* Aubl. (Clusiaceae) to take nectar from its flowers, and peccaries passing far below on the forest floor.

One of my most exciting arboreal experiences was a nocturnal climb into a large tree adjacent to a *Hymenaea courbaril* L. (Fabaceae/Caesalpinioideae) tree to observe animals visiting its flowers. Because the flowers open shortly before sundown, and are reported in the literature to be visited by bats (Howell & Burch, 1974), I expected to see bats taking nectar from the flowers. The climb was difficult because the lower part of the trunk was at the upper limit of my climbing spikes, and the trunk expanded and became cracked at 10 meters, perhaps as the result of a lightning strike. The first evening I scaled the tree, I reached 20 meters when a ferocious squall hit, causing this large tree to sway back and forth. Deluged by torrents of rain, I made a nervous and hasty retreat to the safety of the ground. The next evening the weather was clear as I climbed toward the canopy at 4:30 p.m. and settled into position to observe the flowers as they opened. At first I only saw a few hummingbirds visiting the flowers, but by 5:00 p.m. hundreds of them, along with perching birds, swarmed over the canopy showing flashes of brilliant color as they fed on nectar. I was thrilled to see such an impressive sight even before the fall of darkness. About an hour after sunset, I heard the sounds of animals moving through the canopy, and when they entered the *Hymenaea* tree I saw three pairs of eyes moving up the branches toward the flowers. I directed my flashlight beam towards one of the animals and saw a kinkajou staring back at me. Although I did not actually witness them visiting the flowers, I suspect that they had come into the tree to feast on the abundant nectar. I did not end up seeing any bats that evening, but they probably also visit and pollinate the flowers of *H. courbaril*, as has been reported several times (Geiselman et al., 2001).

Preparing specimens. Each specimen added to a herbarium incurs both immediate and continuing expenses. Anderson (1996) estimated that mounting a single specimen and incorporating it into the herbarium of the University of Michigan costs approximately $5.00. His calculation did not include other costs such as printing labels from a database, sending duplicates to specialists for identification, managing duplicates sent on exchange to other herbaria, changing the location of specimens in the herbarium when there are name changes, and keeping the names on the actual specimens in sync with their records in the institution's database. Even beyond that, there are the costs of long term specimen storage, which includes keeping them free of insect and fungal damage, and the expense of making the specimens available for botanists to study. Thus, the costs are much higher than that estimated by Anderson and probably reach $30 per collection at NYBG (Kallunki, pers. comm., 2009), and that excludes the costs of collection, long term storage, and use of the specimens for botanical study. In the face of these high expenses, cost/benefit ratios dictate that only well-prepared, informative specimens should be archived in herbaria.

The best way to make high quality specimens and to record collection data accurately is to enter information in the field collection book, and put the specimens in a field press, when the collections are gathered. Descriptions of bark are almost impossible to remember at the end of the collecting day and should, therefore, be recorded in immediate proximity of the tree to insure accurate and detailed recording of the bark features (Figs. 4-3, 4-11, 4-12). The most time-efficient way to handle specimens is to arrange them carefully as they are put into the field press (and then make only minor adjustments when they are being positioned in the drying press at the end of the day. Field-pressing specimens rather than placing them in plastic bags to be pressed in the evening ensures better herbarium specimens in three ways. First, specimen data are more accurately recorded when the features of the plant are fresh in mind. Next, the collections have not yet wilted, and leaves, flowers, and fruits are still attached to the stems. Finally, the probability of creating mixed collections, for example pressing the leaves of one species with the flowers of another species, is significantly reduced. Mixed collections cause problems for all researchers who subsequently study them. For example, Aublet, who collected plants in French Guiana from 1754 to 1756, described a number of species based on mixed collections, including one based on the flowers of *Lecythis grandiflora* Aubl. (a synonym of *Eschweilera grandiflora* [Aubl.] Sandwith) and the fruits of *L. zabucajo*. The mixed collection was treated as *L. zabucajo* in his *Histoire des Plantes* (Aublet, 1775; Howard, 1983).

After a long day of collecting, it is difficult to spend hours separating collections that have become compacted together, sometimes in many separate bags, and accurately record data and press specimens with the care needed. Under these conditions, the pleasure of collecting turns into a tedious and tiring job that leads to mistakes in numbering, poor quality specimens, and a diminished quality in field notes. The objective of plant collecting is to make the best possible collections, accompanied by the most accurate and informative data; a goal best reached by taking notes and pressing as the plants are collected.

Preparing plant speci-

Fig. 4-11. *The author is recording bark characteristics, such as the outer bark appearance (e.g., smooth, scalloped, or fissured) and the thickness and color of the inner and outer barks, which are useful in plant identification. Bark features are observed by making a slash and this should not be done in nature reserves. Bark characters are very useful, but they are seldom recorded on herbarium specimens.. Drawing by Michael Rothman.*

mens in the rain is difficult, if not impossible, unless the collector is well prepared. I recommend using a field press of water resistant canvas (available from the Herbarium Supply Company (705 Bridger Dr., Unit D; Bozeman, MT 59715-2292; http://www.herbariumsupply.com/nu_dynamicIndex.asp) to keep the specimens dry. In addition, I sometimes carry a very light tarp with ropes tied to the grommets, and at the first sign of rain, I set it up quickly by tying the ropes to adjacent trees. If there is room to attach it to my pack, I often carry a small folding field chair in the rainy season to make it more comfortable while pressing and, then, while waiting for the rain to stop. This enables me to protect specimens from the rain as I prepare them and put them into the water resistant field press, which is, in

Fig. 4-12. The author pressing plants in the field. The best collections are made by taking notes and pressing the plants onsite just after they have been collected. Drawing by Michael Rothman.

turn, placed into a plastic bag. If the field press gets too full, I tie specimens in bundles and place the bundles in plastic bags and leave them along the trail to pick up on the way back to camp. Even with all this care, the newspapers often get wet and have to be replaced back in camp.

It is better to dry herbarium specimens in a field dryer (Fig. 4-13), rather than to treat them with a liquid preservative and dry them after the field trip. For shorter trips, specimens can be held in bundles for two days without markedly reducing the quality of the resulting herbarium sheets. Significant deterioration begins on the third day, and after three days the

Fig. 4-13. The author (left) and Brian Boom (right) attending to drying presses. Specimens that are dried in the field are more aesthetic and informative than those that are put in alcohol and dried after returning to the herbarium. Drawing by Michael Rothman.

specimens begin to lose color, shed leaves and flowers, and fungus starts to grow on them, resulting in specimens of such poor quality that they are of little use.

When it is not possible to dry specimens within three days of collection, they can be moistened with a 50-70% alcohol solution and preserved in sealed plastic bags. This is best done by spraying the specimens in their individual sheets, tying the moistened sheets into bundles, placing the bundles in plastic bags, and sealing the bags. Alternatively, the collector can first put the bundles in bags and then pour the alcohol solution into the bag, but it is imperative to use enough liquid to moisten the entire contents of the bag. If the center of the bundle remains dry, the specimens will lose their leaves and eventually rot. With enough liquid, specimens can remain in the bags for several months. Other preservatives, such as formalin, can be used, but they are more caustic and unpleasant to work with and, thus, should be avoided. Herbarium labels should indicate if specimens have been liquid-preserved prior to drying, because specimens preserved in this way are not suitable for chemical or molecular studies.

Specimen images. Herbarium specimens provide a great deal of information, but photographs taken in the field markedly increases their value, making it well worth the extra time expended. This is especially true for succulent plants such as Cactaceae, very large plants such as palms, plants with fleshy flowers such as Lecythidaceae, and plants with fruits and seeds that have characters difficult to study once they have been dried (e.g., the position and shape of the hilar scar in Sapotaceae; their seeds must be removed from the fruit before they are dried in order to see the features). In addition, bole form and bark characteristics are more effectively communicated through images. Digital images are especially informative because they can be attached to specimen records in databases and, in turn, displayed on web pages.

   Although images without voucher specimens can be useful, the images become much more valuable when they are documented with herbarium specimens. In the Brazil nut family, photographs of the two best-known species, the Brazil nut (*Bertholletia excelsa* Bonpl.) and the cannon ball tree (*Couroupita guianensis* Aubl.), have some value without specimen vouchers because they are relatively easy to identify based on images alone. However, most other species of the family are more difficult to identify and, thus, herbarium vouchers are needed to confirm the identity of the species represented by images.

   Careful photographic documentation is time-consuming and, thus, there is often not enough time for general collectors to make field images for all of their collections. However, images are so valuable that monographers and floristicians should make every effort to photograph as many of their collections as possible. It is especially critical for monographers to photograph the characters that are used in classification. For example, in the Lecythidaceae the flowers possess features that are lost when they are dried, especially over excessive heat, and the seeds have arils that are not easy to observe when the fruits are dry because dry fruits are difficult to open, and the arils are modified by the drying process. In addition, there are many bark features of Lecythidaceae that go unrecorded on herbarium specimens. Nearly all plant families have features that can be more accurately communicated in photographs than in herbarium collections alone—ideally collectors should learn what these features are and document them with images. See Box 4-1 for the photographic protocol I use in imaging collections of Lecythidaceae.

   Although adequate pictures can be taken with a handheld point-

Box 4-1. Protocol for Images of Lecythidaceae
by Scott A. Mori

Each plant family has a suite of characters used in classification. The following photographic protocol was developed to insure consistent imaging of the features used in the classification of Lecythidaceae, and to encourage general collectors to follow this protocol when photographing collections. Specialists of other families are encouraged to develop protocols appropriate for the groups they study.

1. **Bark:** Images should give an idea of the general appearance of the bark. In the Lecythidaceae, there are sometimes striking differences in the barks of closely-related species. For example, the fissured bark of *Corythophora rimosa* is very different from the scalloped bark of *C. alta* (Fig. 4-3). To best demonstrate the relationship between the thickness and color of the inner and outer barks, make and photograph an oblique cut of the bark called a slash or blaze. The first bark image should include a small paper ruler with the collector's name and number written on it. The final image of the collection includes the same label so that all images of the same plant are bracketed; this reduces the possibility of subsequently confusing the identity of images. The best bark pictures are taken under natural light with the camera held on a tripod.
2. **Trunk:** In Lecythidaceae, the most important trunk difference is that some species are cylindrical at the base and others are buttressed in various ways. The best trunk pictures are also taken under natural light with the camera held on a tripod,
3. **Specimens:** Once the collected branch is on the ground, a piece of stem with attached leaves, and preferably flowers or fruit, is placed on a black velvet cloth along with the paper ruler showing the collector's name and number. At least one leaf should be turned to show the abaxial surface, because some species are characterized by leaves having surfaces with different colors. Fallen empty fruits should be included if they are present. This photograph provides a colored image similar to the herbarium specimen that will be prepared. If the specimens are lost, a print of this image can used as a substitute for the herbarium sheet.
4. **Stems:** This image should show the bases of the petioles and the apex of the stem. If young stems differ from older stems, both should be photographed.
5. **Petiole cross section:** To standardize position, the cross section is made at the base of the petiole (i.e., at the attachment of the petiole and the stem). To demonstrate the petiole's orientation in relation to the leaf blade, the base of the blade should be visible.
6. **Inflorescence:** The inflorescence is usually included in the specimen image taken on black velvet. Sometimes additional detailed images are needed to show its branching pattern, the orientation of the rachis (straight or zig-

zag), the shape and persistence of bracts and bracteoles, and the presence or absence of pedicels.

7. **Flowers:** Apical, lateral, and basal views of flowers show the pedicel (when present), the orientation and imbrication of the calyx-lobes, and petal orientation. In Lecythidaceae, a medial longitudinal section of the flower shows the nature of the androecial hood (e.g., flat or coiled, and presence or absence of appendages or a protective external flap), the nature of the style (well-defined or difficult to distinguish from the summit of the ovary) and its orientation (erect or oblique). Medial and cross sections of the ovary or of very young fruits show the type of placentation, the number of locules, and the presence or absence of mucilaginous ducts. Single-edged razor blades are useful for making the sections.

8. **Fruits:** Apical, lateral, and basal views of the fruits show the position of the calycine ring, the orientation of the infra- and supra-calycine zones, and the persistence of the pedicel. In addition, the operculum is photographed to show the presence or absence of an umbo (the style remnant) and columella (a projection from the operculum into the fruit).

9. **Seeds:** Photographs of fruit with mature seeds *in situ* are taken to demonstrate seed arrangement, and then seeds are removed and photographed. Lateral and basal views of the seeds show seed coat surface, venation pattern, and presence or absence of an aril and, when present, the position of the aril.

and-shoot camera (especially for display on the Internet), the best images are obtained with an SLR digital camera. The smallest aperture available in most point-and-shoot cameras is f8, whereas most SLR cameras have stops up to f32. Higher f-stops allow greater depth of field which, in turn, gives a clearer image of the object photographed. Shooting pictures with a tripod reduces camera shake, allows images to be taken without a flash (under certain conditions), and permits shooting at slower shutter speeds. The colors of images captured without flash are more true to life, and this seems to be especially true of trunk and bark features. However, under the low light conditions of tropical forests, a flash is usually needed and, in those situations, I have taken satisfactory photographs using a ring flash for close-ups and the built-in flash for overall trunk and bark images (decreasing the intensity of the flash by placing a thin piece of white tissue over the flash will, however, improve the color rendition). A tripod improves even images captured under flash because it insures that the camera doesn't shift in between the time that the image is composed and the moment that the shutter is released. The plant holder described in Box 4-2 will secure flowers while they are being photographed (Fig. 4-14). Clamps available in most hardware stores can also be used to hold specimens in

Box 4-2. Construction of a Plant Holder for Photography

by Scott A. Mori

Materials

1. Vinyl coated copper wire, 9 gauge: approximately 20 in (50 cm) long.
2. Alligator clip with the screw removed.
3. Epoxy glue.

Procedure

1. Remove 0.75 in (19 mm) of the vinyl coating from one end of the wire.
2. Insert the stripped end of the wire into the tube at the base of the alligator clip.
3. Glue the alligator clip to the wire.

Notes

1. When the alligator clip is glued to the wire, it should be oriented so that the plant specimen will be in the best possible position for photography.
2. To use this plant holder, wrap the free end of the wire around a support such as a small tree, or coil it at and place it on the ground. In the latter case, if the sample being photographed is heavy, weight the base of the plant holder with a rock or other heavy object.
3. The sample (petiole, flower, fruit etc.) or the stem bearing the sample is secured in the alligator clip. This frees both hands for camera operation.
4. The best pictures are obtained when a tripod is used to ensure that the camera does not move.
5. A more flexible 8 or 9 gauge aluminum wire can be substituted for the copper wire because it is more easily wrapped around the stem of a small tree. On the other hand, the more rigid copper wire works better when the plant holder is placed directly on the ground.
6. Clamps that are commercially available in hardware stores (Fig. 4-14) are also useful for attaching branches with flowers or fruits to small trees.

place (Fig. 4-14).

Because each image takes time to process, I do not keep those that duplicate other images. This may seem like a trivial point, but archiving only those images that show unique features saves time, both in image management and for those consulting websites where the images are posted (see Naming Image Files in Chapter 5).

Fig. 4-14. Top: *A plant holder used for securing flowers and other plant parts to facili-tate photography. This plant holder can also be bent into loops at the end of the rubber coated wire and placed on a flat surface. Bottom left: Two different sized clamps that can also be used to secure parts of plants for photography. These are inexpensive and can be purchased in any hardware store. Bottom right: a plant of Oxalis stricta at-tached to a branch with a clamp.*

Drying specimens. Drying specimens requires a drying frame (Figs. 4-13, 4-15, 4-16, 4-17 and Boxes 4-3 & 4-4), a heat source, and a drying press. The best heat is provided by light bulbs, hot plates, or forced air space heat-ers. Forced air heaters are probably the best, but most of the models that are commercially available automatically turn off when they are positioned with the air blowing upwards, the direction that will most efficiently dry plant specimens. Thus, most forced air heaters need to be modified by (1) deactivating the switch that turns the heater off when placed in unusual positions, (2) making a stand that holds the heater and directs its output upward, and (3) deactivating the thermostat so that the heater does not shut down when used with the drying frame. A modified forced air space heater must then be used with extreme caution to reduce the risk of fire.

On expeditions where electricity is not available, propane gas is the cleanest and most easily handled heat source. Propane is available in 5.5 and 24.5 kilogram sizes (measured when tanks are full). A 5.5 kilogram bottle burns at medium intensity for approximately 33 hours. Because the burning time varies under different conditions, I keep a running record of burn time by writing the number of hours of each drying session with a magic marker on the bottle. The drying frame for using propane heat sources is described in Box 4-3.

Specimens dry best over a steady, gentle heat. When specimens are dried slowly, they are more attractive, and are more informative because they retain color better and don't lose the structure of their reproductive parts. For example, if the fleshy flowers of the Brazil nut family (Lecythidaceae) are dried on extremely hot dryers, they become brittle and do not regain their form even if they are reconstituted in boiling water. The best way to dry flowers of this and other groups with fleshy flowers is to put them in a paper bag and place them on the coolest part of the dryer or, if the weather permits, in direct sunlight. The paper bags should not be flattened.

A traditional drying press consists of a series of specimens, each of which is carefully arranged within a folded newspaper sheet, intercalated between two "sandwiches." A "sandwich" consists of a blotter, an aluminum or cardboard corrugate, and another blotter. The blotters absorb moisture from the plant specimens, and the corrugate allows air to pass between the blotters and hasten the drying. First, a press end board is laid down, and then a "sandwich," followed by a specimen, followed by another "sandwich" until all of the specimens in the field press have been transferred to the drying press, then another press end board. Periodically alternating the direction of the newspaper fold to make sure that thick parts of the specimens counterbalance one another prevents the drying press from becoming lopsided. Although the drying press can be tied with ropes, it is much more efficient to hold and tighten the press using nylon webbing straps with hand-tightened parachute buckles (available through the Herbarium Supply Co.).

Box 4-3. Construction of a Standard Wooden Propane-Driven Drying Frame

by Scott A. Mori

A plant drying frame is a rectangular box slightly narrower than the width of the plant press and as long as the average number of specimens one intends to dry per day. This drying frame will hold about twice the number of collections as the mini-dryer described in Box 4-4. The height of the box depends on the heat source—the press should always be at least 14 in (35.5 cm) from open flames. The frame described here was designed to accommodate two 5½ kg propane bottles attached to screw-in burners. If the heat source is a short stove connected to a propane bottle that remains outside of the drying frame (Fig. 4-13), the height of the frame needs to be reduced. A non-propane driven drying frame is shown in Fig. 4-15 to give the reader an idea of the parts that make up a drying frame constructed with wood—all that differs from the frame illustrated in that figure and the one described in this box are the size and height of the dryer.

Materials

1. Two plywood end boards (= the width of the press): 0.25 in (6 mm) thick × 17 in (43 cm) wide × 24 in (61 cm) tall.
2. Two plywood side boards (= the length needed to accommodate specimens collected during a day of collecting): ¼ in (6 mm) thick × 24 in (61 cm) long × 24 in (61 cm) tall.
3. Two railings for the tops of the side boards: 1 in (2.5) thick × 2 in (5 cm) wide × 22.5 in (57 cm) long. If weight is a concern the railings can be eliminated.
4. Four wooden legs: 1 in (2.5 cm) thick × 2 in (5 cm) wide × 26 in (66 cm) long
5. Ten bolts with wing nuts (this includes two spares): 0.25 in (6 mm) diameter × 1.5 in (4 cm) long, wing nuts to fit.
6. Sixty finishing nails: 1.25 in (3.8 cm) long.

Procedure

1. Nail the railings to the tops of the side boards, leaving 0.75 in (1.8 cm) spaces at both ends to accommodate the legs.

text continues on next page

Box 4-3 continued

2. Nail the side boards to the 1 in (2.5 cm) side of the legs. Note that the end boards will be bolted to the side boards when the frame is assembled.
3. Temporarily nail the end boards to the legs of the side boards to facilitate drilling the holes that will accommodate the nuts and bolts that hold the frame together.
4. Drill two holes large enough to accommodate the bolts that pass through the end boards and the legs. One hole should be 3 in (7.5 cm) from the bottom of the side board and the other 3 in (7.5 cm) from the top of the side board.
5. Drill two extra holes in the railings. These holes do not need to be in any particular place because they will hold the spare wing nuts and bolts.
6. Remove the nails used to temporarily hold the frame together.

Notes

1. The exact measurements of the thickness and width of the railings and legs are not important as long as they are strong enough to support the press.
2. Metric measurements are to the nearest centimeter.
3. The legs are 2 in (5 cm) longer that the side and end boards to allow air to enter under the bottom of the drying frame and then pass upward through the press.
4. A mesh wire can be placed on top of the dryer (Fig. 4-14) to keep the press strap and other inflammable material from falling onto the flame and causing a press fire.
5. Burner heads that attach directly to the top of the gas bottle should be acquired in the country where the gas is purchased to ensure that they are compatible with the bottle's screw threads.
6. A 5.5 kg propane gas bottle will burn for approximately thirty-three hours. Either one or two bottles can be used, depending upon the size of the drying press on a given day. If a day's collections do not completely fill the press, the dryer can be run on alternate days.
7. The open flame of the heat source should always be at least 14 in (35.5 cm) from the bottom of the plant press. The dryer should be closely monitored, especially the first time it is used, to make sure that it does not overheat and cause a fire. If the drying protocol is modified in any way, the dryer should also be closely monitored following the change. Fire is a constant risk when plants are dried over open flames; thus, it is preferable to use an electric heat source, such as a hotplate, whenever possible. Even with a hotplate it is important to be aware of the possibility of fire.
8. The dryer should not be used in the place where expedition members sleep, both to avoid the possibility of injury in case of a fire or the release of fumes (especially if the stove goes out before the tank is empty).

Fig. 4-15. A wooden hotplate driven mini plant drying frame. The dimensions for this frame are provided in Box 4-4 (i.e., the aluminum mini frame) and are different from the dimensions for the standard propane-driven drying frame provided in Box 4-3. The drying press shown here differs from the standard frame only in the dimensions of the frame. This and the aluminum mini frame are the same dimensions. A mini frame has the ends wider than the sides because it is designed to accommodate fewer collections. Top: The pieces needed to make a wooden plant drying frame. The length of the frame side is determined by the number of collections to be dried and the height of the frame is determined by the source of the heat, this frame is short because a hotplate provides more moderate heat. Bottom left: A protective screen keeps debris and the drying press straps from contact with the heat source. Bottom right: The drying frame, protective screen, and drying press in place. This drying set-up is inexpensive, easy to make, and fast to set up and take down. Photos by the author.

Box 4-4. Construction of an Aluminum Hotplate-Driven Mini Plant Drying Frame

by Scott A. Mori

This mini plant dryer is designed for specialty collecting of relatively few specimens per day. A larger dryer is needed for general collecting (see Box 4-3). The mini dryer is handy because it can be dissembled and carried in a suitcase or backpack and relatively safe because the metal frame and welder blanket make it fire resistant. On the other hand, it is relatively expensive and time-consuming to build, takes about 20 minutes to assemble, and requires a lot of nuts and bolts to hold it together—and these are easy to lose. The heat source is a small hotplate; thus, if a different heat source is used (e.g., a larger hot plate or propane gas), the size and height of the frame given here will have to be modified to ensure adequate space between the heat source and press. A mini dryer can also be made from wood by modifying the sizes for the frame described in Box 4-3 (Fig. 4-15).

The mini-dryer accommodates a plant press with 65 cardboard corrugates (without specimens included). As specimens are added, the number of corrugates that end up in the drying press depends on the number of specimens collected and their thickness. If the specimens are very thin, the press can hold 10 collections, with three duplicates of each. With thicker specimens, it is usually possible to dry five collections, with three to six duplicates of each, per day. Using a single burner hot plate set on high, easily dried plants (such as grasses and understory shrubs) dry in seven hours, and most plants can be removed after nine hours. Large, fleshy flowers; large, fleshy or woody fruits; and other fleshy plant parts are more effectively dried in bags placed on top of the drying press. This mini-dryer is not effective for drying large or fleshy plants such as aroids, cacti, and palms.

Before attempting to construct this dryer, study Figs. 4-15 and 4-16 and become familiar with what is meant by the ends and sides of the drying frame and the plant press.

Materials

1. Two upper end angle irons or aluminums, henceforth called angles, these are slightly longer than the side of the frame: 17 in (47.5 cm) long.
2. Two upper side angles (these accommodate the length of the press and determine the number of specimens the dryer can support): 12.5 in (31.8 cm)

long.

3. Four angle legs: 15.5 in (39.4 cm) long.

4. Two end braces: 18 in (45.7 cm) long.

5. Two side braces: 12.5 in (31.8 cm) long.

6. Fire resistant welding blanket: 63.5 in (162 cm) long × 14.5 in (36.8 cm) wide.

7. Welding blankets can be purchased on the Internet by searching for "welding blankets and curtains." The margins of the blanket should be hemmed by the manufacturer to prevent them from fraying.

8. Hole punch or knife. The punch or knife is needed to make holes in the welding blanket to attach it to the frame.

9. Eighteen bolts with wing nuts (this includes two spares), to attach the end and side angles to the legs, the lower braces to the legs, and the welding blanket to the top of the drying frame: 0.25 in (6 mm) diameter × 1 in (25 mm) long.

Procedure

1. Drill bolt holes in the legs, end angles, side angles, and braces (see Figs. 4-16, 4-17). The holes to attach the braces to the legs should be made 1 in (2.54 cm) from the bottoms of the legs. Other holes can be drilled in the middle of each of the top and bottom angles and braces of the four sides for attaching the fire-resistant cloth more tightly but I do not find this necessary with such a small drying frame).

2. Attach the end and side angles to the legs with the bolts and wing nuts. Place the wing nuts on the outside of the frame.

3. Attach the braces to the legs.

4. Use the punch or knife to make holes in the welding blanket. The blanket will attach to the same nuts that hold the frame together, therefore the holes should be made at the points where the blanket passes the nuts holding the angles to the legs, i.e., two at each upper corner. Grommets can be used to line the holes but they often fall out with use. A more permanent way to prevent the holes from fraying is to use strong thread and make a button stitch to line the holes (Fig. 4-17).

5. Secure the welding blanket to the frame. The lower margin of the blanket only needs to be attached at the bottom of the leg where the blanket overlaps. In all, the blanket is attached in nine places, twice at the top of each side of the four legs and once at the bottom of the leg where the blanket overlaps.

6. Tighten the frame.

Notes

1. Four bolts are needed at each corner of the frame. One to attach the end angle to the leg, one to attach the side angle to the leg, one to attach a brace

text continues on next page

Box 4-4 continued

to the end side of a leg, and one to attach a brace to the side of a leg.

2. A 1 in (2.54 cm) opening must be left around the bottom of the dryer to allow air to enter from below, be heated by the heat source, and move up through the press (therefore the width of the welding blanket is shorter than the total dryer height). For this reason, the braces should be placed one inch (2.54 cm) from the bases of the legs.

3. Forced air heaters are a better heat source than a hot plate, but they usually need to be modified to allow them to blow air upward without shutting off automatically (as a safety feature). The height of the dryer has to be adjusted according to the height of the heat source.

4. A drying frame built with iron angles weighs 6 pounds (2.7 kg), and one built with aluminum angles weighs 3.5 pounds (1.6 kg). The iron angles come with holes, but they do not always line up with one another, and may have to be expanded with a drill or a file so that they can be properly aligned. The aluminum angles do not come with holes, so they have to be drilled *de novo*. For larger drying presses, aluminum angles and braces must be used with caution because they may not be strong enough to support the weight of heavier plant presses.

5. In this system, the angles face downward (Fig. 4-17, top right) and this allows the plant press to be easily rested on their margins. If the angles face upward, the plant press has to be perfectly aligned within the angles, and this causes unnecessary work.

I sometimes eliminate the blotters altogether, and use only high quality double-sided cardboard corrugates in the drying press. In most cases, I have not been able to detect any difference in the quality of specimens dried only with cardboard corrugates and those dried between blotters as described above. In the early years of botanical exploration, when plants were dried without an artificial heat source, blotters were more important because warm blotters absorbed the moisture from the specimens. The blotters were replaced periodically with others that had been dried in the sun or by another heat source. For many plants, the reliable and gentle heat sources available today allow for the preparation of beautiful specimens without the use of blotters. Heat that escapes from the plant dryer without passing upward through the corrugates decreases the efficiency of the drying process. If the press does not completely cover the space on the drying frame, heat should be prevented from escaping through the open areas. Extra corrugates can be added to both ends of the press as "filler," so that all of the hot air will be forced to move through the press. In the past, if the drying press did not cover the entire length of the drying frame, I

Fig. 4-16. An aluminum mini plant drying set-up. Top: From left to right: the drying frame, the hot plate heat source, and the drying press. Bottom left: A view of how the angle aluminums of the press are assembled. Note that aluminum angles are used to reduce the weight of the frame but angle irons can also be used. Bottom right: An assembled plant dryer with the drying press in place (See Box 4-4). This drying frame is used when weight and space are an important factor and when few plants will be collected each day. Photos by Scott A. Mori.

covered the open ends with a few extra corrugates at the ends of the drying press (Fig. 4-13). Now, I actually add corrugates to the inside of the drying press to make it as long as the entire length of the drying frame. Even small batches of specimens in the center of the press will dry as efficiently as when the press is entirely filled with specimens.

Fig. 4-17. Additional notes on the mini-plant drying frame shown in Fig. 4-16. Top left: A view of how the legs are attached to the frame. Note that the angle aluminums have their outer edges facing up, which makes it more difficult to place the plant press on the drying frame. Top right: The place on the frame where the fire-resistant cloth overlaps is where the two ends of the cloth are attached. Note that the angle aluminums in this image have their outer edges facing down which makes it easier to place the drying press on the drying rack. Bottom left: A hole lined in the cloth with a grommet. Grommets tend to fall out when they line this kind of cloth. Bottom right: A hole in the cloth which is lined with strong thread with a button hitch.

A wooden drying frame (Box 4-3) 24 inches (61 centimeters) long will accommodate 100 sandwiches, and costs about $30.00 to make. To maximize space in the press, several specimens that dry easily (e.g., Cyperaceae and Poaceae) can be accommodated between sandwiches. When the sandwiches consist only of cardboard corrugates without blotters, more drying space is created and more specimens can be accommodated. Aluminum corrugates allow plants to dry faster but they are much more expensive and heavier to transport than cardboard corrugates. In wet areas, such as lowland rain forest during the rainy season or cloud forest, aluminum corrugates may be necessary, but in dryer areas cardboard corrugates are adequate for drying most plant specimens. Thus, the choice of aluminum versus cardboard corrugates depends on the habitat visited, as well as the kinds of plants collected. For general collecting in the tropics, the best solution is to bring a combination of cardboard and aluminum corrugates.

Most specimens, except those of fleshy plants such as Araceae and Cyclanthaceae, will dry in eight hours on a propane gas dryer. I usually place the press on the drying frame just before I retire to my hammock at around 9:00 p.m., and by 5:30 a.m. most of the specimens are ready to take out of the press. Generally speaking, it does not harm the collections to leave them on the dryer for an extra hour or two, especially if the heat is not extreme (specimens dried over very hot heat become brittle and lose color). For some specimens, the straps may have to be tightened several times during the drying process to prevent wrinkling, but this will not always be necessary.

There is always the possibility of a press fire. I have had three over the course of my career, but all were caught in time to avoid major damage to specimens or property. The closest call came at the old ORSTOM research station in Saül, French Guiana. The collecting team placed the press on the dryer without properly adjusting the heat source, and left to spend the day in the field. Fortunately, Louise Raymond, an avid bird watcher who had left for an early birding expedition, returned to the station at mid-morning and noticed smoke pouring out of the drying area. She ran to the drying frame just as flames were erupting, knocked the smoldering press onto the ground, and doused it with water. If Louise had arrived a few minutes later, the entire ORSTOM station would have gone up in flames.

To prevent press fires, it is important to watch the press carefully during the first hours of operation at the beginning of an expedition; especially if there has been any change in the heat source or if someone new is controlling the heat. A screen mesh can be placed over the top of the drying frame to prevent press straps and debris from falling and reaching the heat source, this is especially important when using propane gas heat sources (Fig. 4-13). Press strap ends should always be visible on top of the press and tucked into the tightened straps, especially if a protective screen (Fig. 4-15) is not used. In addition, avoid using straps or ropes that might melt or ignite under excessive heat.

As the specimens are removed from the press, the numbers on the sheets should be checked against the numbers in the field notebook to make sure that there are no errors. The dried specimens are generally placed in numerical order, tied into a bundle, and placed in a plastic bag with several other bundles. A few mothballs are added to the bag to discourage insects from eating the specimens, and the bag is tied shut. In remote areas, the specimens often have to be carried out on the backs of the collectors in boxes attached to pack frames with diamond hitches (Fig. 4-18). Some botanists have suggested that prolonged use of mothballs may deteriorate DNA; thus, I recommend that use of mothballs should be indicated on labels.

If there are too many mothballs in the bags they will have a strong aroma, and if the specimens are transported on a small plane, the overpowering smell may cause the pilots to refuse to fly the plane! I was once returning to New York from French Guiana, when the pilot of the small plane we were taking between Martinique and Haiti entered the cabin and asked who owned the boxes marked "New York Botanical Garden." I identified myself, and he told me that for the comfort of the passengers the boxes would have to be removed and sent to New York on a plane with a separate cargo hold. He assured me that they would arrive without problem—but I never saw them again. Fortunately, I was able to study the duplicates that we had deposited in the Cayenne Herbarium or else the work of the entire expedition would have been lost. Too many mothballs may also cause postal inspectors or custom agents to open the specimens even though all of the permits are in order.

When the fieldwork is complete, duplicates of each collection are separated, put in numerical order, and left at the host herbarium to be mounted when the labels arrive. The newspapers holding the remaining duplicates are marked "minus the name of the herbarium where they are left" (e.g., -CAY for species left at the Herbier de Guyane in Cayenne). This is done so that duplicates will not be sent twice to a herbarium where the collections are already deposited. If time permits, specimens are identified in the herbarium of the host institution. It is often easier to identify them there than it is in larger herbaria that include many specimens of species not found in the area where the collections were made.

Fig. 4-18. After the specimens have been dried, they are placed in plastic bags with several moth balls, the bags are tightly packed into cardboard boxes, and the specimens are transported back to the herbarium. If the locality is remote, the boxes of specimens are tied to the pack frame with diamond hitches and often transported on the backs of the collectors. Drawing by Michael Rothman.

Tips for Collecting in Neotropical Countries

Botanical expeditions to neotropical countries place demands on the time and resources of the local botanists hosting the visits. My year as a resident botanist in Panama (1974–1975) and two years in Bahia, Brazil (1978–1980), along with the experiences of Danish botanist Lauritz Holm-Nielsen in Ecuador prompted us to write guidelines for visiting botanists (Mori & Holm-Nielsen, 1981). The following discussion is based on those guidelines.

Before the Arrival

A. Host botanists have their own research interests, teaching assignments, field trip plans, and family obligations. Therefore, visits that require their assistance should be arranged well in advance to accommodate their schedules. Moreover, the areas to be explored and the purpose of the field work should be established with the concurrence of the host prior to the arrival of the visiting botanist.

B. Visiting botanists should always arrive with the documents needed for plant collecting, including permits to collect material for molecular studies (now the most closely regulated part of plant collecting). In some countries, it is against the law for local botanists to help visitors who do not have the correct collecting permits. In many neotropical countries, permits must be obtained before the visiting botanist arrives in the host country. A research permit may be needed even if the visiting botanist is only going to work in local herbaria. Some neotropical countries have posted instructions for obtaining permits on the Internet, and at least part of the application process can be completed online. There are usually fees for research or collecting permits, as well as export permits, so collectors should remember to include these costs in their grant applications.

Efforts to obtain permission to collect in a foreign country may take considerable time and aren't always successful. In 2006 I started trying to obtain permission to visit the coastal forests of northern Venezuela to collect herbarium specimens and leaf samples for molecular study of the Brazil nut family; at the same time, I initiated a similar application to study Lecythidaceae in the Atlantic coastal forests of Brazil. As of 2010, neither of these permits had been issued.

It is not clear why tropical countries are so reluctant to issue permits, but it may have something do to with the misconception that systematic research may bestow economic benefits to the collector that properly belong to the host country. However, it is highly unlikely that the sequencing of the *ndhF*, *trn*L-F, and ITS genes of Lecythidaceae, for example, would ever have any direct economic value because these genes are not known to code for proteins of economic importance.

C. The collection numbering system should be established before the onset of the trip. Most herbarium curators are avid plant collectors themselves, and more than likely will want to make the general collections under their number series, while allowing visitors to number the specimens in the group that they study. The same plant should never be collected in two different number series, because that doubles the work and cost of those who subsequently handle the collections, and no additional information is gained by making more than one collection of a given species from the same location on the same day. The visitor should never collect specimens for other botanists unless groups outside the specialty of the visitor are included in his/her collecting permit.

D. The host and the visitor should make an agreement about the number of duplicates that will be made for each plant, and how they will be distributed. Collecting regulations generally require that one or more duplicates be left in the host country. If a single collection is made it should remain in the local herbarium; thus, if the visiting collector wants to study a unicate specimen, it must be done while the collector is still at the host herbarium, or the specimen must be sent on loan. For unicate collections of Lecythidaceae, I take extensive notes, photograph the specimen, and attach the image to my database record of the collection. When new species are discovered, many countries require that the holotype be deposited in the host herbarium or in the national herbarium of the host country, and this should be done without fail.

E. It is important to establish what kinds of material will be collected, and make sure that the appropriate permits are obtained for each. For example, if pollinators will be collected, permits for their collection must be requested. Many countries require a separate permit from a different agency if the collected material will be used in molecular studies. Additional permits are usually required to collect in national parks or other protected areas.

F. The financing of the expedition should be established at the planning stage. In some cases, the host country provides the means of transport and the visitor pays for the cost of running the vehicle and paying the driver if he/she is not a member of the field team. All expendable field equipment that the visitor will need, such as plastic bottles, preserving fluids, etc., should be provided by the visitor. Payment for food and lodging should also be agreed upon for the host country participants before the start of the trip.

During the Visit

A. The visiting botanist should make his or her way to the host institution without relying on the host for transportation. Moreover, changes in flight times and last minute changes of plans by the visiting botanists cause major inconveniences for the host; thus it is important to notify the host about any travel changes.

B. Latin American herbaria are seldom open on evenings and weekends. Visitors should ask for permission to work outside of normal work hours only if it will not inconvenience the host.

C. Visiting botanists should remember that their visit is just one in a continuous stream of visits by foreign botanists. These visits, although profitable and interesting to the resident botanist, can be disruptive of his or her family and social life. Consequently, visitors should be prepared to be on their own while the resident botanist fulfills other obligations.

D. Shortly after arriving, the local botanist and the visitor should discuss and agree on final field trip plans. However, the visitor should understand that the resident botanist may suddenly have to postpone a trip because other obligations arise. This happens infrequently, but, nonetheless, the visitor should be prepared for such an eventuality.

E. Visiting botanists should make every effort to inform local scientists about their research and the plant groups they are studying, either in formal seminars, in a casual manner during field work, and or by including local students on field trips. The latter also enables the visiting botanist to observe the abilities of local students and perhaps recruit them to study for an advanced degree at the visitor's home institution. Although Latin American botanists are well-trained at in-country universities, overseas study broadens the student's education and facilitates mastery of a foreign language.

F. Visitors should take the time to annotate specimens in the herbarium of the host institution. This is the most useful way for visiting botanists to improve the collections of local herbaria.

Before the Departure

A. Visiting botanists should budget time to dry and ship their own collections. These time-consuming tasks should not be left to the host, nor should the visitor expect the host institution to pay for the cost of shipment. When the visitor collects only specimens of his or her specialty, the number of specimens is usually not excessive and can be taken home as baggage, as long as the collector arrives at a point of entry designated in his or her collecting permit.

A permit from the Animal and Plant Health Inspection Service (APHIS) of the United States Department of Agriculture is needed to import plant specimens into the United States. The permit is accompanied by shipping labels to a stipulated port of entry to which the specimens can be either sent or hand carried. A copy of the permit and the contact information for the collector need to be placed inside each box of specimens. If plants protected by the Convention on International Trade in Endangered Species of Wild Fauna and Flora (CITES) are included in the shipment, a CITES permit must be included in each of the boxes. It also helps to include a list of the species collected, usually as a copy of the field notebook, in each box of specimens. It is important to periodically review the regulations because they can change from one field trip to the next. To make sure that all regulations required by the host country are fulfilled, it is best to send the specimens from the host herbarium.

B. The visitor, in collaboration with the host, should separate the duplicates to be left in the host herbarium. The specimens should be placed in numerical order, tied in bundles, and placed in plastic bags with several mothballs. Each bag should be clearly marked with the name of the collector and the number series of the collections in the bag.

C. Any extra equipment, such as plastic vials, drying materials, silica gel, etc. can be donated to the local herbarium.

After the Expedition

A. As soon as possible, labels for all collections should be sent to the host herbarium, and the host herbarium should be subsequently informed of new determinations. If the visiting botanist maintains an online database, host herbaria can check for new determinations online, thereby eliminating the task of sending determinations to the host institution. Space in tropical herbaria is often limited, so bags of specimens without labels not only get in the way, but are a constant reminder that the visiting botanist has not yet met the obligation of sending labels. Moreover, the longer the specimens are left outside of herbarium cabinets the greater the chance that insects will damage them.

B. As previously noted, some countries require by law that the holotype of a new species be left in a herbarium of the host country; however, with or without such a law, every effort should be made to deposit duplicates of type collections in host country herbaria.

C. Field work often stimulates continued collaboration between the host and the visiting botanist. This may result in joint publications, which can be enhanced by additional collecting and research after the visitor has left. For example, a new species collected in flower can be gathered in fruit by the local botanist. Publications that result from the field work should acknowledge the contribution of the host and other collaborators at the host institution. This is important to the local herbarium because it demonstrates to its administration that the institution's cooperation with visiting botanists produces tangible results, and gives the institution an international reputation. Reprints of publications based on collections made while based at the host institution (as well as others related to the local flora) should be sent to the host institution.

Conclusions

This chapter discussed the process of collecting plant specimens in the field. The major themes are that specimens should be well-prepared and information-bearing before they are accepted and deposited in herbaria. The following two chapters outline how specimens are handled in herbaria

(Chapter 5) and how data associated with collections are made available on the Internet (Chapter 6).

To insure that specimens merit the cost of including them in herbaria, each collection: (1) should add information that is not provided by the specimens already in herbaria, (2) should be aesthetically prepared, and (3) should include value added data and material, such as labels with information about the plant that cannot be obtained from the herbarium specimen, e.g., flower color; digital images that capture all morphological structures of the stages of a plant's life cycle; accurate locality information; liquid preserved material for use in understanding the anatomy and morphology of the species; and silica-dried leaves for molecular study. If collectors adhere to the following recommendations, it will enhance their contributions to the knowledge of the plant species they collect.

1. Before collecting a specimen, the collector should ask "Will this collection add new information about this species?" If the answer is no, the specimen should not be collected because the time and cost of collecting, handling, and archiving it will not be justified by the data that researchers will be able to extract from it.

2. Collectors should take notes and place specimens in the field press at the time plants are collected. This ensures that the greatest amount of error-free data will be documented, and that the specimens will be of the highest quality possible.

3. Collectors should arrange specimens as attractively as possible, because aesthetically pleasing specimens also provide the greatest amount of information. Both sides of the leaf should be shown, and if extra flowers are added for dissections it will facilitate study of the taxonomic features of the species collected.

4. Collectors should add value to their collections by taking digital images; collecting silica-preserved leaf samples for molecular study; and pickling flowers, fruits, and vegetative material for anatomical and morphological research.

5. Collectors should avoid making sterile collections unless there are compelling reasons for doing so. Sterile specimens cost as much as fertile collections to collect, handle, and archive in herbaria, but they provide a fraction of the information provided by fertile specimens. Sterile specimens are sometimes justified because they are needed to voucher scientific studies.

6. Collectors should dry specimens with gentle heat to ensure

that the specimens provide the best possible material for study. Rapid drying of plants under extreme heat often results in delicate structures, such as flowers, becoming so brittle that their structure is impossible to reconstitute for future study.

Chapter 5
Into the Herbarium

by Scott A. Mori

Collections of plants gathered in the field are integrated into museums of plant specimens (Figs. 5-1, 5-2) called herbaria. Most herbaria have as their goal the accumulation of specimens representing all species of the plant or fungal groups found in the areas of expertise of the herbaria, which may range from regions as small a county of a state in the United States or as large as the world. In addition, if the curators of a given herbarium are given the mission of preparing monographs of plant and fungal groups another goal is to gather specimens of the groups throughout their geographic distributions.

Information about the herbaria of the world is provided in Index Herbariorum (IH). This important reference to herbaria is described in the online version (Thiers, 1997 onward) as follows:

> "For the past three centuries, scientists have documented the earth's plant and fungal diversity through dried reference specimens maintained in collections known as herbaria. There are approximately 3,990 herbaria in the world today, with approximately 10,000 associated curators and biodiversity specialists. Collectively the world's herbaria contain an estimated 350,000,000 specimens that document the earth's vegetation for the past 400 years. Index Herbariorum is a guide to this crucial resource for biodiversity science and conservation.
>
> The Index Herbariorum (IH) entry for an herbarium includes its physical location, Web address, contents (e.g., number and type of specimens), history, and names, contact information and areas of expertise of associated staff. Only those collections that are permanent scientific repositories are included in IH. New registrants must demonstrate that their collection is large (usually 5,000 specimens minimum), accessible

to scientists, and actively managed. Each institution is assigned a permanent unique identifier in the form of a four to eight letter code, a practice
that dates from the founding of IH in 1935.

The first six editions of Index Herbariorum were published by the
International Association for Plant Taxonomy in the Netherlands (1952–
1974). Dr. Patricia Holmgren, then Director of The New York Botanical
Garden [Herbarium], served as co-editor of edition 6, and subsequently
became the senior editor of the of IH. She oversaw the compilation of
hard copy volumes 7 and 8, and Dr. Noel Holmgren, a scientist on the
NYBG staff, oversaw the development of the IH database, which became
available on-line in 1997. In September 2008, Dr. Barbara M. Thiers, Director of the NYBG Herbarium, became the editor of IH."

The permanent, unique identifier mentioned above is called an
acronym; for example, the acronyms of several large herbaria are: COL for
the Universidad Nacional de Colombia; K for the Royal Botanic Gardens,
Kew; NY for The New York Botanical Garden; P for the Muséum National
d'Histoire Naturelle in Paris. Acronyms are used by systematic botanists
throughout the world to indicate where specimens that they cite in their
monographs and Floras are archived.

An especially useful feature of IH is the ability to search the database to learn the names and addresses of specialists working on different plant groups and different floras of the world. Because it is often very
difficult to identify specimens, duplicates of a collection are often sent to
specialists for identification. For example, one can query the IH database
for "Sapotaceae" and find out that 21 different botanists have interest in
this family. From that list, T. D. Pennington at K, who wrote the definitive
monograph of Sapotaceae for Flora Neotropica (Pennington, 1990), has
identified all of the author's collections of this dominant family of lowland
neotropical trees over the last 35 years. The contribution of specialists to
the work of other biologists is invaluable (Mori & Heald, 1997)—but its
importance is seldom recognized by scientists in other fields of research.

Herbaria are not restricted to collections of mounted herbarium
specimens (Fig. 5-1) but also include fruit collections, pickled collections
of flowers and fruits, silica-dried specimens for molecular study; and field
images attached to herbarium records of specimens that document images.

This chapter adds the observations of the author to previously
published manuals of collecting techniques and herbarium management
(see Chapter 4 for a list of some of these publications). The ideas presented
in this chapter have not been discussed at all or, in the opinion of the author, were not emphasized sufficiently.

Figure 5-1. Herbarium collections. Top left: Sheranza Alli glues a specimen onto a herbarium sheet. Specimen mounters at NY prepare at least 50 specimens a day. Top right: A herbarium sheet of Eschweilera longipedicellata S. A. Mori. In order to keep track of all objects in a herbarium, each should have a bar-code (lower left on this specimen) and each bar-coded object should have a separate record in the herbarium's database. Bottom left: Images, associated with a collection, provide more information than given by the specimens alone. The medial longitudinal section of this flower of Lecythis tuyrana shows the complicated androecial structure of the species, a feature that is very difficult to observe in dried flowers. Lower right: A fruit and seeds of Lecythis tuyrana. Note the basal aril of this species is a feature that is difficult to see in dried specimens, especially if the fruits have not been opened. Delicate fruits are stored in boxes and less delicate ones in plastic bags. Note that two important parts of most modern collections, flowers and fruits preserved in a liquid preservative and dried leaves used for molecular studies (stored at low temparatures in plastic bags with silica gel) are not shown in this figure.

Herbarium Management

This section is based on my experiences as curator of the Summit Herbarium in Panama and the Cocoa Research Institute in Bahia, Brazil (see Chapter 1) as well as observations I have made as a research curator at NYBG. The discussion is directed toward curators of herbaria with fewer than 100,000 collections and may or may not be the curatorial policy of any of the herbaria with which I have been associated. Some of the recommendations might not be practical for very large herbaria because the management of them is so demanding that the additional work resulting from following these recommendations would be too much to expect of staff already overwhelmed by work.

Geographical focus. It is important for herbaria to determine the geographical focus of their collections. With the advent of digital photography, the scanning of specimens, and easier and less expensive travel, it is not as important for local herbaria to have collections from areas larger than their immediate area of interest because large collections are available in herbaria with collections focusing on states, countries, and the world, and many of those herbaria make their collection data and specimen images available online. By limiting geographical focus, curators of local herbaria can more efficiently focus on documenting the plants of the region in which they are located. Collections of species previously unknown from their area can be identified by using hard copy publications, consulting images on web sites, and by visiting larger herbaria to compare unknowns with specimens in the larger collections. It is such an enormous task to document the plants of large geographic areas that an attempt to do this by curators of local herbaria will make it difficult for them to produce a Flora of their area, make conservation recommendations based on their collections, or educate local people about plant diversity.

An example of a well-managed local herbarium is that of the Herbier de Guyane in Cayenne (CAY) which specializes in the flora of French Guiana. The curators of CAY send duplicates of their collections to non-resident specialists for identification and to the Muséum National d'Histoire Naturelle (P) in Paris as a secondary repository for its collections. For the most part it is impractical, expensive, and time-consuming to maintain a collection of plants from all of the Neotropics or even for the other two Guianan countries (Guyana and Surinam) at CAY. Moreover, maintaining a large herbarium without a geographical focus on the plants of French Guiana would cost so much money and consume so much time that it could interfere with the mission to document the flora of French

Guiana.

Failure to recognize the roles of herbaria with different scales of geographic focus results in duplication of effort that increases the overall costs of maintaining herbaria. For example, if a country is small, there may not be enough financial support to maintain more than one national herbarium. The herbarium of the University of Utrecht (U) in the Netherlands provided an example of duplication of herbarium facilities, but not research programs. The program at U made very important contributions to neotropical flowering plant systematics as the result of its studies of the flora of Suriname and the Guianas, and its emphasis on monographic work in the Annonaceae, Cannaceae, Cecropiaceae, Costaceae, Haemodoraceae, Moraceae, Rutaceae Violaceae and saprophytic flowering plants (Gentianaceae, Burmanniaceae, and Triuridaceae). The U group has been responsible for the publication of 11 Flora Neotropica Monographs. In addition, U played a major role in the organization of plant systematics on an international scale by promoting the establishment of the International Association for Plant Taxonomy (IAPT). IAPT has advanced the cause of systematic botany in countless ways, including the publication of the journal *Taxon* and the establishment of an international list of herbaria called Index Herbariorum.

There are two other large herbaria in the Netherlands, one in Leiden with a focus on the flora of SE Asia and another in Wageningen with an emphasis on the African flora. These three Dutch herbaria joined in 1999 to form the Nationaal Herbarium Nederland which was maintained in three separate localities until June 2008. In order to reduce repetition of herbarium infrastructure in the relatively small geographic area of the Netherlands, the three herbaria have been centralized in Leiden as part of the Netherlands Centre for Biodiversity (NCB), which will include all natural history collections in the Netherlands. At the time of this writing (November 2009), the ownership of the U herbarium has been transferred to the state and the specimens, including the large wood and spirit collections, have been temporarily housed in the Leiden Herbarium until the new facilities of the NCB are available.

As disheartening as it has been for the international community of neotropical botanists to see U closed, it probably makes economic sense to avoid duplication of herbarium infrastructure in three separate places in such a small country. Fortunately, the transfer of the U herbarium to Leiden has been successful, and Dutch botanists will be able to continue with their tradition of making major advances in neotropical botany.

Label making. In the past, labels were made after returning from the field.

However, the ability to work remotely has made it possible to enter the label data into the home institution's database, download the data onto a laptop while still in the host country, print out the labels, insert the labels in the duplicates left for the host herbarium, and deliver the duplicates to the specimen preparers before leaving the country. On a field trip to central Amazonian Brazil in 2008 to collect Lecythidaceae, I was able to enter the label data of the 85 collections made directly into the NYBG database, attach images to the records, and generate the labels while still in Brazil. Use of this technology reduces the work of the staff of the local herbarium, nearly eliminates the chance of losing specimens left in host herbaria, and allows the collector to make productive use of his or her time between field trips. If a fast enough Internet connection is not available, the data can be entered into a field version of the home institution's database and imported into the institution's database when the collector returns home.

This procedure, however, would be difficult for anything other than specialty collections, because it is time-consuming to prepare labels for the large number of specimens gathered as part of a general collecting expedition. Thus, the remainder of this discussion assumes that the labels are made after the collector returns to the home institution.

After returning from a field trip, labels are made and added to the duplicates at the home institution, and labels for duplicates left at the host herbarium are sent to its curator. Next, duplicates are sent to specialists as gifts for determination. This is indicated on the newspaper holding the remaining duplicates (e.g., by writing "-MO" on the newspaper holding specimens of Araceae to indicate that a duplicate was sent to Tom Croat at the Missouri Botanical Garden as a gift for determination) as well as noted in the field book in order to avoid sending another duplicate to the same specialist at some time in the future. Unfortunately, there are fewer and fewer monographers of tropical families and it is more difficult to obtain determinations from specialists than it has been in the past (Mori, 1992c).

When a determination is returned, it is either added directly to the label file before the labels are printed or separate annotation slips are added to the duplicate specimens if the labels have already been printed. The specimen is then mounted and archived in the collector's herbarium, and the remaining duplicates are used as exchange with other herbaria. The exchange of duplicates allows the collector's herbarium to acquire additional specimens, it distributes duplicates of collections in herbaria throughout the world for others to study, and it insures that specimens are still available in the case of a disaster, such as fire, at the collector's herbarium. Many herbaria image their collections, especially the types, so that much of the information associated with specimens is available on the

---

Box 5-1. An Example of a Collection Label
by Scott A. Mori

THE NEW YORK BOTANICAL GARDEN
PLANTS OF FRENCH GUIANA

Lecythidaceae
Lecythis zabucajo Aubl.
Det. S. A. Mori, 1995

Saül, Route de Belizón, 3 kms. N from the intersection with the airport road. 200 m alt., 3°37'N, 53° 12' W.

Non-flooded moist forest, on well-drained soil.

Tree, 50 m x 100 cm. Bark fissured. Flowers with purple-tinged petals, yellow androecial hood. Fruits 15 cm diam., empty, collected from ground.

S. A. Mori, C. A. Gracie & J. D. Mitchell 20100
20 October 1995

Expedition funded by the Fund for Neotropical Plant Research of NYBG

Internet. Nevertheless, an image does not provide all of the information available by studying the specimen, so actual specimens sometimes still have to be studied by visiting the herbaria where they are located or by borrowing them.

It is important to present the label data in a way that maximizes readability. The plant name and determiner, the locality, the habitat, the description of the plant, the collector and number, and the date of collection should be set off in easy-to-read blocks, as this makes it much easier to enter the data. Placing the name of the collector and number in the lower left hand corner of the label allows for faster location of the collection when one searches for it in the herbarium. The placement of barcodes, which aid in keeping track of specimens, should also be on the bottom left edge of the herbarium sheet, thus facilitating rapid reading of the codes when specimens are being sent out as, or returned from, loans. In Box 5-1, I provide an example of a label in which the information is arranged in easy-to-read blocks.

If the expedition has been financially supported by a foundation or other source, it is appropriate to acknowledge the organization at the bottom of the label. For example, the National Geographic Society has supported many botanical and zoological expeditions and should be given credit for the important role it has played in biodiversity studies.

Cataloging. Modern databases generate specimen labels from the specimen record. Thus, one of the most important steps in producing a Flora or a monograph, the preparation of a checklist vouchered by herbarium specimens, is accomplished by the entry of the name of the species to the specimen record in the database as part of the process of producing the labels. When this is accomplished a checklist can then be generated from the database.

There are, however, specimens in herbaria relevant to floristic or monographic projects that have to be entered retroactively into databases. Ideally, information on a herbarium specimen should be entered into the database exactly as it appears on the label. For example, *P. Acevedo 4500* may be on the label and even if it is known that this is P. Acevedo-Rodriguez (a specialist in Sapindaceae and writer of Floras of the Caribbean at the Smithsonian Institution), the name and number for that particular collection should be entered into the database as *P. Acevedo 4500* (i.e., the way it appears on the label and not P. Acevedo-Rodríguez as it has been standardized in some authority files). It is important to follow this rule because the principal reason a collector and number are provided is to allow others to identify a collection based on a name and number combination which is unique. For example, if there is a collection of DNA marked with *P. Acevedo 4500* and vouchered by an herbarium collection with the same name and number, there will always be some doubt about whether it is the same collection if one finds it recorded in the database as *P. Acevedo-Rodriguez 4500*. This problem is especially common with collectors having Spanish and Portuguese names because collectors with several family names may not be consistent in the use of their names. Standardizing data is something that database developers strive to achieve, but in these kinds of situations the name should probably not be standardized, i.e., changed from the way it is written on the label. Unfortunately, this dictates maintaining all permutations of a collector's name in database authority files. Most database managers would not agree with this recommendation because they feel that it is less confusing for interpreting collections if one collector is recognized by one name—however, this is nearly impossible because of the numerous permutations that a collector may have used for his or her name as mentioned in Chapter 4.

Entering fruiting phenology brings with it a problem caused by the presence of fruit at all stages of development over long periods of time. The only stage of phenological importance is that in which mature seeds are ready for dispersal. Thus, the registration of a plant as being "in fruit" when it is in its developmental stages or long after the seeds have been dispersed conveys little biological information. This is especially a problem with species that have woody fruits, such as those of some species of the Brazil nut family. Its fruits are conspicuous on the tree long before they are ripe and may persist on the tree or on the ground for many months after seeds have been dispersed. For this reason, the database should allow for distinguishing among immature, mature (i.e., fruits with mature seed), and old fruits without seeds.

A problem with English language databases for collections of neotropical plants is that many of the labels are written in Spanish, Portuguese, or French. The diacritical marks of these languages increase the time needed to add collection data to databases but, nonetheless, they must be added because foreign words without them represent misspellings. The date should be standardized in the form day/month/year with the month spelled out or unequivocally abbreviated (i.e., 4 October 2005 or 4 Oct 2005) because in this form the date cannot be misinterpreted. Dates entered with the month given as a number should be avoided because it is not clear which number represents the month and which the day if the collection is made before the 13th day of the month (e.g., 10/4/1955 can be interpreted as either 4 Oct 1955 or 10 Apr 1955).

The best way to facilitate accurate data entry is to participate in the training of those who enter data. Although I have a great deal of experience in tropical botany and have a vested interest in making sure that the data for my projects are properly entered, I catch many mistakes that I have made entering data. Thus, I appreciate how easy it is to make errors and understand the ways they are made. The number of errors committed by those who enter data for my projects is markedly reduced if I take the time to give them further training in my preferred techniques after they have been given general training in database management. This advanced training consists of me adding specimen records in the presence of the trainee, checking the initial records entered by the trainee, making the trainee feel comfortable in asking questions, and periodically checking the accuracy of the data.

How many records are needed? Because specimens are often collected in duplicates and each duplicate has the same basic information there is question about the need to keep a specimen record for each duplicate. The purpose of this section is to contrast the difference in making separate records for each duplicate or making a single record that represents all duplicates of a collection.

Databases designed for keeping track of all specimens in herbaria require the creation of a separate record for each object in that herbarium. Because of the large size of the leaves, inflorescences, and fruits of palms, their collections usually consist of multiple sheets and/or boxed specimens; i.e., there may be separate herbarium sheets with the base, middle, and apex of the leaf; the base and apex of the inflorescence; and a box of separate fruits representing a single collection number in a single herbarium. In a given herbarium, data from the first sheet is entered in the database and the remaining sheets and separate fruit collections are dittoed from the first record entered. When the records are dittoed, all records in the database for a given collection have the same information except that the barcode, the type of collection (i.e., whether it is a record for a herbarium sheet or a fruit), the type of type (e. g., holotype, isotype, etc.) if it is a type collection, and the attached images may differ for each record representing a duplicate of a collection. Preparing a record for each duplicate of a collection and assigning barcodes for each duplicate is the only way that herbaria can unequivocally keep track of their holdings.

As mentioned earlier, collectors usually make more than one duplicate of their collections and distribute these duplicates to other herbaria. Some collectors may gather as many as 12 or more duplicates of a collection depending upon how interesting they judge a collection to be. For example, large numbers of duplicates of a species previously not seen by a collector, especially one judged to possibly represent a new species, are often collected. In addition, collectors gathering specimens for exchange with other herbaria tend to gather large numbers of duplicates. These collections have the same labels and, thus, one duplicate bears essentially the same information as the other duplicates. In this case, the multiple duplicates were gathered not to show different aspects of the plant but to get a wider representation of species archived in different herbaria (see previous section entitled Collection Numbers and Duplicates).

In the case of duplicates in different herbaria, monographers and floristicians sometimes decide NOT to make separate records for each duplicate of a collection and note the differences among duplicates in a single specimen record. In this philosophy, images of all duplicates are attached to a single record and notes about other duplicate differences, such as the herbaria in which specimens are archived, are entered into that record. The reason given for maintaining a single database record to represent all duplicates is that it eliminates the additional work of making duplicate records because this work may not compensated for by the information obtained from the duplicate records.

In addition, making a single record eliminates entering slightly different information for duplicates of the same collection, a problem that is reduced if the multiple records are dittoed. Differences in data entry for duplicates of a collection are especially pronounced when the data for the different duplicates are entered by different people, many of whom do not read the language that the label is written in. For example, Guyane is the shortened French name for Guyane Française (= French Guiana), an overseas Department of France located in NE South America. This name is easily mistaken for Guyana, a neighboring country in NE South America. As a result, English speakers sometimes mistakenly enter Guyana when the collection is actually from French Guiana and this may create duplicates of the same collections being reported as coming from two countries.

Another frequent cause of differential recording of label data comes from the interpretation of the phenological state of collections. One sheet in one herbarium may be sterile and another sheet of the same collection in another herbarium may be in flower, or specimens in the same phenological state can be easily interpreted as different phenological states by different data entry personnel. This is especially true for fruiting collections which are easily scored as mature when they are actually immature. In addition, see the section below for a discussion of errors in geo-referencing duplicates.

Differences in data entry are especially problematic when records are imported from the database of one herbarium into the database of another. In my monographic work of Lecythidaceae, I take a very conservative view toward importing data from other sources because of the large numbers of errors caused when independently created databases are imported into my Lecythidaceae database. Correcting errors in records imported from other herbaria creates so much work that it may be easier to visit other herbaria and personally enter the data from the specimens, many of which can be dittoed because they have already been entered into the user's database from other duplicates.

In summary, if the goal of a database is to keep track of all of the objects archived in a herbarium then separate records have to be made for all of the duplicates of all of the collections archived in that herbarium. On the other hand, if the goal of the database is to keep track of the information associated with collections for monographic and floristic research only one record for each collection may be sufficient. Keeping track of all duplicates of all collections in all herbaria involves so much work and consumes so much time that there is little time left for doing research. In addition, so little additional information is obtained by creating separate records for duplicates, and it is so difficult to keep records maintained by different herbaria congruent, that the work involved may not be justified by the information that can be extracted from the duplicate records.

Geo-referencing. Coordinates are placed on specimens by collectors because they allow for the (1) production of distribution maps, (2) relocation of individuals of species of interest, (3) development of models of where species might occur based on the ecological parameters of where they currently or at least have occurred in the past, and (4) determination of the ecological factors required by a species to grow where it grows.

The production of distribution maps is a mandatory part of monographs because it is useful to know where a species does or has occurred. The location of individuals in the field using coordinates and map descriptions is useful because it allows monographers and floristicians to find and collect species that are poorly known. For example, it facilitates collecting the flowers of a species previously known only from fruits. Of particular importance is recollecting species from a type locality, and this is critical if the species is poorly understood because the type is fragmentary. The development of models to show where a species might occur provides data for interpreting phytogeographic and evolutionary relationships, but this requires sophisticated analysis not addressed in this essay. Finally, determining the ecological requirements of species provides insight into the evolutionary relationships of species, for example determining if two closely related species occupy different niches. These studies are usually undertaken by ecologists and often require maps at a finer scale than normally presented in monographs and Floras. Because the last two items are not normally presented in monographs and Floras, these uses of distribution maps are not discussed. The use of distribution maps is also addressed in the "Conservation" section below.

The accuracy of collection coordinates varies according to the methods used to obtain them; for example a GPS reading is generally more accurate than retroactively added coordinates derived from label data and maps. For this reason, there are usually fields in databases to indicate how the coordinates were obtained and to give an idea of how precise the coordinates are. In general, label coordinates give a good approximation of where a species occurs or at least has occurred in the past. This, however, does not guarantee that the species will still be found there. For example, if a collection represents a forest tree and examining the area with Google Earth indicates that the collection site is now part of urban sprawl, it is most likely that the tree will no longer be found there.

Geo-referencing has several caveats. In the first place, the accuracy of distribution maps depends upon the accuracy of specimen identifications. There are a number of widespread species that may actually be represented in herbaria by several species. Thus, a map of all specimens of such "species" is actually a map of more than one species; therefore, the distribution of the mapped "species" is meaningless. In the second place, earlier collections often do not have GPS coordinates and have to be geo-referenced retrospectively based on label data that are not precise enough to allow the coordinates to be determined with the precision needed for some types of studies. The latter problem is not of much concern when the purpose of the map is to provide an overall idea of where the species occurs or has occurred in the past. However, it does become a problem for more precise ecological studies. For example, locality coordinates are usually not precise enough to determine that a species prefers moist soils, especially because many moist habitats may be so small that they cannot be mapped at the scales of most distribution maps.

Because coordinates are recorded and then entered into databases by humans they are subject to error by the collector that originally registered them in the field as well as by those entering them into the database. As a result, it is not surprising that some collecting localities are mapped in impossible places, such as trees mapped as occurring in the ocean. The Lecythidaceae database includes records imported from other sources and this sometimes results in discrepancies among the imported records and records independently entered of the same collection in the NYBG database because data entry is subject to different errors by different data entry personnel. Unfortunately, there is seldom a one-to-one correspondence in the way data are entered from duplicates unless the data are dittoed from the same original entry.

Another problem is that the original coordinates on the labels may not be accurate enough to pinpoint the exact locality of a collection. For example, my collections from Central French Guiana are entered with the coordinates of the village of Saül (3° 57' N, 52° 12' W). These coordinates indicate that the species represented by the collection occurs somewhere around the village, and that information provides a general idea of the location of the species. However, if one wants to find a given species collected by Mori et al. in the Saül area, the description of the locality on the label must be consulted. For example, to increase the chance of locating a plant of *Banisteriopsis carolina* W. R. Anderson (type = *Mori et al. 22751*), one has to walk seven kilometers N of Saül to Eaux Claires and search in the area of the Allinckx family gravesite. Good geo-referencing protocol now dictates that readings be taken as close to the collection site as possible and that the description of the locality be based on more permanent geographic localities. Unfortunately, relocating the plant from which *Mori et al. 22751* was made is not as simple as using the coordinates on the label to walk directly to the plant.

Geo-referencing errors surface in completely unexpected ways. For example, an imported duplicate specimen record from MO of *Cariniana rubra* Gardner ex Miers (*Plowman et al. 8621*) from the state of Pará, Brazil was mapped at 8° 18'S, 49° 33'W but the NY specimen label showed the coordinates as 8° 03' S, 50° 10' W and that specimen was mapped at that locality As a result, duplicates of the same collection were mapped in two localities. After some detective work, it was determined that the reason for the discrepancy was because the coordinates had been changed, presumably by the collector, and the corrected coordinates had not been entered in the field book. The data had been entered directly from the field book at MO into the database and not from the specimen as was done at NYBG. This explained how two specimens, both with the same coordinates on the labels, had different coordinates in their respective records. Ideally, data should be entered directly from specimens and coordinates on specimens should not be changed unless there is good justification for doing so. If they are changed, the reason for the change should be noted in the specimen record and that information should be part of the online record display.

Collections of cultivated plants should not be mapped unless they are clearly indicated on the map as being cultivated. The Brazil nut is native to the Amazon basin, and a map showing collections in Africa, Hawaii, Malaysia, and Trinidad has no phytogeographic significance. Nevertheless, maps of the distribution of cultivated plants give an idea of what kind of ecological conditions are needed for it to grow. Both types of distributions can be shown because biodiversity databases usually have a field for stipulating if a collection represents a native or cultivated plant. Thus, separate maps can be generated showing native and cultivated distributions, or a single map using different symbols for the two types of occurences can be displayed.

Because there are many possible sources of error in geo-referencing it is important to verify the points added to distribution maps to eliminate mistakes before the labels are printed. Some errors, such as symbols placed in the ocean for a collection of a rain forest tree or a collection mapped in Venezuela when it is from Brazil, are easily spotted and corrected. Programs have been developed that will catch the most obvious errors, but more subtle errors still depend on careful human monitoring of the maps. Fortunately, many database systems allow checking the distribution of a species on a Google map immediately after the coordinates have been added.

Naming image files. One of the most effective ways to communicate information about plants is to show their features in scanned herbarium specimens and in field images vouchered by specimens. It is paramount that the identity of species represented by images can be verified by examining the herbarium specimens that voucher them. Online databases allow images to be seen by the entire world, and, therefore, the handling of images has become an important part of herbarium management.

An available, but often untapped, source of information about plant features are slides vouchered by herbarium specimens. In the past, some slides were printed and the prints were glued to herbarium sheets. Now, slides can be scanned and the images attached to specimen records in databases. The combination of displaying scanned herbarium specimens and specimen vouchered field images on the Internet has revolutionized systematic botany and provides an ideal way for herbaria to communicate information about their collections and the species represented by them. File names for scans of herbarium specimens and field images should be as simple as possible and should follow standardized protocols for the naming of electronic files. For scanned specimens, most of the data that one might put into a file name is also available from the specimen record as well as directly from the scanned label so a file name for a scanned specimen only has to provide the information needed to facilitate attaching the image to the specimen record. Handling images is very time-consuming so, although I describe the way I name image files, I recommend that the reader develop the shortest and easiest way of naming image files needed to achieve his or her goals.

The purpose of the image's file name is to provide a unique identifier of the image which, in turn, facilitates its attachment to specimen records. For images of herbarium specimens, the most complete file names include a code for the taxon, the name of the collector, the collection number, and a method of keeping track of specimen duplicates. As an example, take the file name for an image of a herbarium sheet of *Wendt & Montero 4468* of *Eschweilera mexicana* T. Wendt et al. I use the code EschMex for the species, so the file name for the image is EschMex_Wendt4468.jpeg. If there is more than one duplicate of the specimen in one or more herbaria, they need to be identified with unique file names. For example, if there was a single duplicate of the specimen at the Missouri Botanical Garden and another at NYBG, they can be distinguished by the file names EschMex_Wendt4468_MO.jpeg and EschMex_Wendt4468_NY.jpeg, respectively. If there were more than one duplicate in a given herbarium that should also be apparent in the file name. In the case of two specimens of *Wendt & Montero 4468* at NY they would be named EschMex_Wendt4468_NY01 and EschMex_Wendt4468_NY02. Note that a zero is placed before single digit numbers so that the images are arranged in numerical order when a list is generated. It is important to remember that a period should appear in the name only in front of the file type, for example in front of "jpeg." If a period is placed elsewhere in a file name, software programs or operating systems may not be able to read the file type correctly. Finally, it is not necessary to provide too much detail in the file name of an image of a herbarium specimen (e.g., the initials of the collectors or the names of other collectors) because that information is available by opening the image file).

For file names of images taken in the field, it is often useful to indicate what part of the plant is represented in the images. For example, images of the bark, flower, and fruit of *E. mexicana* would be named EschMex_Wendt4468_Bark, EschMex_Wendt4468_Fl, and EschMex_Wendt4468_Fr, respectively. If there is more than one photograph representing different aspects of a plant collection, each should be discriminated with a number or a letter, as in: EschMex_Wendt4468_flA, EschMex_Wendt4468_flB, etc.

As an example of the process of scanning slides and making the images available on the Internet, I will use the placement of images of French Guianan plants online as part of a project to produce an e-Flora of French Guiana. Most of my collections from French Guiana had already been databased as part of research projects starting from my first trip there in 1971. Because there are very few non-databased specimens, and because it is too time-consuming to pull specimens that have not been databased, add them to the database, and then attach the images to the records, I usually only scan slides vouchered by specimens in the database or add a specimen's data to the database if the image is of special importance. The first requirement for adding an image to the taxon database is that there has to be a specimen record (i.e., the voucher) to which the image can be attached. Another requirement is to avoid scanning images if another image based on the same collection shows more-or-less the same feature(s). The rationale for this policy is that it takes time to name and attach images to the database, duplicate images add no additional information, excessive images do not justify the storage space they take up on servers, and repetitive images unnecessarily take up the time of the viewer.

The process of retroactively converting slides of French Guianan plants to electronic files to be placed on the Internet includes: (1) removing the slides family by family from the slide collection, (2) checking the database to see if the collection has a record, (3) selecting the images making sure that unique features are represented by each slide, (4) organizing the slides in bundles with as few families as possible in each bundle, (5) instructing the operator of the scanner to either place the images scanned from each bundle in a separate folder or to name the images derived from each bundle with a letter identifier, (6) cropping, adjusting the brightness and contrast, and sharpening the scanned images as needed, (7) labeling the image file as described above (8) saving a version of the image at a pixel width that optimizes screen display, (9) opening the database record for the voucher collection, (10) attaching the image to the record, and (11) adding metadata to the image record (e.g., species name, what the image shows (habit, bark, slash, leaves, flowers, fruits, etc.), collector and number of the voucher, site of the voucher collection, creator of the image, and keywords).

Once named, images should be displayed to take advantage of the most commonly used computer screens. I use a 700 to 800 pixel wide display for field images and a 1000 pixel width for herbarium specimens. The latter is larger because herbarium specimens often have difficult-to-read text and the slightly larger size facilitates reading it. Ideally, the image should fill most of the screen and the description of the image should be visible with the image or by scrolling down only slightly to see it. While looking at the image, the user should be able to read a description that at least includes what the online image represents, the name of the species, the voucher information, and the locality of the voucher.

More advanced viewers allow for more flexibility in the size of the images displayed such that larger images can be attached to records because they are automatically adjusted by these programs for best screen display. In addition, images can be managed with programs such as Photoshop® and Lightbox®, which make it possible to adjust image size and add some metadata with actions and batch files. It is important, thereover, to ask your Herbarium Information Managers if there are ways in which image handling tasks can be more efficiently carried out. This is extremely important because automated image handling will save countless hours of repetitive and monotonous work!

Use of space. The easiest, and perhaps most enjoyable, part of making herbarium specimens is collecting them. More tedious or difficult tasks are making the labels, identifying the species represented by the collections, making sure that the duplicates are distributed, supervising the mounting and archiving of the home institution's duplicates, and attaching new determinations to the specimens and entering them into the database. Specimens that are not labeled and entered into the herbarium can not be used for scientific study and take up space that could be used for more productive purposes such as the storage of mounted specimens. If the specimens are stored in bundles or boxes outside of herbarium cabinets the herbarium starts to look disorganized, and a messy herbarium is one of the first indicators that it is not being properly managed. In addition, specimens stored outside of herbarium cabinets are more prone to insect damage. Thus, a well-managed herbarium is one that makes the best use of its space by integrating specimens into the herbarium in a timely fashion.

In some plant families, such as the Lecythidaceae, the fruits are large and need to be archived in a way that economizes herbarium space. With most specimens having large fruits, it is more conservative of space to store the fruits separately from the herbarium sheets in boxes or bags in herbarium cabinet cubbyholes at the end of their respective families. The number of fruits needed should be decided by a specialist or, in the absence of a specialist, by the curator. For example, it is not necessary to keep 30 fruits of a collection if variation in size and form are well-represented by six. If the fruits are not fragile, they can be stored in plastic bags rather than in boxes to save space; fragile fruits, however, need to be protected in boxes just large enough to accommodate them. Another way to save space is to cut large, woody fruits in half if the fruit halves are mirror images of one another. Fruits of species such as *Lecythis ampla* Miers, which can be as large as 25 centimeters long by 20 centimeters wide, will not fit into the compartment of a standard herbarium cabinet unless cut in half. Very large fruits that can not be cut in half are archived in special herbarium cabinets.

Bark specimens should be cut just large enough to be sewn onto a herbarium sheet. If the salient features of the bark are not shown in a specimen that size, then bark collections can be stored with the fruit collection of the family. Additional smaller pieces of bark can be added to a packet for use in anatomical study. Images and detailed descriptions of bark are especially useful for helping users to understand bark features.

Specimen preparers should be trained to alternate the positions where fruits and bark are attached to herbarium sheets. Herbarium sheets with these bulky items placed consistently in one corner take up more space than if the positions of the fruit and bark are changed from one sheet

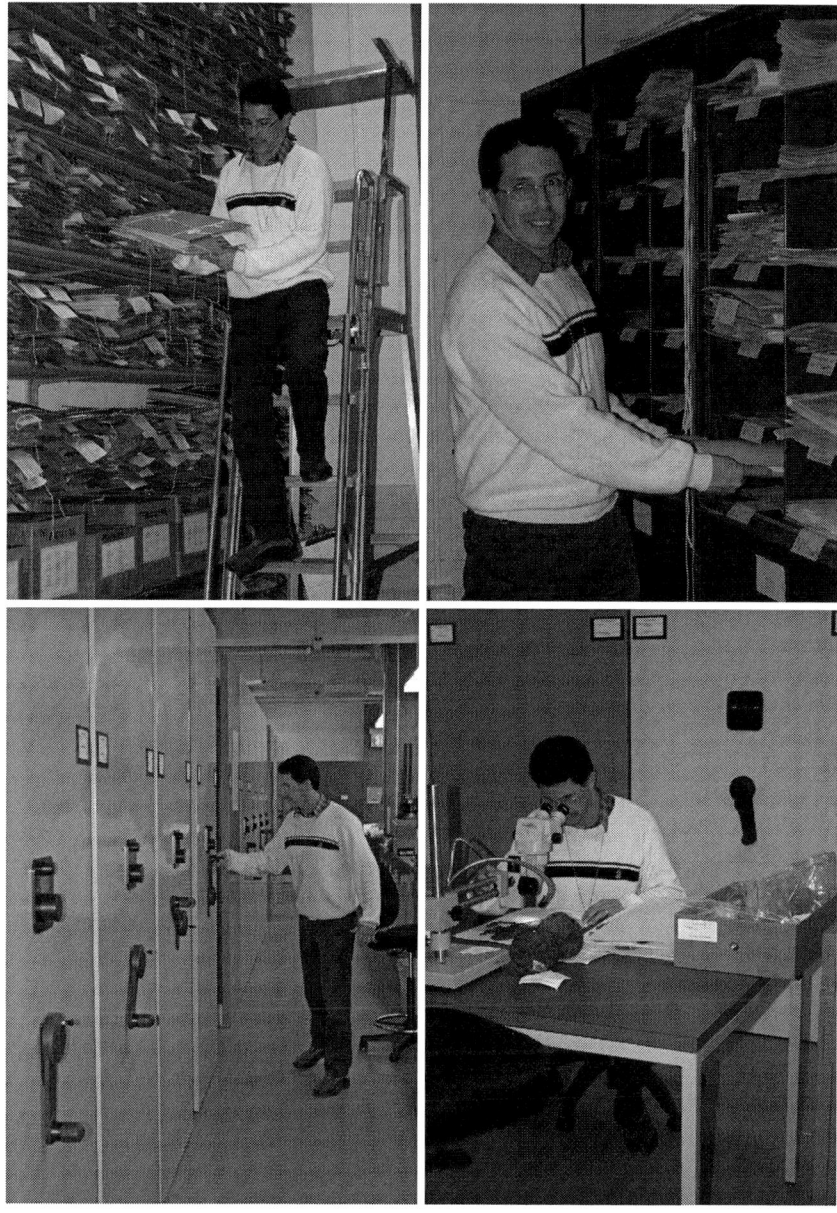

Fig. 5-2. After the collection is determined. Top left: The specimens are removed from storage in the cold room where low temperatures protect them from insect damage. Top right: A specimen is kept for NYBG and others are used as exchange with other herbaria. Bottom left: The specimens are stored in airtight herbarium cabinets and only removed when they are being studied at NYBG or in other herbarium which request them on loan. Bottom right: Specimens of the Brazil nut being studied at NYBG. The person in the images is Xavier Cornejo. Photos by Scott A. Mori.

to the next.

The retention of more than one exact duplicate of a collection represents poor use of herbarium space. If a duplicate does not provide new information then it probably should be used for exchange with other herbaria. In other words, there should be a specific reason for maintaining more than one duplicate of a collection in a given herbarium. For example, some species of *Pourouma* possess two morphologically different leaves. The vegetative sprouts have divided leaves and the fertile branches have simple leaves (Kincaid et al., 1998) so, in this case, the retention of two sheets of some collections of *Pourouma* is justified.

It is especially important to have duplicates of type specimens, but the duplicates generally should be archived in different herbaria for the obvious reason that some duplicate types will still be available for study if those in a given herbarium are destroyed. The classic example is the destruction of important types in the Berlin herbarium caused by bombing in Word War II. Many of those types were not represented by duplicates in other herbaria (Hiepko, 1987) and, thus, it is now difficult to interpret species concepts because the type specimens that those species are based on have been destroyed.

Uncontrolled growth of herbaria may not be possible due to economic constraints, so curators should exercise common sense over which specimens are archived in their herbaria. This may be impractical in large herbaria, but in smaller herbaria it should be possible to limit the number of specimens of a given species to only those that add new information to the herbarium. For example, if a local herbarium already has collections of a species at all stages of its life cycle, from the various habitats where it grows, and from throughout its range, the addition of new collections of that species probably should be allowed only if they provide information not found in extant collections or if the new collections are vouchers for scientific studies.

Ideally, voucher specimens should not be added to the herbarium until the study for which they provide vouchers has been published. Botanists are sometimes momentarily enthusiastic about a given project but for one reason or another do not carry the study to completion. On one of my expeditions, I was asked to collect 50 collection numbers of a species to document what was planned to be a detailed study of within population molecular variation. The study was never done so the collections were not mounted and not filed in the general herbarium. However, the specimens still take up space in a herbarium cabinet in my office and will probably eventually have to be discarded. Because it is extremely difficult for curators to keep track of when papers are published, the responsibility of submitting specimens for deposit in a herbarium should be with the authors of papers in which the specimens are cited.

Specimen annotation. An efficient way to obtain accurate identifications of specimens is to send duplicate specimens to specialists for determination. When the identifications are returned, the proper protocol is for the names to be entered into the collector's notebook, an annotation slip attached to the herbarium sheet, and the name entered into the institution's database. If there has been a change in the identification of the collection, the specimen has to be relocated to where other collections identified as the species are stored. Adding the determination to the database but not annotating the specimen creates discordance between the name on the specimen and the name in the database, and may result in different species being filed under the same name, or the same species filed under different names, in the herbarium. Obviously, having collections of different species filed under the same species makes it extremely difficult to use the herbarium for identifying unknown collections. Ideally, the new names should also be sent to other herbaria holding duplicates of the collection but there is seldom the man power needed to do this.

Another way to keep names in herbaria up-to-date is to consult the lists of *exsiccatae* published in monographs and to use the keys and descriptions in the monographs to identify specimens that have not been cited. If, after using the monograph, there are still specimens that have not been identified; the curator can ask the monographer if the unidentified specimens can be sent as a loan for identification. As more-and-more electronic monographs become available, this process will become somewhat easier because online images of the specimen can sometimes be consulted instead of sending the actual specimens.

Visits to local herbaria by specialists provide an opportunity to have an expert put the most up-to-date identifications on specimens in his or her group. If funds are available, a good use of them is to pay the costs for specialists to visit local herbaria. If the funds are not available, many specialists are willing to pay their own expenses if their herbarium visits are facilitated, and if they are given the opportunity to see some of the plants they study in the field as part of their visit if the herbaria visited are in an area where the plants occur.

Herbarium mission statements and collection policies. A mission statement and a policy establishing criteria for archiving specimens is the first step in controlling the quality of specimens in herbaria. However, the main reason that there is usually not more control over specimens that enter herbaria is that there is generally not enough qualified staff to manage acquisitions on a sheet-by-sheet basis.

Mission statements for herbaria should include statements about (1) the area of geographic focus, (2) the groups of plants and fungi of interest, (3) the types of studies promoted, and (4) the mode of dissemination of information derived from the collections. Collection policies should define: (1) the quality standard for archived specimens, (2) the ba-

---

Box 5-2. Mission and Collecting Policy
by Scott A. Mori

National Herbarium of La Botanica (NHLB)

The mission and collecting policy described here are based on the national herbarium of a fictitious country in the New World tropics called La Botanica. In this space, a description of a herbarium for which this document is being written should be added. A statement about the research goals of the herbarium (e.g., the publication of Floras and monographs, making conservation recommendations, identifying weeds that invade agricultural fields, ect.) should be included.

Mission

The mission of the National Herbarium of La Botanica (NHLB) is to document the plant and fungal diversity of La Botanica with collections showing (1) all stages of the plant or fungus life cycle, (2) morphological variation, (3) habitat preference, (4) phenological changes throughout the year, and (5) geographical distribution. In addition, the NHLB's mission includes serving as a repository for specimens cited to voucher data reported in publications about the species of plants and fungi of La Botanica.

Policy

1. Specimens added to the Herbarium must be carefully prepared and, thus, aesthetically pleasing. The Director of the NHLB reserves the right to reject poorly prepared specimens.
2. Only collections that add new information to that available from collections currently archived will be accepted for archiving in the NHLB.

3. Collectors wishing to archive their collections in the NHLB must submit a written request to the Director to do so. In the letter of request, the collector must state the justification for archiving the collections in the NHLB.
4. Large numbers of sterile collections will be accepted for archiving in the NHLB only if the reason for doing so is justified in a letter by the collector and found to be valid by the Director. If sterile specimens of the collection are already archived in other herbaria, the reason for duplicating them at the NHLB must be explained in this application.
5. Sterile collections vouchering species treated in scientific publications will be accepted only after the paper has been accepted for publication. It is the responsibility of the author(s) of the paper to submit the sterile specimens for archiving after a letter of acceptance of the paper has been received. The NHLB is not responsible for holding specimens pending acceptance of the paper unless previous arrangements have been made.
6. An understanding of how the cost of archiving specimens in the NHLB will be apportioned must be reached between the collector and the Director of the NHLB. If the addition of collections falls within the mission of the NHLB it may choose to bear the archiving cost. If not, the cost of archiving the collections will be paid by the collector. The NHLB encourages collectors to seek herbarium support in their grant proposals.
7. Collections from outside La Botanica will not be archived in the NHLB unless they are of special interest. For example, associates of the NHLB who are specialists in given groups of plants are encouraged to archive specimens of their group from throughout their distribution range.
8. The NHLB will collaborate to maintain duplicates of specimens from La Botanica in other herbaria as a measure to avoid loss of information in the event of destruction of specimens at the NHLB. Excessive duplication of specimens, however, will be avoided.

sis for acceptance of sterile specimens, (3) the acceptance of specimens for vouchering specimens cited in publications, and (4) who pays the costs of integrating and maintaining collections in the herbarium.

A model mission and collection policy developed for the imaginary Central American country of Botanica can be found in Box 5-2.

Conclusions

1. Collectors and herbarium managers should make certain that collections are handled as rapidly as possible by producing labels soon after returning from the field; by sending the labels, in timely fashion, to the herbarium in which duplicates were left; by routing

duplicates to specialists for determination soon after the labels are prepared; and by having the identified specimens mounted and deposited in herbaria without excessive delay.

2. Collectors and herbarium managers should keep identifications up-to-date by recording changes in the database, attaching annotation slips to specimens, and moving specimens with changed determinations to their new locations in the herbarium. They should also forward determinations to herbaria where duplicates of the collection have been archived.

3. Herbarium managers are encouraged to database their collections because databases offer the best possible way to communicate information about a herbarium's collections. In addition, databases facilitate the Floras and monographs that are the main goals of the curators of many local herbaria.

4. Images of herbarium specimens and specimen-vouchered field images attached to herbarium specimen records offer an unparalleled opportunity for local herbaria to communicate information about the plants of their area to other botanists and to the general public, and for monographers to more effectively illustrate the features of the plant groups they study.

5. Ideally, herbarium managers and their staff should develop mission statements and collection policies for their herbaria. Specific goals provide direction for a herbarium's research program and this will result in increased productivity.

Chapter 6
Onto the Internet
by Scott A. Mori

The previous two chapters dealt with the processes of gathering collections in the field and managing them in herbaria. This chapter focuses on how information about specimens and species can be presented over the Internet.

This discussion refers only to electronic monographs and Floras that are database-driven. I do not include the conversion of hardcopy monographs to html or PDF files, which are then placed on the Internet, because they do not make full use of database technology. The goal of this chapter is to emphasize how powerful database-driven, specimen-based, illustrated, online studies are for presenting data previously available only in hard copy monographs and Floras.

Technology that enables botanists to manage botanical collections electronically, and to prepare electronic species pages and identification keys, has been available for several decades. User-friendly software, that allows botanists without extensive computer expertise to prepare electronic monographs and Floras is, however, a fairly recent innovation. I personally use two of the improved software packages, (KE Emu, an object-oriented data management system for museums [KE Software, 2009] and Lucid, an electronic key program [Lucid, 2009]) to prepare and maintain an electronic monograph of the Brazil nut family (Lecythidaceae), and electronic floras of French Guiana, the Osa Peninsula in Costa Rica, and the island of Saba in the Netherlands Antilles. Never before have botanists had the ability to present data derived from field and herbarium studies in such an efficient, effective, and collaborative way. Moreover, electronic publications can reach a much wider audience than traditional hard copy publications.

Database-driven Internet dissemination of monographs and Floras is so superior that in the future, traditional hard copy publications will probably only be printed if they are derived from database-driven websites. Electronic publication is preferable because it makes images available at a much lower cost, facilitates collaboration with other researchers, provides identification keys that are more efficient and easier to use, and the information can be instantaneously accessed almost anywhere in the world (Farr, 2006). In addition, the sites are easy to develop and update, and research is markedly improved because electronic publications can be immediately and continually critiqued by a worldwide user-group.

Although I use the two programs cited above, other useful programs include Atrium (Botanical Research Institute of Texas, 2009), BioOffice (Tiroler Landsmuseen, 2009), Brahms (Department of Plant Sciences, University of Oxford, 2009), DELTA (Dallwitz et al., 1999), DeltaAccess (Hagedorn, 2002), EOL (Encyclopedia of Life, 2009), Linnaeus II (ETI Bioinformatics, 2004), Mx (Yoder et al., 2006), SLIKS (Guala, 2006), Scratchpads (Scratchpads, 2010), Specifiy (Biodiversity Research Center, 2009), TAXIS (Meyke, 2004), Winclada (Nixon, 2009), and 31 Interactive Keys and Taxonomic Databases (Dmitriev, 2003). Botanists can also use off-the-shelf programs such as Access and FileMaker Pro to develop electronic monographs and Floras, or even write their own programs using software that is downloaded free from the Internet. All of these programs have their advantages and disadvantages; thus, the reader should read this chapter not as an endorsement for any particular program, but as a general description of what can be accomplished with these programs as a class.

The Encyclopedia of Life (EOL) is an ambitious attempt to make electronic species pages available for as many species inhabiting the planet as possible (Encyclopedia of Life, 2009). As part of this effort, the EOL is developing LifeDesk, which is designed to make sharing and managing biodiversity data as simple as possible. The Encyclopedia of Life and LifeDesk can be considered by anyone wishing to undertake e-monographs and e-Floras, especially researchers not associated with institutions that have appropriate database systems and the IT support needed to manage them. The Scratchpads project (Scratchpads, 2010) is similar to the EOL effort.

I chose to use KE Emu to prepare my species pages and other electronic publications for three reasons. First, I can use computer programs but am not capable of resolving problems that arise with them, so I could not develop a system on my own or even use a system without IT support. Second, technical support is available to me from both NYBG Information Managers and indirectly from the KE EMu support service. Third, all of the authority files, for example the lists of collectors and herbaria, are already part of the NYBG database system and, thus, were immediately available for my use. For systematists without advanced computer skills, the most important consideration in software selection is to make sure that the selected program will minimize the time required to learn and manage the system. The research leading to monographs and Floras is so labor-intensive that most botanists do not have the time to effectively manage unnecessarily complex databases!

The KE EMu system is the database program used by some of the major herbaria in the world, including NYBG, the British Museum, the Field Museum of Natural History, and the Smithsonian Institution (US). KE EMu allows a researcher to (1) manage specimen data; (2) attach multimedia such as images, PDF files, and links; (3) manage bibliographic information; (4) prepare glossaries; (5) handle names, including homotypic and heterotypic synonyms; (6) prepare species pages; and (7) export data and images in a straightforward and inexpensive manner so they can be shared among collaborators, and moved from antiquated to improved systems. KE EMu, along with most other modern database systems, uses the Darwin Core format (Biodiversity Information Facility, 2010) to facilitate export into collaborative projects such as the GBIF Portal (Global Biodiversity Information Facility, 2010). These are the tasks that must be handled by database systems designed to manage collection data, and from which electronic monographs and Floras are produced. The programs used today will change as the technology improves. What will not change is the kind of information that will be communicated to other scientists and the general public. Monographs and Floras lend themselves to database-driven, image-rich, electronic publication, and herbaria and systematic botanists should make every effort to present their data using this technology.

Electronic Monographs

As mentioned earlier in this book (see Box 1-1), a monograph is a treatment of a given lineage of plants or animals throughout its entire distribution range. In general, monographs include basic information about the included taxa (e.g., taxonomic history, anatomy, morphology, distribution, ecology, and conservation status), keys to genera and species, descriptions of the species and infraspecific taxa, as many illustrations as needed, distribution maps, and references to the collections studied. For example, Flora Neotropica monographs synthesize all information about the New World plants and fungi that is known at the time of publication. In a Flora, all of the species of plants found in a specific geographic area are described. For example, the electronic Flora of the Osa Peninsula (Aguilar et al., 2008) is a project dedicated to discovering and describing all vascular plants known from that part of Costa Rica. Monographs, especially in conjunction with phylogenetic analysis, enable botanists to explore evolutionary and ecological relationships within and among lineages, and contribute to a stable system of classification and nomenclature. Floras are practical guides that facilitate the identification of plants found in a particular region, and can be simple or even monographic in their content.

It would be easier to prepare Floras if monographs were available for all of the plant groups that occur within the geographic confines of the floristic project; otherwise it is difficult for floristicians to interpret the morphological variation in the context of species circumscription because what they observe is small compared to the variation found throughout a species' geographic range. Species are therefore more accurately circumscribed from a monographic, rather than a floristic, perspective. Ideally, after the monographs are completed, species descriptions could be excerpted from them and used to prepare Floras. Unfortunately, this strategy is not practical because, given the relatively few botanists and the great species richness of plants, it would take an unrealistic amount of time to complete all of the necessary monographs before Floras could even be started. For example, several authors (Mori, 1992c; Thomas, 1999) have calculated that, working at the current rate of monograph production, it would take hundreds of years to complete monographic treatments for the flowering plants of the New World tropics.

Important components of an electronic monograph. Electronic monographs can be very elaborate; however, a database that drives even the most basic monograph should contain the following modules: specimens, taxonomy, muiltimedia (e.g., images, videos, and links to DNA sequences), and literature. In addition, keys and a glossary can be linked to these data. These will enhance the value of an e-monograph to professionals and students, and especially to members of the general public, who may not be familiar with certain taxa or terminology. There are many database-driven e-monographs available over the internet; one example is The Lecythidaceae Pages (http://sweetgum.nybg.org/lp/index.html; Mori et al., 2010). This work in progress uses KE Emu and Lucid, and thus far has only been completed for the Lecythidaceae of Mesoamerica, the Guianas, and Eastern Brazil. Nevertheless there has been enough progress that I will use it as an example to illustrate the process involved in using databases to produce an e-monograph.

On the Lecythidaceae Pages (LP), there are links to the six database categories mentioned above (specimens, taxonomy, images [= multimedia], keys, literature, and glossary) as well as to other information including a phylogenetic review of the family. In KE Emu, these database categories are called modules (e.g., the specimens category = the Specimen Module), and these modules are linked together into a much larger database. When visiting each of these modules on the LP, a search box enables the user to easily locate information stored within the category (e.g., data about a specific plant specimen will be stored in the Specimen Module). Of course, other database systems can be used to manage the data in these categories. The most important points are that electronic database-driven e-monographs should include the data in these categories, and the system should manage the information in an efficient way that allows users to access, link, organize, and export the data onto Web sites and into other database systems. The six categories are described in more detail below.

Specimens. Physical plant specimens—in particular the type specimens—serve as the anchors for species concepts in systematic botany, and as the basis for both hard copy and e-monographs. In a hard copy monograph, a list of specimens studied is presented as a list of *exsiccatae*. In an e-monograph, the database table that holds and provides access to information about specimens (here referred to as the Specimen Module) is the equivalent to the list of *exsiccatae*. A complete specimen record at NY provides the label data, an image of the specimen, a Google map of the collection locality, and other images associated with the specimens. In addition, field images, herbarium images, images of anatomical and morphological features, and DNA sequences can be anchored directly to the collection upon which they are based (e.g., images without vouchers are next to worthless and should not be used in scientific monographs). These supplementary materials enable others to understand the author's species circumscriptions because they can access information on the referenced specimens, and it allows curators to put names on duplicates that are archived in their herbaria, but which have not been annotated by the monographer. Other researchers can examine specimen images over the internet, and provide feedback that enhances the accuracy of the data. Images attached to specimen records reduce the number of physical specimens that have to be sent to researchers at other institutions as loans. This reduces the shipping costs and wear-and-tear on the specimens, and eliminates the possibility that specimens might be lost in transit, or for some other reason never returned to the home institution.

Prior to the advent of electronic databases, lists of *exsiccatae* were usually generated from index cards. Electronic management of specimens has very distinct advantages: although database management still takes a great deal of work, the task is easier and more accurate, and records are more easily modified. In addition, once data from all of the specimens of a group or region have been entered in the database, it is possible to generate preliminary specimen-based checklists for any country or major political subdivision throughout the geographic distribution of the taxa included.

Multimedia. Multimedia, including images along with other files (such as PDF documents) and links to related websites, is managed in a section of the database called the Multimedia Module. A multimedia record can be associated to records in more than one module. For instance, an image of the flower from a vouchered plant specimen might be accessed through the specimen record in the Specimen Module, and also through a database-driven species page. Identification keys include links to records in the Multimedia Module to illustrate the characters used to distinguish taxa as well as to images of the taxa (see Keys below).

Plant images should include captions that indicate what the image shows, the photographer, the associated voucher specimen, the collection's geographic locality, and an explanation of image significance. For the most part, an image should not be used if it is not associated with a voucher specimen, but exceptions are often made for images of species that can be identified with a high degree of certainty by images alone.

A species page includes a representative sample of the multimedia available for a particular species. For example, the most informative species pages on the LP include botanical line drawings; field images of the trunk, bark, flowers, fruits, and seeds; the protologue; images of type specimens; and links to any DNA sequences available through GenBank. Although each species page contains a selection of images, the Multimedia Module generally contains additional images attached to specimens that give an idea of intraspecific variation.

Taxonomy. The Taxonomy category of the LP (here referred to as the Taxonomy Module) contains all of the names of neotropical Lecythidaceae, indicates which names are accepted and which are synonyms, and identifies which synonyms are homotypic (= based on the same type) and heterotypic (= based on different types). Species and genus descriptions (i.e., the species and genus pages) are also driven by this module. The fields used in the LP include the same kinds of information found in hard copy monographs, and are listed and explained in Box 6-1. The Taxonomy Module allows users to search for names (either accepted or in synonymy) and to view the pages and data associated with those names. In addition, an advanced search option allows the user to produce a list of all species known to occur in a given geographic region.

Species pages on the LP are arranged in the following order: (1) scientific name (2) literature citation, (3) status of the name (e.g., accepted or a synonym), (4) list of synonyms, (5) link to type specimens at NY (images of types at other herbaria are provided as multimedia attachments at the bottom of the species page), (6) species description (see Box 6-1 for details of the contents of the description), (7) link to a Google map of the species distribution based on all specimens with coordinates, (8) a link that executes a search for all specimens in the database determined with that name, (9) a link that executes a search for all multimedia records associated with the species, and (10) the multimedia that have been selected to represent the species.

Box 6-1. Species Page Content

by Scott A. Mori

This example is modeled after the species page for Lecythis tuyrana on the Lecythidaceae Pages (http://sweetgum.nybg.org/lp).

Author: The author(s) of the species page.

Type: The citation of the type collection in the following sequence: Country. Country subdivision: specific locality; date of collection as day/month (three letter abbreviation)/year followed by fertility state(s) (e.g., st, fl, fr, seeds); collector, as name appears on label, collection number; herbarium of holotype or lectotype; isotypes or isolectotypes (in parentheses following the collector's name and number); Additional notes. Example. Panama. Darién: forests around Pinogana, 16-21 Apr 1941 (fl, fr, seeds), *Pittier 6557* (holotype, US – herb. no 716630; isotypes, C, F, GH, NY, US). Notes: Additional numbers of type collections (e.g., the herbarium accession number or the barcode number) should be added if known. In the case of lectotypes, the literature citation of where it was first published should be provided (e.g., lectotype, NY, designated by G. T. Prance, 1979), All references cited within the species page (e.g., G. T. Prance, 1979) should be included in the literature module of the Website.

Description: The morphological description, as well as any other information

text continues on next page

Box 6-1 continued

available, such as chromosome numbers, etc.

Common names: The country-by-country common names as indicated from herbarium specimens or from the literature. The name and collector of the herbarium specimen and/or the literature reference should be cited. Example. Panama: *coco, kula wala* (Cuna name for tree), *sapisuro guala* (Cuna name for fruit). Colombia: *coco de mono, ohetón* (*Callejas et al. 4937*). Notes: Other information about the name is provided in parentheses after the name.

Distribution: A written description of the overall distribution of the species. Example. *Lecythis tuyrana* ranges from eastern Panama into northwestern Colombia. Although expected to occur there, this species is not known from the Colombian Chocó or from Ecuador. Previous reports from the latter country are based on misidentifications.

Ecology: A description of the canopy position, frequency, and vegetation type of the species. This information is usually incomplete because these data are seldom reported in sufficient detail on specimen labels and in the literature. Example. This species is a canopy or emergent tree of lowland, wet forests.

Phenology: A list of the months in which the species has been collected in flower or fruit, and details from the literature of any phenology studies made on the species. The latter should be documented with the literature reference(s) included in the literature module of the database. Example. In the Darién of Panama, flowers have been collected in Apr and Oct and seeds in Jul and in Colombia flowers have been collected in Mar, May, and Jun. Notes: In the Lecythidaceae, the presence of empty fruits provides little phenological information because the woody fruits persist for many months under the tree. The most important fruiting event is when mature seeds are available for dispersal and that is seldom observed.

Pollination: A report of pollination observations obtained from herbarium specimen labels and the literature. Notes: Suggestions of pollinators can be made based on information available from species with similar flower structure and color, but these suggestions must be clearly indicated as such. Lack of information is recorded as "No observations recorded."

Dispersal: A report of dispersal observations based on data obtained from herbarium specimen labels and the literature. Suggestions of dispersal agents can be made based on information from species with similar fruit and seed structures, but these suggestions must be clearly indicated as such. Notes: Lack of information is recorded as "No observations recorded."

Predation: A report of predation on leaves, flowers, fruits, or seeds based on observations obtained from herbarium specimens and the literature. Notes: Lack of information is recorded as "No observations recorded."

Field characters: A summary of the characters used in field identification. Example. *Lecythis tuyrana* is recognized in the field by its large size; cylindrical trunk; rough bark; oblong, large leaf blades; androecium with a single, incipient coil; yellow petals and androecium; relatively large fruits with an erect supracalycine zone and muricate opercular rim; and seeds with a basal aril that ascends for a short distance up the side of the seed.

Taxonomic notes: Provide a description of the features that indicate relationships with other species. Example. Based on molecular studies (Huang, 2010; Mori et al., 2007), *Lecythis tuyrana* forms a clade with *Lecythis ollaria* (type of the genus) and *L. minor*. These three species possess single-coiled androecial hoods with the vestigial stamens restricted to the outside of the coil. In *L. tuyrana*, the coil is much shorter than it is in *L. ollaria* and *L. minor* and is, therefore, described as an incipient coil.

Uses: Provide observations obtained from herbarium specimen labels and the literature. Example. The Cuna Indians use the fruits and seeds in their folk medicine. It is a handsome tree with a wide-spreading crown when grown in the open and has frequently been planted as an ornamental. The empty fruits are used as containers to store various domestic articles (Carrasquilla R., 2005).

Etymology: The derivation of the species epithet. Example. According to Dr. Mireya Correa, this species was probably named after the Tuira River in Darién. The type locality is given as Quebrada Honda, in the vicinity of the Tuira River.

Conservation: The classification of the species according to the *IUCN Red List of Endangered Species* and *Plantas Raras do Brasil*, if the species occurs in Brazil. Additional comments about the endangered status of a species can be added from other sources.

Source: The publication upon which the species page is based. Example. Based on Mori in Mori & Prance, 1990. Notes: If the description is new, "New description" should be placed in this field. An Internet description of a species new to science does not, however, constitute valid publication.

Literature. An e-monograph will be particularly useful if it includes as many references as possible about the all aspects of a group. This makes life easier for students, professionals, and members of the general public because they can search a single database compiled by a specialist, instead of searching for citations in scattered sources such as the Internet, journals, and library databases. If permission is granted, or there are no copyright restrictions, important works included in the literature category (here referred to as the Literature Module) can be uploaded to the Multimedia Module (e.g., as a PDF) and are then available to download for free.

By making references and PDFs widely available, the Literature Module facilitates collaboration, and significantly reduces the burden on specialists to provide literature upon request. A recent paper on the phylogeny of Lecythidaceae (Mori et al., 2007b) includes insights into the classification of the family; thus, reprints of it are periodically requested. Because the *American Journal of Botany* allows authors to make PDFs of their articles freely available online, to honor a reprint request I need only to provide the URL to its entry in the Literature Module.

My relationship with Alex Popovkin is one example of how collaboration is enhanced through the Literature Module of the LP. Alex is a Russian-born editor who works remotely from his small farm in Bahia (Brazil). Alex has a life-long passion for plants and studies the local flora, making observations almost every day. His photographic archive, which includes several thousand plant images, is still growing and is available online (search for "Popovkin Lecythidaceae"). He has begun to collect plants and deposit his collections in the herbarium of the Universidade Estadual de Feira de Santana. In October, 2007, Alex noticed that trees of *Eschweilera ovata* (Cambess.) Mart. ex Miers periodically flushed new leaves. He wrote to me "Maybe I should start a phenological study of local Lecythidaceae. Could you give me some guidelines? Or a source for such?" In the past, a request such as this would have required a great deal of my time, because it would have been necessary to consult the appropriate literature, copy selected papers, and respond via snail mail. With the help of the LP, this process took minutes; I simply directed Alex to the Literature Module and instructed him make an "Advanced Search" for "Phenology." I then suggested that he read the attached PDFs of the papers that provided the relevant information. This communication, which would have taken hours in the past, took less than 10 minutes because of information available in the Literature Module of the LP.

All plant studies are based on previous studies, and relevant literature involving classification may extend back for hundreds of years. Protologues are the original publications of plant names. These must be consulted to determine correct species names, and are the most obvious application of past literature to present classification. The *International Code of Botanical Nomenclature* is based on this concept (McNeill et al., 2006). The LP, by making protologues and other important literature immediately available to the user, eliminates the need to search for the original references each time they are needed. Another invaluable source for older literature in systematics is the Biodiversity Heritage Library (BHL, Biodiversity Heritage Library, 2010).

Electronic monographs and e-Floras are not the only electronic vehicles for the dissemination of information about organisms; there are also open access journals. ZooKeys (ZooKeys, 2009) publishes original research in animal systematics, and PhytoKeys, a journal dedicated to plant systematics, will soon publish its first addition. These journals allow open online access to their publications. Electronic journal access is usually so expensive that only large universities, zoos, and botanical gardens have the resources to pay for subscriptions, but open access journals are available free of charge, which makes them accessible to smaller institutions and individuals. Even journals that are not entirely open access sometimes provide authors with the option to make their papers open access after the payment of fees; when an author has the financial resources, purchase of these rights make his or her publication available to others free-of-charge.

Glossaries. It is difficult, if not impossible, to perceive the differences among similar plants without understanding the terms describing the characteristics that distinguish them. Botanical terminology is so highly specialized that it creates a major obstacle to communication between professional botanists and general science students or plant enthusiasts. In monographs, botanical terms are generally defined at the beginning of the morphology section. Because many terms are needed to describe morphology within the diverse group of plants included in a Flora, a glossary is often appended at its end (e.g., see Mori et al., 1997, 2002a). In addition, many hard copy glossaries have been published for plant terminology (e.g., Ellis et al., 2009; Ferri & Menezes, 1990; Font Quer, 1985; Hickey & King, 2000; Harris & Harris, 1994; Jackson, 1965; Kiger & Porter, 2001; Stern, 1992), and many online glossaries can be found by simply searching for "botanical glossaries.

Botanists now have the ability to integrate electronic glossaries into e-monographs and e-Floras, and while this ability is just beginning to be exploited, it has opened a new era of botanical communication. E-glossaries are particularly informative because they can be accompanied by numerous images, or even videos, that more accurately communicate the meaning of terms than do words alone. Terms stored in a glossary database can now be directly associated with text included in other parts of the database (e.g., a species description), which allows users to see definitions of terms simply by hovering over or clicking on a term mentioned on a Web page. On the LP, a glossary record includes the term, its general definition, a definition of the term as it is applied to the Lecythidaceae, and multimedia attachments that illustrate or provide additional information.

Keys. There are several electronic programs that can be used to generate identification keys (e.g., Dallwitz et al., 1999; Guala, 2006; Lucid, 2009; Nixon, 2009). Electronic keys are generally easier to use than hard copy keys because the user can begin the identification process using whichever characteristics can be seen in the unknown specimen. In contrast, in a hard copy key, a specimen will be very difficult to identify if the character described in the first couplet is absent or difficult to see. Electronic keys, such as those generated by Lucid and used on the LP, can include links to glossary entries, thereby providing easy access to definitions and images that elucidate obscure terms. Identifications can be verified by clicking on the taxon name, which is linked to the descriptions and images on a species or genus page—and if the identification does not appear to be correct, it is easy to backtrack and follow an alternate path through the key.

Electronic Floras

Most of the information provided in the preceding discussion of e-monographs also applies to e-Floras. The major exception is that floristic treatments generally do not provide the detailed information included in the species descriptions found in monographs. The goal of a Flora is to document the species found in a given geographic area and to provide keys and descriptions that facilitate their identification. Because the species found in the area covered by a Flora usually display less variation than is found throughout their entire range, it will be easier to identify them if descriptions are based on specimens that occur in the area of the Flora. Drawings should ideally be based on collections from the area of the Flora and should include flower, fruit, and seedling collections derived from the same population, if not the same plant, because this reduces the chance of making mixed-species illustrations.

Before preparing a Flora, botanists must explore the region at all times of the year so that they can prepare a comprehensive checklist. It is also necessary to visit herbaria that have large numbers of collections from the Flora's area, and enter specimen data into the Flora database. In addition to field images of freshly collected plants, herbarium specimen images facilitate species identification, and therefore representative, well-prepared collections should be selected and photographed. After specimen labels have been prepared for any new collections, and data from existing collections have been entered into the database, a specimen-driven, illustrated checklist can be generated. That checklist serves as the basis for communication among the collaborators working on the Flora.

The best strategy for preparing Floras is for monographers and floristicians to work together. Experienced monographers are more likely to know names that should be applied to the plants in their focal groups. On the other hand, monographers frequently depend on firsthand information from field botanists. Hammel and Zamora (2005) provide a convincing argument for the added value of including resident botanists in floristic projects. For instance, botanical exploration carried out as an integral part of the *Manual de Plantas de Costa Rica* project added nearly 1000 species that were either new to science, or new to Costa Rica as of 2005 (Hammel & Zamora, 2005); this represents nearly 10% of the vascular plant flora of the country (Zamora et al., 2004). Just as important, the resident botanists trained as part of this project have made countless additional contributions to the knowledge and conservation of plants of Costa Rica.

Now that field botanists have the ability to communicate using the Internet, especially with images, providing information about the characters that distinguish plant species is much more efficient. When monographers and floristicians work together the result is synergistic— their collaboration yields the most accurate and complete Floras and, at the same time, provides information that markedly improves monographic research.

The steps and time needed to place collection data and the images they voucher online are described in Box 6-2. This process represents the first stage in the preparation of both e-monographs and e-Floras and in itself is time-consuming.

Box 6-2. Specimens from the Field to the Internet
by Scott A. Mori

This box provides a step-by-step breakdown of the time involved in collecting plant specimens, archiving them in the herbarium, and making the data and images associated with the collections available on the Internet. This breakdown does not include passive processes such as the time required to dry the specimens. Nor does it include the time required for various "downstream" processes: specimen identification, indefinite maintainance in the herbarium and the database, or the study that would be needed to incorporate specimen-based data into Floras, monographs, and other publications.

text continues on next page

Box 6-2 continued

   To document the process, the time it took to handle 45 collections made on a trip to the Adirondacks, New York was recorded. To reduce our workload, we made a single duplicate of each collection—if more duplicates had been collected it would have been longer to process the collection. In this case, there was no need to gather duplicates because we were collecting species from a well-known flora and were able to make positive identifications in the field, or knew we would be able to use the literature and the herbarium to identify them when we returned to NYBG. In some cases a specialist identified specimens for us, e.g., the Cyperaceae. The specimens were collected to voucher images taken of the plants as part of a NYBG project to image and describe the plants of the northeastern United States. At the time, not a single one of these species was represented by online, specimen-vouchered images from the Adirondacks and, thus, they all provided new information for the NY Virtual Herbarium. As we collected the plant specimens, we photographed them to show features useful in their identification. Of the numerous images initially taken, we selected 132 to include in the database. We discarded redundant images to save time and conserve server space.

   The procedure we followed in processing these collections is outlined below. The time given for each step includes the handling of all 45 collections.

1. Notes on the collecting locality and specimen data were entered in the field-book at the collection site. Each collection was given a unique number in sequential order in the collection series of S. A. Mori (Mori et al. 27179 through Mori et al. 27223).
2. The plants were photographed.
3. Each collection was immediately pressed in newspaper, the collection number was added to the newspaper, and the specimen was placed in a field press. An effort was made to make the best possible specimen layout, so that we would not need to rearrange specimens before placing them in the drying press. Steps 1, 2, and 3 = 5 hours.
4. Specimens were placed in the drying press. Step 4 = 0.25 hours.
5. After eight hours, the specimens were taken from the press and placed in numerical order. Step 5 = 0.5 hours.
6. The dried specimens were tied into a bundle and the bundle was placed in a plastic bag. Step 6 = 0.1 hours.
7. Upon returning home, the collection data were entered in the institutional database. Step 7 = 4 hours.
8. The labels were proofed, corrected in the database, printed, cut, put in numerical order, and added to the specimens. Step 8 = 1.5 hours
9. The images were adjusted in PhotoShop®. This included deleting some images, cropping, adjusting exposure, sharpening, and sizing the images to

800 pixels in width. Time = 4 hours.

10. The images were attached to the specimen records in the database. This included adding the following information: name of the photographer, title of the image, description of the image, and key words. The title of the image includes the part of the plant photographed, the name of the species, and the name of the photographer. The description adds the name of the collector and the collection number, the collection locality, and any special information about the plant. Key words include at least the name of the plant family; the name of the species; the region, country, state or other major political subdivision of the country where the plant was collected; and what the image shows, i.e., flower, fruit, seedling, etc. Time = 4 hours.

11. Specimens were mounted onto herbarium sheets and stamped as part of the collection of NYBG. Time = 6 hours. (This time is estimated; plant specimen preparers at NY are expected to produce 50 mounted collections per seven-hour day).

12. Specimens were barcoded. The barcode was added to the herbarium specimens (15 minutes), the specimen record searched in the database, and the barcode number added to the database (30 minutes). Time = 0.75 hours.

13. The mounted herbarium specimens were scanned, and the images were added to the specimen records. Information was added to each image record, and that record was attached to the catalog record of the specimen. Time = 2 hours.

14. Finally, the specimens were filed in the general herbarium of NYBG. Time = 3 hours. Usually specimens from different projects are sorted into families and then filed in the herbarium one family at a time. The time presented here is for the relatively unusual situation in which the collections of a single project were all filed at the same time.

In summary, making field collections (steps 1–3) and drying the specimens (steps 4–6) took 5.85 hours. Subsequent specimen handling, including databasing and labeling (steps 7 and 8), mounting and barcoding (steps 11 and 12), and filing (step 14) took 15.25 hours. Image management, including photo editing, scanning herbarium vouchers, and entering images into the database (steps 9, 10, and 13) took 10 hours. The process of collecting 45 unicate specimens documented by 132 images took a total of 31.1 hours, or approximately 41 minutes per collection. This process would have taken longer if they had been done by less experienced botanists.

The records for these collections can be seen by (1) accessing the C. V. Starr Virtual Herbarium of NYBG at http://sciweb.nybg.org/science2/VirtualHerbarium.asp, (2) scrolling down to "Online Specimen Catalog," (3) clicking on "All Vascular Plant Collections," and (4) doing an "Advanced Search" by typing "mori" in the collector field and ">27178<27224" in the collector number field. This search yields a summary list of the collections and clicking on each collection provides herbarium label data, field images (when available), and an image of the scanned specimen.

Citation of e-monographs and e-Floras

It is important to make clear who the authors of electronic publications are because without knowing this, it is difficult to have confidence in on-line information. This is especially true for scientific publications which, in contrast to other types of pages such as those published by conservation organizations, have more reliability if the author is cited. In addition, knowing who the author is allows the user to contact a person with questions about the online content or to provide information that might be of interest to the authors. For these reasons, a scientific web page should not be attributed exclusively to an organization because people, not organizations, write electronic monographs and Floras in same sense that people, not organizations, collect plants (see Chapter 4).

In order to make it clear how to cite websites, the home page should have a link entitled "How to Cite" that leads to a page that provides the following information: the authors, the year established, name, URL, and organization publishing the website. There should be no blind pages in the website, i.e., pages that do not have a link back to the home page because it will be difficult for a user to find out how to cite the website. If different pages of the website have different authors, those pages should have separate "How to Cite" pages.

If there is a major change in the website, for example when a re-searcher retires and the pages are maintained by someone else, that change should be registered in the "How to Cite" pages. If the pages have become static, that should be mentioned in the "How to Cite" pages by stating that the website is no longer being actively maintained.

When preparing the citation for an e-monograph, if only the starting date is known, the date should be followed by "onward" and the date that the website was accessed should be added to the end of the cita-tion followed by "accessed." In cases in which the publication date is not cited, the date that the website was accessed should be cited in place of the publication date, e.g. as "2010 accessed." See the "Literature Cited" at the end of this book for examples of how to cite websites.

Problems Remaining to be Addressed

Although current databases accomplish most of the tasks needed to pro-duce e-monographs and e-Floras, there are usually several features that have not been integrated into most of the systems. For example, for those using KE EMu, electronic keys must be generated from stand-alone programs such as DELTA, Lucid, or WinClada. Thus, data used for constructing keys is entered into a data matrix associated with the key program and not in KE EMu. As a result, when new information about a plant species is obtained, two separate databases must be updated and this may result in the KE Emu database becoming out-of-sync with the key matrix. This problem would be eliminated if the keys were based directly on information provided in a KE Emu database module.

Ideally, the information for electronic species pages would be derived directly from data recorded in specimen databases. Species pages would then be immediately updated if new data were added to the database. For example, a species page lacking information on fruit size would immediately be updated as soon as new fruit measurements were recorded in the specimen module. It would also make species descriptions parallel (i.e., with all of the information about each species appearing in the same order), and, if the descriptions were linked to the keys, ensure that information included in both was concordant.

Database-driven descriptions would be especially useful when taxonomic changes are made; for instance when what is initially recognized as a single plant species is subsequently separated into two. If the original species description was database-driven, to generate the new descriptions the monographer would simply change the names of the specimen records representing each of the segregated species. On the other hand, when the original descriptions are not database-driven, descriptions for both of the segregated species need to be rewritten. The ability to link information from Specimen Modules to species pages in database-driven e-monographs or e-Floras represents a major advance in the storage and dissemination of botanical knowledge.

Unfortunately, it would be nearly impossible to have all species pages derived from specimen databases because of the wealth of information that has already been accumulated about plants, but is no longer associated with particular specimens. Thus, the information for species pages would have to be: (1) recorded from specimens into databases *de novo* or (2) derived from literature but treated in a similar way as data derived from specimens. In the later case, the bibliographic source of the information would be recorded, but that would not guarantee that the description could be traced to the specimens from which the data were originally obtained.

Another obstacle to the creation of species pages for all species on the planet, as envisioned by the Encyclopedia of Life project, is that contributors do not necessarily agree on species limits. For instance, taxonomic *lumpers* and *splitters* have divergent philosophies; *lumpers* tend to circumscribe species with considerable morphological variation, while *splitters* tend to describe species separated by more subtle differences. These philosophical differences are difficult to resolve because it is a matter of opinion regarding exactly how different plant populations need to be to merit recognition as distinct species. This oversimplifies the situation because species delimitations are defined by a much wider range and complexity of species concepts than those of simply *lumpers* and *splitters* (Luckow, 1995)... this issue is not simply academic, because it affects legal recognition of plants and interpretations of their conservation status.

Interpretation of morphological variation in plants sometimes gets more complicated as more information becomes available. In the very early stages of plant exploration, nearly all collections represented "new species." However, as exploration intensified, new collections were gathered and morphological intermediates were found linking populations that were previously thought to represent separate species. Some researchers assume that molecular data will help resolve problems with species limits, and while that will undoubtedly be the case for some taxa, molecular data is not a panacea that will resolve all problems. The same problems that arise when defining species on the basis of morphological features are likely to arise with molecular data (particularly when collected from a small sample of plants). For instance, as more and more populations are sampled, genetic differences that were originally thought to distinguish species might turn out to be part of a gradient or network. Regardless of the data source, almost all species descriptions are based on a small subset of what is essentially an infinite number of individuals making up plant populations; thus, further sampling will reveal more variation.

Although great strides have been made to enhance the quality and reproducibility of botanical data and plant classification, systematic botany is not an exact science (Burger, 2006). Therefore different botanists will most likely continue to have different interpretations of species limits, no matter how complete the sampling or how sophisticated the techniques used to distinguish species become. The changes in plant nomenclature that accompany a better understanding of their evolutionary relationships are a considerable source of frustration for non-specialists, but users of taxonomic data must understand that botanists are striving to achieve a stable classification that indicates evolutionary relationships. Up-to-date information that is easily accessible via internet species pages will help end-users better cope with changes in plant names.

Because of these and other problems, it is not realistic to expect that all species can be represented by universally accepted species pages.

It is already apparent that non-vetted, wiki-created species pages will be difficult to maintain because different contributors will sometimes fail to agree on the species delimitations, or even on the page content. For example, I know a professional botanist who, after publishing a monograph, updated the Wikipedia entry for a genus in the family he studies. Soon after he uploaded the page it was modified by another contributor, who removed much of the detail that my colleague had provided. Although there is a "History" tab in Wikipedia that allows one to trace changes in a page, this is a difficult process, and does not necessarily enable the user to clearly identify the authors of different versions, especially when they use pseudonyms. Wikipedia can be a useful source of information for some disciplines, but it may not be appropriate for maintaining species pages. Rather than investing time and energy in creating detailed species pages for Wikipedia, professional botanists would be better advised to spot-check entries on "their" groups for accuracy, correct the errors, and insert links that direct users to species pages under their own control. I have done this in Wikipedia and also placed information for non-scientists on the Encyclopedia of Earth (Mori & Swarthout, 2007, 2008a, 2008b). The links inserted into these pages bring users directly to the LP where they can obtain more detailed information about Lecythidaceae. The links markedly increased traffic to the LP and, thus, made the pages more highly ranked than they were previously.

Species are not described exclusively by professional botanists associated with botanical institutions or universities; anyone may publish species descriptions as long as they follow the rules set forth in the *International Code of Botanical Nomenclature* (McNeill et al., 2006). One of the tenets of the *Code* is that species descriptions are published by the distribution of printed matter to the interested public; currently, most new species are published in peer-reviewed botanical journals. This ensures that the rules of the *Code* and other basic botanical protocols are followed, but it does not guarantee that a species is a "good" species. It is, thus, up to those who use botanical names to determine which validly-published names should be accepted.

According to the *Code*, when a contemporary botanist changes the classification of a plant species that was described by someone else, that change is documented by a concomitant change in authorship. In other words, established traditions prohibit botanists from changing species names or descriptions without leaving a clear trail about the author of the change. In open-access species pages, such as those created on wikis, there is no equivalent form of oversight. Google, in Google Knol, recognizes that wikis do not protect authors' rights and, thus, allows authors to respond to essays with which they do not agree in separate essays. For example, if someone does not agree with the information presented in a web page, then he or she can write a separate page on the topic and users can select the concept they wish to follow (Manbar, 2007). This is essentially the same protocol that has traditionally been followed in hard copy publication, i.e., authors sometimes publish conflicting interpretations, and it is up to the users to select which author to follow.

There are sometimes good reasons to prepare more than one species page, even in the absence of conflicts over species concepts or species page content. For example, monographic species pages need to be as complete as possible; including the protologue (or reference to it), images of types (or links to them), synonyms, reference to specimens representing the species from its entire range, descriptions that apply to the species throughout it range, etc. This could be too much information for a species page in an e-Flora, and could potentially hinder rather than facilitate identification. It might therefore be appropriate to have one species page for an e-monograph, and a different page for the same species in an e-Flora.

A system for efficiently documenting biodiversity should include tools to carry out all of the functions of systematic botany, and incorporate flexibility that allows for individual creativity and control over content. It should be available to all systematists (or other users) and provide them with an efficient support system. There are databases that already have all, or nearly all, of these features. Ideally, one of those systems should be selected and developed as the world's biodiversity database system, and made available to systematists that want to use it. There should be dedicated servers to archive the data, and the software could be accessed from a content management system using "cloud" technology. This would allow systematists to access their e-monographs and e-Floras from practically anywhere in the world, including field stations in the most remote parts of the planet. System administrators would allow qualified botanists to post e-monographs and e-Floras even if another is already online and, in the case of differences in concepts among websites, users would determine which author to follow.

As mentioned earlier in this chapter, the Encyclopedia of Life provides a system that could be universally employed for making species pages, especially when its LifeDesks tools are improved and become available to systematists other than beta-testers (Encyclopedia of Life, 2009). However, professional biologists would be more inclined to use EOL if they were able to create species pages that would remain fully under their control. Open-access contributions to species pages should be discouraged, because very few contributors have the expertise needed to distinguish among species, especially in difficult groups. This is especially important when images are posted without being vouchered by specimens. Although most people would agree that a picture of a polar bear represents *Ursus maritimus*, many plant images do not convey enough information to support positive identifications. If multiple, uninformed contributors add unvouchered images to species pages; it increases the chances that those species pages will end up including images of more than one species. This problem will only be rectified if specialists act as gatekeepers, and oversee contributions to species pages. Even Wikipedia has recognized that non-vetted revisions—at least to some entries—need to be controlled (Cohen, 2009).

The degree of confidence that a user has in a species page is related to the experience and reputation of its author, and many species pages online today do not provide this information. Authorships should always be clearly indicated, as should instructions for appropriate citation. When available, the date that the site was created should be included in the citation, and the date that the page was consulted should also be included (e.g., accessed 4 January 2010). To make website citations as easy and accurate as possible, authors can include a "How to cite" link at the bottom of each page. Unless circumstances make it impossible, the authors of electronic publications should be cited—not the organizations that sponsor these works; because people, not organizations, do the work and they should receive the credit (or be held responsible if a mistake is found!).

Although I am a strong advocate for e-monographs and e-Floras, these are valuable supplements to—not replacements for—hard copy publication. The "moving target" nature of electronic publication makes it difficult to know if what was cited one day will still be valid the next. Thus, it is essential that new species and new taxonomic combinations be published in hard copy to provide a static point of reference for names. In addition, it might be advisable to develop a system for periodically publishing hard copy versions of ongoing e-monographs and e-Floras. Hard copy publication provides periodic summaries of the state of knowledge in groups of plants (monographs), and plant diversity (Floras). It also acts as insurance against total loss of information should there be drastic changes in electronic media technology, or should home institutions fail to support database-driven Websites that authors no longer maintain.

Conservation

Monographs and Floras are essential conservation tools because they enable conservationists to determine what species grow where, how wide their ranges are, and if their populations are likely to be threatened due to habitat destruction. Most important, they provide the tools (keys, descriptions, and images) needed to identify species. It is logistically—and legally—impossible to protect a species without knowing its name. Because systematists are so aware of the complex interactions among plants, animals, and their environments, they are often avid conservationists (See Chapter 5). After all, if they are not willing to apply their knowledge to protect the organisms and natural ecosystems they know so well, who will? Conservation planning is informed by knowledge of species distributions; therefore distribution maps play an important role in developing strategies for protecting biodiversity. These maps pinpoint areas of high endemism and indicate where conservation areas should be established to protect the greatest number of species.

For instance, our tree inventory in the Atlantic coastal forests of eastern Brazil demonstrated that 53.5% of the tree species studied were endemic to that region, and another 11.8% were restricted to the coastal forest plus some part of the *Planalto* of Brazil (Mori et al., 1981). To qualify as a hotspot, an area must harbor endemic species at least equal to 0.5% of all vascular plant species worldwide, and be threatened with loss of habitat (Myers et al., 2000). The Atlantic coastal forests clearly make up a "biodiversity hotspot."

Recognizing areas that are rich in endemics—and simultaneously under threat—as in biodiversity hotspots aids in setting conservation priorities (Myers, 1988); this recognition is based, to a large extent, on mapping species distributions to reveal areas of high endemism. Mesoamerica is another of the world's 34 hotspots, with 5000 of its estimated 24,000 vascular plants endemic to the area (Mittermeir et al., 2005; Myers et al., 2000). The idea behind the hotspot concept is to make the best use of the limited resources available for conservation by establishing reserves in regions where the greatest number of endemic species can be protected. Areas rich in endemic plants are likely to harbor endemic animals as well! Proponents of the hotspot strategy estimate that 44% of all vascular plants, along with 35% of all mammals, birds, reptiles, and amphibians, can be protected by conserving the 1.4% of the earth's land surface covered by hotspots (Myers et al., 2000). An updated review of the world's hotspots is found in Mittermeir et al. (2005).

In another study that generated conservation recommendations, all species of the Brazil nut family (Lecythidaceae) found in central French Guiana were mapped. Of the 53 species of this family known to occur in the Guianas, 30 were found in central French Guiana (Mori, 1991). We concluded that a large biological reserve in central French Guiana would protect 57% of all of the Guianan species of this family, and that the addition of a second large reserve in central Guyana would add another suite of species and increase the percentage of protected Guianan species to 83%. To further increase the number of protected species it would be necessary to protect special habitats, such as the savannas of southwestern Guyana and adjacent Brazil, where *Lecythis brancoensis* (R. Knuth) S. A. Mori and *L. schomburgkii* O. Berg are endemic, and the periodically flooded forests, which harbor species such as the endemic *L. pneumatophora* S. A. Mori.

Nearly 35 years ago, French botanist Jean-Jacques de Granville proposed a system of forest reserves in French Guiana (Granville, 1975). One of the proposed reserves surrounded the village of Saül in central French Guiana. Although this reserve was never established, in 2007 the French government created an enormous national park, the Parc Amazonien de Guyane. It encompasses 33,900,000 hectares (33,900 square kilometers), and includes the region in central French Guiana that de Granville proposed. The southern boundary of the French park abuts the Parque Nacional das Montanhas do Tumucumaque in the Brazilian state of Amapá. Together, these two parks form one of the largest terrestrial protected zones in the world, and their establishment represents a major victory in rain forest conservation (Ministère de l'Ecologie, de l'Energie, du Développement Durable et de la Mer, 2009).

Electronic maps that show plant distributions represent a valuable new resource for conservation. Species pages in e-monographs and e-Floras usually allow the generation of Google maps, which use geo-referenced collections to show where specimens have been collected. With the satellite view, a user can determine if a collection locality (or dot) on the map occurs in an area still covered by natural vegetation, or whether the area is now occupied by shopping malls or agricultural fields. These maps are not quite as useful for conservation as they might be, because the satellite maps are usually not in real time. The technology does exist to map collection localities onto real-time LandSat images, so it is possible—at least in theory—to closely monitor land use in areas where populations of endangered species exist. In simple terms, if the pixels close to mapped populations of endangered species turned from green (forest intact) to brown (deforested), an alarm system could alert local conservation authorities to intervene before vulnerable populations of endangered plants were lost. For species of special interest, GIS data (e.g., vegetation cover,

238 / *Tropical Plant Collecting: From the Field to the Internet*

total rainfall, periodicity of rain fall, etc.) can be used in conjunction with locality data to create maps to predict areas of suitable habitat (species distribution models). It is then possible to determine how much, if any, suitable habitat exists within protected areas, or pinpoint other areas that might be appropriate for protecting species of conservation concern.

Conclusions

1. When possible, monographers and floristicians should present specimen and taxon information in e-monographs and e-Floras. Even researchers without sophisticated computer skills can use the programs that are currently available to make specimen databases, species pages, and multiple entry electronic keys.

2. Although electronic publication has many advantages, hard copy publication is still needed because some data needs to remain static, for example the description of new species. Hard copy publication is also needed to guarantee that electronic information in e-monographs and e-Floras is not lost in transfer between generations of electronic media.

3. The new era of e-monographs and e-Floras presents problems that have not yet been adequately addressed. The most important of these are the failure to recognize that a single species page may not be possible for many of the world's species; that non-vetted Wiki technology may not be appropriate for the production of all species pages; and that haphazard data entry by untrained personnel can lead to chaos.

4. Electronic monographs and Floras support important conservation goals by providing the keys, descriptions, and images needed to identify plants, including threatened species of conservation concern. In addition, electronic distribution maps provide the information needed to justify conservation planning, and to determine where biological reserves are most effectively established. The Parc Amazonien de Guyane serves as an example of how French biologists, working in collaboration with conservationists and governmental organizations over a period of decades, finally achieved major success in rain forest conservation in the overseas Department of French Guiana.

Chapter 7

Rain Forests of Tropical America:
Is there Hope for Their Future?

by Scott A. Mori

This chapter is based on the following beliefs: (1) tropical rain forests are complex, and this complexity is caused to a large extent, but not exclusively, by the interactions between plants and animals; (2) plant/animal interactions are the result of long, co-evolutionary processes in which human beings have played an inconsequential role; and (3) biodiversity and plant/animal interactions are almost always compromised by human activities. The last belief was summarized by John Robinson (1993) in the following quote:

> "Sustainable use is very appropriate in certain circumstances, but it is not appropriate in all. It will almost always lower biological diversity, whether one considers individual species or entire biological communities, and if sustainable use is our only goal, our world will be the poorer for it."

Tropical Rain Forest

Tropical rain forests are loosely defined as forests within the tropics in which most trees are seldom entirely leafless and, therefore, the canopies are considered "closed." Within this broad classification, ecologists recognize numerous forest types – many of which are not actually rain forest in the strict sense. These distinctions, however, are not important to the ideas conveyed in this chapter.

The tropics are that portion of the earth's surface lying between the Tropic of Cancer (23° 27'N) and the Tropic of Capricorn (23° 27'S), which, respectively, mark the northern and southern limits of the passage of the sun. This band encompasses about 40% of the earth's land surface (Longman & Jenik, 1987) and possesses an amazing diversity of vegetation types, ranging from the desert-like *caatinga* (Fig. 1-9 in Chapter 1) of northeastern Brazil to one of the wettest regions of the world, the Chocó of the Pacific coast of Colombia, where 10,000 millimeters (ca. 400 inches) of rain fall each year. Mountain ranges, such as the Andes, extend into the tropics and the ensuing altitudinal variation creates conditions for additional vegetation types, and the species that inhabit them. For example, high elevation *paramos* harbor a completely different set of plants and animals than occur in lower elevation cloud and rain forests. Thus, geological history, coupled with climatic patterns, has resulted in wide variation in topography, soils, and rainfall. Speciation, associated with both environmental heterogeneity and plant/animal co-evolutionary processes, has left an impressive legacy of biodiversity, and New World tropical forests arguably harbor the richest concentrations of species of plants on the planet.

Although the tropical forests of the world occupy only 7% of the earth's surface (Longman & Jenik, 1987), they may possess as much as one-half to two-thirds of the world's plants and animals. Systematists (scientists that, among other things, define and classify biodiversity) argue over the exact numbers—but nobody disputes the claim that tropical forests are home to a disproportionate share of the biodiversity of the earth.

The tropics of the world are divided into two realms, those of the New World (the Neotropics) and those of the Old World (the Paleotropics). The latter refers, collectively, to tropical areas in both Africa and Southeast Asia. The organisms of these different tropical regions are distinct from one another, and the sociological and conservation problems within each of the three areas differ. Nevertheless, the underlying biological principles that govern the evolution and ecology of the plants and animals in the Paleotropics and the Neotropics are essentially the same.

One of the reasons I chose to dedicate my career (Chapter 1) to studying plants of the Neotropics is that it is not only home to the greatest number of flowering plant species, but also a region in which additional botanical exploration still remains to be undertaken. Tropical America is where approximately 30% of the known flowering plant genera and species on earth live (Smith et al., 2004). The estimated number of seed plant species, previously accepted at a global total of 250,000 to 270,000 species (Mabberley, 1997; Raven, 1988), has been under debate and estimated to be as high as 300,000 to 420,000 (Bramwell, 2002; Govaerts, 2001; Prance et al., 2000) or as low as 223,300 species (Scotland & Wortley, 2003; Wortley & Scotland, 2004). If we accept the assumptions that two-thirds of the world's plants are tropical, and that one-half of these reside in the

Neotropics (Raven, 1988), the estimated number of neotropical flowering plant species ranges from a minimum of 78,800 (Smith et al., 2004) to a maximum of 140,000 (Bramwell, 2002; Govaerts, 2001). This flora is so incompletely known that, at the current rate of publication, it will take hundreds of years to complete Flora Neotropica (an international effort to inventory all plant species in the region) (Forero & Mori, 1995; Thomas, 1995, 2005). It is clear that a much more basic inventory must be accomplished before the diversity of neotropical plants can be accurately estimated, and their relationships with animal pollinators, seed dispersers, and predators are even less understood.

Biodiverstiy

My home state of Wisconsin has approximately 101 species of trees, including trees of all size classes (Elias, 1980). In contrast, the forests of Latin America are strikingly more diverse. In central French Guiana, my colleague Brian Boom and I studied a sample of 800 trees with a diameter breast height (DBH) greater than or equal to 10 centimeters (ca. 4 inches). There were 295 species of trees in the sample (Mori & Boom, 1987)—a different species for every second or third tree we recorded! The maximum number of species of trees of this size class is that reported for a forest in Amazonian Peru, where Al Gentry (1988) found 300 species in one hectare (a hectare is 100 × 100 meters, or 2.47 acres). In his study, every second tree represented a different species. Gentry (1988) hypothesized that the diversity of trees increases relative to total rainfall up to about 4000 millimeters (158 inches) per year. However, my colleague, Alexandre A. de Oliveira, and I conducted an inventory of trees with a DBH >10 cm in central Amazonian Brazil, and found 285 species per hectare in a region where rainfall is only 1500 millimeters (59 inches) per year; thus, factors in addition to rainfall account for the high tree diversity in the Neotropics (Oliveira & Mori, 1999). Some neotropical forests are so diverse that in an area only slightly smaller than two and one-half American football fields, nearly three times as many tree species can be found as occur in Wisconsin! This extremely high tree diversity has only been found in central and western Amazonia. Forests further to the east and forests in suboptimal habitats, such as those on nutrient-poor white sand soils (Davis & Richards, 1933, 1934) or in periodically flooded forests (Campbell et al., 1986), harbor fewer species. The Amazon is not only home to the greatest diversity of trees and lianas (woody vines) found in the Neotropics (Gentry, 1982), but it may harbor the greatest number of tree species per hectare found anywhere in the world—botanists studying the flora of SE Asia would, however, dispute this suggestion.

The same pattern of diversity (a latitudinal gradient of species richness) exists in many other groups of organisms. For example, there is a single species of hummingbird (the ruby-throated) with an established summer range in the northeastern United States (Sibley, 2000), whereas 117 hummingbird species are listed in the *Birds of Colombia* (Hilty & Brown, 1986). Bats number only 20 species in the United States (Tuttle, 1988), and most of these are insectivorous. In contrast, 105 bat species are known from French Guiana (Lobova et al., 2009), with 78 species documented at a single locality (Simmons & Voss, 1998)—and these include insectivores, leaf eaters, nectar feeders, fruit consumers, carnivores, and blood-feeders.

Because of the great diversity of beetles that Smithsonian entomologist Terry Erwin found in the forest canopy, he estimated that there might be as many as 41,000 different species of insects in a single hectare of tropical forest, and as many as 30–50 million species of insects in the world (Erwin, 1982, 1997). The overall number of tropical insects may be lower than Erwin estimated because neotropical beetles probably have broader niches—or lower host plant and stratum specificity—than he assumed (Berkov & Tavakilian, 1999). Nevertheless, the diversity of insects in tropical forests is astronomical, to the point that many captured insects can not be identified to species because they are new to science.

Some authors argue that adequate species richness can be protected by protecting the secondary forests that arise after primary forests are disturbed, but this has been negated in studies involving plants (Oliveira & Mori, 1999), bats (Lobova et al., 2009), birds (Barlow et al., 2007a), butterflies (Barlow et al., 2007b), dung beetles (Nichols et al., 2008), and reptiles (Luja et al., 2008). In general, secondary forests have lower plant and animal diversity compared to undisturbed primary forests. For the most part, primary forests have more species but fewer individuals of each species, whereas secondary forests have fewer species but more individuals of each. Thus, the only way to preserve maximum biodiversity in tropical ecosystems is to protect large areas of primary forest from human impact!

Why are there so many Species in the Neotropics?

This question has been pondered by biologists ever since the earliest natural history explorers visited South America. The answer is not simple, and probably is the result of a combination of many factors including optimal growing conditions; great habitat diversity caused by differential rainfall patterns and altitude, climatic changes in the Pleistocene; the geological history of Central and South America (e.g., uplift of the Andes and connection of North America and South America via the Isthmus of Panama); and the great number of plant/animal interactions in the tropics. Because I believe that an understanding of the co-evolution of plants and animals is essential if there is to be any hope of conserving biodiversity in the Neotropics, I will emphasize this aspect of biodiversity evolution. For the most part, I will discuss the relationships between the animals that pollinate the flowers and disperse the seeds of flowering plants, especially members of the Brazil nut family (Mori et al., 2010).

Plants and animals in the Neotropics interact in such a way that the diversity of both plants and animals is enhanced. For example, plants that are pollinated by hummingbirds are often characterized by a syndrome of features that attract hummingbirds and reward their visits. In return, the hummingbirds aid in plant reproduction by transporting pollen from one plant to another. Hummingbird flowers are usually red in color, positioned such that they are only easily visited by hovering animals, open during the day, do not emit an aroma (most birds have a poorly developed sense of smell), produce nectar with a dilute sugar concentration dominated by sucrose, and have their petals fused into a tube that serves as a container in which the nectar accumulates. Perching birds also pollinate plants. For example, *Symphonia globulifera* L. f. (Clusiaceae) is visited by a guild of some of the showiest perching birds in tropical America, including the black-faced dacnis, blue dacnis, green honeycreeper, purple honeycreeper, red-legged honeycreeper, fulvous-crested tanager, paradise tanager, silver-beaked tanager, bananaquit, and even the waved woodpecker. Although the flowers of this species are red, they differ from those of a typical hummingbird plant because they are easily reached by birds perched on the branches, and have hexose-dominant nectar. Hummingbirds may occasionally visit the flowers of *Symphonia globulifera*, but the visits of perching birds are far more frequent (Gill et al., 1998).

Flowers pollinated by bats are open during the night, often white in color, suspended or otherwise positioned so that they are away from the foliage, may emit a musty aroma, and produce nectar with a dilute sugar concentration dominated by hexose. The interactions between plants and hummingbirds, perching birds, and bats have resulted in the evolution of species of plants characterized by different combinations of attributes (syndromes) and this co-evolution has contributed to a significant increase in biodiversity.

Much of my research in tropical forests has been dedicated to understanding the evolution and ecology of species of the Brazil nut family (Mori et al., 2010). I will use several examples from this family to illustrate the degree of interdependence between plants and animals in the tropics. *Eschweilera pedicellata* (Rich.) S. A. Mori (Lecythidaceae), a tree species that is widely distributed in northeastern South America, possesses a bilaterally symmetrical flower that provides a nectar reward located at the apex of an inwardly coiled, chamber. The pollinators are numerous species of robust bees with long tongues. Some of these bees, known as euglossine or orchid bees, have the strength to force open the flower and, via the long tongues (Mori & Boeke, 1987), the ability to extract nectar from the apex of the coil. The nectar serves as a source of nutrition to male and female bees alike. The foraging bees inadvertently carry pollen from the flowers of one tree to those of another—ultimately resulting in fertilization and the production of seeds. Although euglossine bees are major pollinators of Lecythidaceae, other groups of bees, such as carpenter and bumblebees, can also pollinate their flowers.

Males, and only males, of these same bees also visit certain species of orchids (Orchidaceae), the African violet family (Gesneriaceae), and the jack-in-the-pulpit family (Araceae). They gather floral aroma compounds which they sequester in specialized slits in their hind legs. Coincidentally, while collecting the aromas, the bees transport pollen from one plant to the other which, again, leads to pollination and the eventual formation of seeds. It appears that the male bees use the floral perfumes, either in the same form as collected, or in some modified manner, to attract female bees as part of their mating ritual (Dressler, 1982).

The Brazil nut (*Bertholletia excelsa*) provides an example of a relationship between a seed dispersal agent and a rain forest tree. In this species, the cannon-ball like fruit falls from the canopy (as far as 45 meters; ca. 150 feet) to the ground when ripe. Each fruit, weighing as much as two kilograms (ca. five pounds), contains from 10 to 25 seeds. The seeds, however, remain trapped inside until they are removed, usually by sharp-toothed neotropical rodents called agoutis (*Dasyprocta* spp.) or, to a lesser extent, by squirrels. These mammals consume some of the seeds and bury others for future retrieval and consumption. However, a few of the seeds they store are never retrieved. These seeds may germinate, as much as a year later and at some distance from the parent tree, in a locality that affords the new plants a better chance of survival because dispersal away from the parent tree increases the likelihood that seeds and seedlings will escape potential predators (Janzen, 1971)

These are only two of the many interactions that have evolved in the Brazil nut family (Mori et al., 2010). Although most species are bee-pollinated, some are pollinated by bats, and others may be pollinated by beetles (Knudsen & Mori, 1996), and the seeds are variously dispersed by monkeys, bats, peccaries, wind, or water (Tsou & Mori, 2002). In addition, all kinds of insects utilize various parts of the plant as their feeding substrates, from lycaenid butterflies that feed on the flowers (Feinstein et al., 2007) to beetles that feed on the wood (Berkov & Tavakilian, 1999).

Louise Emmons (1989) published an eloquent essay describing how plant/animal interactions have resulted in increased biodiversity in the tropics. She classified birds by dietary preference, and then determined the number of species in each class in both a rain forest in Peru and a temperate forest in South Carolina. In each comparison, the Peruvian forest was more diverse, with 25 versus six species of carnivorous birds, 98 versus 27 insectivorous birds, and 84 versus seven birds that eat nectar or fruit. The carnivores and insectivores were only about four times more diverse, whereas the birds that depend on plants were 12 times more diverse. Based on these and other observations, Emmons concluded that the diversity of tropical forests was greatly enhanced by co-evolution between plants and animals. Moreover, she pointed out that the extirpation or extinction of animals, for example by overhunting, will result in the extirpation or extinction of the plants that depend upon them for the pollination of their flowers or the dispersal of their seeds. If there is to be hope for the preservation of tropical forests and the biodiversity they harbor, then the relationships between plants and animals must be understood and appreciated, and measures must be taken to protect all of the species in this complex ecosystem; the loss of a single species can start a cascade of negative impacts affecting the survival of other species.

Economic Importance

One of the justifications for rain forest conservation is the economic potential of the plants found there. Tropical forests are the home to relatively few plants that are part of the global economy, although others may yet be discovered. Economically important native plants of the tropical American forests include seeds of the Brazil nut tree, chocolate from the seeds of *Theobroma*, pineapples, palm hearts, palm fiber, palm fruits, cocaine, various timbers, and rubber (Balick, 1985; Clement, 2008). It will become apparent in this essay that the ecosystem services provided by tropical rain forests may be much more important than any direct economic benefits their plants and animals provide to humans.

Because the tropics lack the severe temperate winters that serve as a "pesticide" to control insect populations (Ewel, 1986), the plants often depend upon chemical defenses to protect themselves from predation by insects and other animals. Man benefits from bioactive defense compounds, resulting from the interactions between plants and herbivorous animals, by using them in arrow poisons and medicines. For example, some Indians of South American rain forests tip their blowgun darts with *curare*, a poison extracted from several species of woody vines, the most important of which are members of the Moonseed (Menispermaceae) and Strychnine (Strychnaceae) families. The poison paralyzes the muscles used in breathing, causing a dart-wounded animal to suffocate. *Curare* is part of the pharmacopoeia that has been used in many different kinds of surgery as a muscle relaxant (Balick & Cox, 1996). The rain forest is a source of many chemical compounds of use to mankind, some of which have yet to be discovered.

It is stated, often without sufficient documentation, that products of the rain forest can be extracted without causing negative impacts on the ecology of the surrounding forest and that, therefore, these products can be harvested sustainably. In the Amazon, Brazil nuts are commonly cited as a non-timber forest product that is harvested from naturally occurring groves of trees without harming its populations. Traditionally, Brazil nuts were gathered during the wet season and rubber trees tapped during the dry season, thereby providing a year-round source of income for the Amazonian people. Today, however, rubber trees are not an important source of income for the Amazonians because of the lower price of rubber produced on plantations in Southeast Asia. Brazil nut gatherers, deprived of rubber as a dry season resource, supplement their incomes by gathering other forest products such as medicinal plants, fruits, and fibers; extracting timber; and farming using slash-and-burn agriculture. This traditional way of living is thought to provide a way of life that does not have a negative impact on the biodiversity needed to maintain the integrity of the forests from which these products are extracted. As a result, large areas called extractive reserves have been set aside to protect the cultures of the inhabitants, as well as the forests upon which they depend.

John Terborgh (1999) pointed out that "sustainable development" may be wishful thinking on the part of conservationists proposing extractive reserves. He adds that it depends upon a series of assumptions about product yields, biological sustainability, the potential for mass production under cultivation, and the potential for intact tropical forests to yield more than if they are cut and put to other uses. He concludes that in the absence of sustainable use of forest products, humans will destroy most tropical forests. He notes, however, that "the converse notion—that sustainable development will lead inexorably to the harmonious coexistence of human-kind and nature—is patent nonsense."

The Brazil nut has been the symbol of sustainable use of a non-timber forest product, and, until recently, it has been accepted as dogma that the seeds can be harvested from Brazil nut groves indefinitely without adverse affects to their continued productivity. However, Carlos Peres and his collaborators (Peres et al., 2003) demonstrated that in frequently harvested groves the populations consist only of large, fruit-producing trees and no young trees. On the other hand, in areas where Brazil nuts were not harvested the populations contained Brazil nut trees of all sizes and ages. These authors concluded that continued harvest of Brazil nuts over long periods of time leads to the eventual senescence of such groves, and, thus, harvesting of Brazil nuts may not be sustainable after all.

Although extractive reserves are hardly a panacea, the clearing of rain forest for agriculture or pastures leads to rapid environmental degradation. When the soils are poor, agricultural fields are abandoned in about three years (Ghazoul & Sheil, 2010), and pastures in around 10 years (Hecht & Cockburn, 1989). It is noteworthy that most Amazonian forests grow on poor soils, with only about 7% of the entire Amazon basin possessing soils suitable for sustainable agriculture (Hecht & Cockburn, 1989). Much of that arable land is found along the margins of periodically flooded rivers, where the soil nutrients are replenished each year when sediments are left by high water.

Relying heavily on extractive reserves for rain forest conservation is unwise because human exploitation in any form reduces biodiversity. In the first place, people living in extractive reserves may not understand the ecological limits of their use of the forest. These people often hunt, cut forest for slash and burn agriculture, extract timber and other products in a non-sustainable way, and mine for gold. As a result of this exploitation, the ecosystems in which extraction takes place are modified, resulting in a loss of the biological diversity present in forests exploited by humans. The activity of modern man in tropical forests, where so many plants and animals have co-evolved, always brings with it the possibility of severe reduction in biodiversity (Robinson, 1993).

In the second place, the world market for rain forest products might not be able to support much of an increase in their production. Moreover, such low intensity of land use is probably not capable of supporting human populations at the level needed to increase the standard of living demanded by more and more people. John J. Ewel, of the University of Florida, has estimated that hunting-gathering and shifting agriculture can only support one person per five square kilometers and one person per square kilometer, respectively (Ewel, 1986). Although the number of people per hectare that can be supported in extractive reserves is greater than the number supported in hunter gather societies, even those populations are not high enough and productive enough to absorb population growth, let along contribute to national economies.

Extractive reserves are preferable to the replacement of rain forest by agriculture or cattle pastures, because forest cover of any kind is able to harbor more plants and animals than are fields or pastureland. However, in order to protect plant and animal diversity to the greatest extent possible, it is important to protect large tracts of representative vegetation types—where human activity is kept at a minimum—throughout tropical America.

Plant Succession

Primary (also called old growth) forest has not experienced natural large-scale (e.g., landslides, hurricanes, large blow downs of trees) or anthropogenic (e.g., logging, agriculture, or ranching) disturbance for long periods, but even in primary forests the formation of smaller disturbed areas caused by limb and tree falls is a very common occurrence. After large scale disturbance occurs, the species of trees that colonize the disturbed land are called secondary species, and the long process of regenerating primary forest via the different stages of secondary forest is an example of plant succession. There are relatively few tree species that are able to thrive in disturbed areas, but the ones that do improve the conditions for other tree species to grow. Thus, as time passes, the number of species of trees in secondary forests increases, and several hundred years later they may again approximate the diversity of the original primary forest.

Secondary forests are extremely important because they are the first step in regenerating biodiversity after large-scale disturbance, and often serve as corridors for plants and animals to move from one area of primary forest to the next. They also provide ecosystem services such as protecting soil, maintaining hydrological cycles, and sequestering carbon. However, secondary forests alone cannot protect the high biodiversity that characterizes primary tropical forests. Studies of plant succession in the tropics show that there are more species of plants in primary forests than there are in relatively recent secondary forests (Chazdon et al., 2007; Guariguata et al., 1997) and this is also true for most, but not all, groups of animals (Barlow et al., 2008a, 2008b; Luja, 2008).

Biologists mostly agree that primary forest continues to be destroyed at a high rate. The United Nations Food and Agriculture Organization (FAO) estimated that just 36% of the world's forests remained relatively untouched in 2005 (Stokstad, 2008) and Asner et al. (2004) have demonstrated that even much of what remains has been modified by selective logging. Recent estimates suggest that only 5-10% (Dirzo & Raven, 2003) to 36% (Wright & Mueller-Landau, 2006) of the original primary forest will be left by 2050 (given the ongoing pressures, the high-end estimate is unrealistic if the FAO assessment is accurate). Even the most optimistic predictions are cause for concern, because it is clear that humans are well along the way to eliminating and severely degrading many of the planet's old growth forests.

Both Fragile and Resilient

Tropical trees are characterized by very shallow root systems, and they are easily uprooted by a variety of natural causes. Because the trees are often connected by lianas, when one tree falls adjacent trees may also be pulled down. The opening in the canopy created when large branches or trees fall is called a gap, and may be small or very large in area (Fig. 7-1). Small gaps are caused by the falling of large limbs or individual trees, and large gaps result from major disturbances such as fire, landslides, hurricanes, localized strong winds, or unusually prolonged flooding. Gap formation is a natural and frequent occurrence in tropical forests—so common, in fact, that the mean time interval between gap occurrences at a given spot in a tropical forest is between 80 to 136 years (Denslow, 1987; Hartshorn, 1980). This is called the turnover time of forests, and the associated ecological and evolutionary aspects of gaps constitute gap dynamics. Many tree species depend on the light made available by the opening of gaps (Denslow, 1987; Hartshorn, 1989) for growth and regeneration. Tropical forests may be viewed as a mosaic of gaps at different stages of plant succession.

Because gaps are a naturally occurring phenomenon in tropical forests, trees have evolved different strategies that promote their survival after damage. For example, rain forest trees have a great capacity for regeneration after damage, and many of them survive snapped trunks by sprouting from below the damaged point. However, fire is often used by man in tropical agriculture, specifically because it usually destroys the trees' capacity to send out new sprouts, and because it makes nutrients from the trees available to support crop growth.

Different tree species typically occur in gaps than are found in

Fig. 7-1. Gap formation. Top: Falling trees may or may not carry other trees with them to form gaps of different sizes. They differ from most man-made gaps by not being exposed to fire. Bottom: Humans at low population densities make gaps similar in size to those caused by non-human forces. Gaps of this size are like gaps caused by natural forces such as wind storms, floods, and landslides but they differ from non man-made gaps by being exposed to fire. Photos by C. A. Gracie (both taken in French Guiana).

Fig. 7-2. Top: The bat, Sturnira lilium, about to land on an infructescence of Solanum rugosum. Bats disperse seeds of this and other species of plants into gaps as well as into other habitats. Bottom: Plants dispersed by bats have begun to colonize a large, man-made gap. The plants on the left foreground are species of Solanum and the plants on the right foreground are a species of Cecropia, both dispersed by bats. Photos by Merlin Tuttle of Bat Conservation International (both taken in French Guiana).

more mature forests. Some species may be adapted to small gaps and others to large gaps, while others may not be gap dependent (Denslow, 1987). There is also evidence that species adapt to the different zones created by gaps—i.e., the upturned root area, the area along the trunk, and the area around the crown (Hartshorn, 1980, 1989). Natural gaps, because they increase the heterogeneity of habitat, promote, rather than decrease, the diversity of tropical forests.

Bats are essential for the colonization of large gaps because they disperse seeds from plants adapted for growth in the full sunlight of large, disturbed areas. The fruits of these early colonizers are usually fleshy and contain numerous, small seeds, or, as in *Cecropia*, the fruits themselves are small and surrounded by fleshy floral parts (Lobova et al., 2003). The bats ingest the *Cecropia* fruits along with the surrounding floral tissue, then digest the floral tissue and expel the unharmed fruits in their feces.

Fruit or seed retention time is usually less than 20 minutes, and the propagules are often defecated while the bats are in flight. As a result, bat-dispersed plants including *Cecropia*, *Piper*, *Solanum*, and *Vismia* are often the first plants to colonize large open areas (Fig. 7-2). A monitored plastic sheet placed in the middle of a large disturbed area reveals that scarcely any propagules arrive during the day, whereas there is a great deal of "seed rain" during the night. Moreover, observation with a night vision scope reveals bats actively consuming the fruits of those plants that first colonize large gaps. These early gap colonists provide perching sites and cover, making conditions more hospitable to other dispersers such as birds and mammals, which bring the fruits or seeds of plants they consume into the area. After about 100 years, a large disturbed area will have regenerated to the point that it is structurally indistinguishable—at least by non-botanists—from pristine areas (Saldarriaga, 1986; Saldarriaga& Uhl, 1991). It would, however, take much longer for the reestablishment of the high species diversity if bats did not initiate the process of fruit and seed dispersal.

Man's Role

Amerindians living at low population densities create agricultural plots that can mimic, but are not identical to, natural gaps in tropical forests. Slash-and-burn (also called swidden) agriculture is the man-made creation of gaps in forests followed by the cultivation of crops such as cassava (*Manihot esculenta* Crantz) and maize (*Zea mays* L.). Maize has been cultivated in the Amazon for at least 6000 years (Myster, 2004), so swidden agriculture is at least that old. Forest is cut down and the debris is allowed to dry until the end of the dry season, and then burned just before the beginning of the wet season to release nutrients. The crops are planted at the beginning of the wet season and grow well because of the increased nutrients in the soil and the abundant water available during the rainy season. The crops do well for about three years, after which the nutrients are depleted and the fields are abandoned to allow secondary species to colonize the area and once again accumulate nutrients. After cultivated crops are harvested there is a fallow period of six or more years, but both fruit collection and hunting around and within the gaps continue to impact the regeneration process for many years after agriculture in a given area has been abandoned (Balée, 1994). During the fallow period nutrients are replenished and pests associated with cultivated plants are reduced (Ewel, 1986).

Gaps created by Amerindians, however, are not identical to gaps produced by natural events such as tree falls, landslides, and long-term flooding. Perhaps the most important difference between natural gaps and those carved into the forest by Amerindians is that in the latter trees are burned after they have been felled. Fire releases the nutrients tied up in the roots, wood, stems, leaves, flowers, and fruits, making them available for assimilation by crops. This virtually eliminates the natural process of decomposition, in which hundreds of different organisms derive nutrients from decomposing plant material. In addition, the fires used to prepare the fields sometimes escape into adjacent destructively logged forests igniting the debris left on the ground, which, in turn, may ignite the remaining trees. If this happens in very dry years, the fires may even burn adjacent unlogged forests (Nepstad et al., 2008; Sanford et al., 1985). Although natural fires do sometimes occur in rain forests during extremely dry periods (Bassini & Becker, 1990; Saldarriaga & West, 1986), fires associated with natural gaps are probably extremely rare. The more intense and extensive the land use, the less a man-made gap resembles a natural gap (Janzen, 1990) (Fig. 7-3).

Balée (1994) suggested that the slash-and-burn agriculture of the Ka'apor Indians, located at the southeastern margin of the Amazon Basin, does not reduce the diversity of trees in the long term. He sampled four hectares of old growth forest, which had not undergone disturbance by man for at least several hundred years (perhaps never), and four hectares of fallow forest that had been used for slash-and-burn agriculture 40 to 130 years prior to his study. The old growth forest hectares had 123, 126, 144, and 145 species of trees; and the regenerated forest hectres after cultivation had 95 (40 years as fallow), 125 (50 years), 141 (120 years), and 157 (at least 100 years) species of trees greater than 10 centimeters

Fig. 7-3. Secondary succession. Top: Several years after disturbance in central French Guiana, this area was dominated by individuals of the bat-dispersed Cecropia obtusa. This secondary forest provides the conditions needed by other plants to get established in the area. Photo by MerlinTuttle of Bat Conservation International. Bottom: When gaps such as this one in French Guiana become too large and two frequent, forest regeneration becomes more difficult. Photo by Carol A. Gracie.

DBH per hectare. From these data, it is clear that swidden agriculture initially reduces tree diversity in favor of cultivated crops, but that diversity increases with fallow age to the point where old fallow and high forest are not distinguishable in the numbers of tree species present. Old fallow fields may possess more diversity than adjacent high forest because they combine residual trees planted or favored by the Ka'apor, and secondary species that invaded at the beginning of the fallow, as well as primary species. The Ka'apor (Balée, 1994) and other Amerindians may not have an explicit conservation ethic, but, when populations are low, their activities mimic ecological processes that have been taking place in lowland forests for millennia.

The Ka'apor or, for that matter, any group of Amerindians, do not create biodiversity—they simply shift it around. Interactions between plants and animals that promote the evolution of new species take place over long periods of time, and have been going on long before man entered the Neotropics (at least 10,000 years B.P., Roosevelt et al., 1996). For instance, fossil flowers from the Late Cretaceous suggest that ancestors to the genus *Clusia* were already producing the resins that attract contemporary pollinators 90 million years ago. Slightly younger amber deposits document the presence of resin-visiting bees, and the specialized predators of the resin bees appear in Dominican amber 25–40 million years old (Melo et al., 2005). References to agri-biodiversity refer to plant domestication, in which humans select for favored characteristics in their crop plants. This can result in the creation of distinct cultivars, but these domesticated plants are not species that persist and retain their characteristics in a natural environment.

Swidden agriculture, which mimics natural plant succession, has been suggested as a possible model for modern agro-ecosystems in the tropics (Ewel, 1986; Hart, 1980), but the swidden system of Amerindians is one in which forest diversity and their activities are in harmony only when Amerindian populations are low and, thus, their needs are modest. When Amerindians assume the ways of colonists their impact on biodiversity increases by orders of magnitude. As Redford (1991) has suggested, the "noble savage" may not be so noble after all. Low environmental impact is most likely an artifact of low population levels, abundant land, and lack of access to market economies. Although swidden agriculture depletes both soil nutrients and game animals in the vicinity of villages, when this happened villages were relocated and its slash-and-burn fields were left to grow back to forest.

Excessive hunting by both Amerindians and colonists has a major impact on forest ecology. Redford (1992) describes how Amerindians take a much greater variety of game than do colonists, but both tend to hunt the largest mammals (agoutis, armadillos, peccaries, tapirs, monkeys, etc.) and birds (guans, toucans, curassows, tinamous, parrots, macaws, etc.). Both subsistence and commercial hunting for game, skins, and feathers has a major impact on animal populations. Moreover, because the preferred game animals also play important ecological roles as seed dispersers and predators, community structure is indirectly modified by overhunting. Under heavy hunting pressure non-primate mammal and bird densities can decrease by as much as 93.7% and 94.6%, respectively, compared to their densities in sparsely hunted forests (Redford, 1992; Robinson and Redford, 1991).

Although Amerindians have had an impact on neotropical lowland forests for at least 10,000, and perhaps for as long as 30,000 to 40,000 years (Goulding et al., 1996), there is still much of lowland South America that probably never experienced slash-and-burn agriculture. In addition, there are some areas that were farmed so long ago that forest regeneration has advanced to the extent that for all practical purposes it is as if people had never been there. The largest populations of Amerindians occurred along the floodplains of major rivers where the soils are enriched each year by floodwaters depositing silt. For example, Marajó Island at the mouth of the Amazon harbored an advanced culture of as many as 100,000 people 1000 years before the arrival of Europeans (Roosevelt, 1992). However, because Amazonian flood plains occupy only two to three percent of the Amazon Basin (Goulding et al., 1996) there were still extensive areas of unoccupied, non-flooded (*terra firme*) forest. Even in areas that supported relatively high human population densities, such as Marajó Island, there were vast areas of undisturbed forest, especially on *terra firme* where soils are much poorer. Moreover, the Amerindians of prehistoric times did not have the technology to cut trees as efficiently as modern Amazonians; thus, pre-Colombian Amazonians should never be equated with modern man in their ability to destroy Amazonian ecosystems.

Our 100 hectare Lecythidaceae plot in central Amazonian Brazil experienced fire at various times from 6,000 to 400 years ago (Bassini & Becker, 1990). It probably was never used for slash-and-burn agriculture, as shown by the lack of phytoliths, which are characteristic artifacts found in previously cultivated soils (Piperno & Becker, 1996). The presence of 38 species of Lecythidaceae, a family of trees symbolic of primary Amazonian forest that are susceptible to fire, indicates that neotropical forests have the capacity to regenerate after disturbance by fire. Today this forest harbors as many as 285 species of trees equal or greater than 10 DBH per hectare (Oliveira & Mori, 1999), which, on our planet, is second only to the species-rich forests of western Amazonia, where 300 species of trees per hectare have been recorded (Gentry, 1988). Forests subjected to natural disturbances, or to disturbances caused by traditional Amerindians at low population densities, represent areas where Amazonian biodiversity, plant/animal interactions, and ecological processes evolved and continue to evolve.

Problems arise, however, when human population densities become too high, when too much productivity is expected from the nutrient-poor soils, or when large areas are converted to other land uses, such as pasture or large-scale farming. At high human population densities, the gaps created for agriculture coalesce and patches of primary forests become so distant that seeds for natural plant succession are no longer efficiently dispersed into the more remote abandoned fields. Moreover, the bats needed to disperse seeds into large deforested areas may have been eliminated, either killed outright by residents who view all bats as harmful, or because their roosting sites were destroyed. If human populations increase too much and swidden sites are no longer given a long enough fallow between use cycles, not only is biodiversity reduced but crop production is diminished. Not enough time passes for nutrients to accumulate in the biomass of the regenerated secondary forests, so that when they are subsequently cut-and-burned, fewer nutrients essential for crop growth are returned to the land, resulting in reduced crop productivity. As a result swiddens are cultivated for shorter periods before they are again left to grow into secondary forest and this causes deforestation in other areas to make new swiddens.

Conversion to other land uses makes it more difficult for forests to regenerate. If swidden agriculture is followed by pasture, succession back to forest is slowed down. Livestock alter the land by trampling vegetation, eating seedlings and saplings, creating hummocks, and compacting soil. Moreover, grasses that were planted or that have invaded pastures may compete with early successional species of native plants, making it difficult for them to become established (Myster, 2004).

Fragmentation is another consequence of modern man's expanding activity in neotropical forests. Sometimes pressure is so intense that only patches of forest remain, surrounded by agricultural fields or pasture. If the patches are too small, plant and animal diversity will be negatively impacted. Studies carried out as part of the Biological Dynamics of Forest Fragments Project (Chapter 1), a joint effort of the Instituto de Pequisas da Amazônia and the Smithsonian Institution (Bierregaard et al., 2001b), documented the many changes that accompany fragmentation. For example, plants along the edges of fragments are susceptible to drought; this stress, along with stronger winds at the edges, results in an increased number of treefalls, and a progressive retreat of forest margins along the fragment borders.

There are many other subtle changes that take place in fragments that impact the composition of the plants and animals within. For example, the fragments may be so small that army ant colonies are reduced in number, and this causes a negative impact on the antbirds that depend on them to flush out the insects they eat (Stouffer & Bierregaard, 1985). In extreme cases, fragments of very small size may harbor only a single army ant colony. When this colony is in the bivouac (quiescent) stage, the birds that follow the army ants will have a reduced food supply and will either have to leave the fragment or starve.

Likewise, if fragments are below the size needed to support top predators, such as jaguars and pumas, the food chain will be disrupted, and prey animals may increase so markedly that they negatively impact the populations of other animals and plants. In order to address problems caused by forest fragmentation, John Terborgh (1999) began a project on the islands created by the Lago Guri reservoir in Venezuela. When he and his students established their research plots four years after the lake was impounded for generating electricity, there were hundreds of islands representing former hilltops that previously were part of a continuous forest. Their first bird censuses revealed that the islands had fewer species than the nearby mainland, but the bird species that remained on the islands often occurred in increased population densities. They discovered that the 11-hectare Lomo Island was surprisingly free of birds, and that this island harbored capuchin monkeys at a density 10–20 times greater than that found on the mainland. When Terborgh and his students placed artificial nests with quail eggs on Lomo Island, they found that 100% of the above ground and 85% of the ground nests were raided and the eggs destroyed, for the most part by capuchin monkeys. Similar experiments on other islands without capuchins were carried out and very few of the eggs were destroyed; moreover, on the mainland, where capuchins were present in normal densities because population sizes were regulated by predation, the loss of eggs was only 28%. Forest fragmentation, whether it is caused by flooding a hilly landscape, or by deforestation that leaves forest patches encircled by agricultural fields or pasture, has a profound influence on the structure and composition of tropical forests. Because most tropical parks and biological reserves are fragments of varying sizes, Terborgh (1999) concludes: "Nearly every park in the Tropics is being degraded, some rapidly, some slowly, but few are immune to the relentlessly growing pressure of human demands on the environment."

In many parts of South America, a major threat to biodiversity is the rapid expansion of the agriculture frontier (Nepstad et al., 2008) similar to that which occurred in Mato Grosso, a state located just south of the Amazon Basin in west central Brazil. A large part of Mato Grosso was once covered by vast expanses of savanna-like vegetation called *cerrado* that, 40 years ago, was considered worthless for agriculture. Unfortunately for the plants and animals that inhabit *cerrado*, it was discovered that the addition of lime and phosphorous to the soil enhanced the growth of soybeans (Mori, 2010). In addition, new varieties of soybeans and other crops that grow well in the modified soils of the *cerrado* were introduced, markedly increasing agricultural productivity at the expense of biodiversity. In Mato Grosso, the year-long mild weather often allows the production of several crops a year on the same land.

Brazil has become one of the world's largest exporters of beef, chickens, orange juice, soybeans, sugar, coffee, and tobacco, and is rapidly becoming known as a bread basket for the world. These agricultural successes make it difficult to argue convincingly for natural habitat protection on the grounds that, because the poor soils will not sustain continued productivity, conversion into agro-ecosystems is not economically worthwhile. Unfortunately for biodiversity, much of the Brazilian *cerrado* vegetation has been replaced by soybean agriculture. Soybean cultivation has even been introduced to the Amazon, where it has the potential to contribute to deforestation and species loss on an even larger scale!

Ecosystem Services

Tropical forests have value far beyond that derived from the sale of rain forest products such as chocolate and Brazil nuts. They are also a source of protein from the fish found in their rivers and lakes, and the animals living in their forests. Furthermore, rain forests are reservoirs of genetic diversity; play an important role in maintaining the stability of the world's atmospheric gases; and help control hydrological cycles on local, regional, and global scales. These far-reaching effects and resources are known collectively as ecosystem services, and their value is enormous, both locally (Pimentel et al., 1997) and globally (Costanza et al., 1997).

Ecosystem services are, in essence, the *pro bono* benefits that humans receive from nature. For example, the water needed to drink, cook, grow plants, and raise animals is provided by the world's ecosystems free-of-charge; thus, the most significant cost involved is that associated with transport to the consumer. Another important economic concept that is seldom considered in developmental schemes is the risk of "externalities," which are defined as costs incurred (negative) or benefits received (positive) by third parties in business transactions (Friedman. 2008). A very important negative externality in rain forest destruction is the release of carbon into the atmosphere when forests are cleared for agriculture and pastures. Individuals benefit from the crops they sell and the meat they market, but the entire world pays for the negative externalities caused by global warming. The services provided by the world's ecosystems, and negative externalities caused by their destruction, are described by Chivian and Bernstein (2008).

A major problem associated with the world's economy is that ecosystem services are not factored into the cost of manufacturing commercial products. For example, the cost of the soil and water needed for trees to grow is not added to the selling price of timber used in house construction. Healthy soils, fresh water, and clean air are valuable commodities, and their cost should be included in the price of anything sold anywhere. In addition, the externality costs of ecosystem service disruption via, for instance, air and water pollution, should be considered in the pricing of all goods and services. Entrepreneurs and businesses with the resources to market products that require clean water and fresh air should pay for those ecosystem services, and, if they contaminate water and air as part of the manufacturing process, they should also pay the costs associated with that contamination. Until the costs of ecosystem services and negative externalities are included in the prices of products derived from ecosystems, there will be little appreciation for the services those ecosystems provide and the negative externalities that arise when ecosystem services are compromised.

Inherent to the concept of paying for ecosystem services is the belief that no human being has the uncontrolled right to exploit soil, water, air, or biodiversity just because that individual has the economic means to do so. For example, residents that live along a river or above an aquifer do not have the right to consume water in unlimited quantities. The needs of others who depend on that river or aquifer for their water supply must also be taken into consideration. People living in a sumptuously landscaped housing development do not have the right to dig deeper wells just because their homes experience a drop in water pressure during summer months, when outrageous amounts of fresh water are expended to irrigate lawns and gardens. There are no "mineral rights" for water and, thus, the water of an aquifer or river belongs to all. The capability to dig deeper water wells, and to exploit ecosystem services in general, does not confer the right to exploit ecosystem services at will. A more ecologically sound strategy would be to reduce the size of lawns, and garden with native plants that do not need to be irrigated. Fresh water and other ecosystem services are too valuable to be squandered; overuse will make them scarce for others in this and future generations.

The periodically flooded forests of the Amazon provide the ecosystem services that enable fish to be harvested from the waters that run past and into the forests during seasons of high water. The number of fish species in the world is estimated at 30,000, of which 13,000 are found in fresh water. The Amazon is home to nearly 3,000 fish species, which amounts to 10% of all of the fish in the world and 23% of the world's fresh water fish diversity. Nearly 75% of the fish in South America are Amazonian (Santos et al., 2008); they provide an important source of protein for the *caboclos* that live along the river. Some of these fish attain remarkable sizes, feeding on fruits that drop from trees into the water. The astounding fruit productivity of Amazonian riverine forests is illustrated by the *jauari* palm (*Astrocaryum jauari* Mart.), which can produce as much as 1.65 tons of fruit per hectare per year (Maia, 2008). Many jauari fruits drop into the water and are eaten by the *tambaqui* (*Colossoma macropomum* Cuvier, 1818)]—one of the most delicious and commercially important fish in world.

When periodically flooded forests are cut down and converted to cattle pasture, regardless of the ecosystem services they provide, it is not likely that the protein value of beef produced in the pastures is equal to that produced by fish that depend on flooded forests for their survival. Another overlooked ecosystem service provided by flooded (and non-flooded) forests is flood control. The negative social and economic externalities placed on *caboclos* driven from their homes by deforestation-exacerbated flooding is not part of the formula for calculating the costs of producing the protein generated by cattle ranching in deforested habitats along major Amazonian rivers.

As an example of the failure to recognize the importance of rain forests in providing ecosystem services to the world, is the case of the Iwokrama International Centre for Rain forest Conservation and Development, a 371,000 hectare (one million acre) reserve in the northeastern South American country of Guyana. This reserve was a gift from the people of Guyana to the Commonwealth of Nations (= an organization of nations, all but two of which, formed part of the British Empire) in 1996. Iwokrama harbors the highest diversity of birds (over 400 species), bats (over 90 species), and fresh water fish (over 400 species) heretofore recorded for any area of similar size in the world. One half of the reserve is earmarked for research on sustainable forestry, and the other half is excluded from human exploitation in order to protect the greatest possible number of species. The mission of Iwokrama is to "promote conservation and the sustainable and equitable use of tropical rain forests in a manner that will lead to lasting ecological, economic and social benefits to the people of Guyana and to the world in general by undertaking research, training and the development and dissemination of technologies" (Iwokrama International Centre for Rain forest Conservation and Development, 2008).

In exchange for receiving this land, the Commonwealth agreed to provide approximately $500,000 per year to maintain the reserve and its research program in sustainable forestry. At the time of this writing, the per-country share for the 53 nations forming the Commonwealth amounted to a paltry $9,433. In September 2007, I was invited to participate in a brainstorming session convened by the government of Guyana, the Iwokrama Centre, and the Commonwealth. The purpose of the meeting was to discuss an anticipated shortfall in funding; the Commonwealth had deemed that it could no longer afford to support Iwokrama and, thus, alternative sources of funding were needed to sustain its conservation and research programs.

The failure of the Commonwealth nations to guarantee continued financial support for this project represents their failure to recognize that intact tropical forests provide ecosystem services for the entire world, especially because of the role they play in the regulation of the planet's climate and mineralogical and hydrological cycles. Neotropical forests sequester much larger amounts of carbon (Golley et al., 1975) than do the alternative landscapes that replace them when they are destroyed. Overall, the destruction of the forests emits more CO^2 into the atmosphere than all of the world's transport systems combined (Royal Botanic Gardens Kew, 2008). The research on sustainable forestry at Iwokrama is helping to develop less-destructive forestry practices—technologies that will decrease the amount of CO^2 released into the atmosphere because improved harvesting methods reduce collateral damage to non-timber trees by as much as 50% (Putz et al., 2008). Surely the $500,000 per year that the Commonwealth currently invests in Iwokrama is much less than the combined value of the carbon sequestered in its forests and the economic value of the forestry practices being developed there.

Although we discussed the possibility of closing the impending budget deficit by selling carbon credits to multinational corporations in exchange for Iwokrama's agreement to protect forests (thereby sequestering carbon in the forest biomass), I do not favor this option. I question the ethics and ecological sustainability of overlooking pollution in one area simply because polluters are able to purchase carbon credits in another. The most sensible way to protect the Iwokrama forests is to place a value on the ecosystem services they provide, and require all those benefiting from these services to pay for them. Failure to set compensable values for the ecosystem services provided by tropical forests will contribute to their disappearance!

Human Population and Consumption

When I was born, in 1941, the world population was somewhere between 2 and 3 billion people. By 2008, it had reached approximately 6.5 billion and, if it continues to grow at current rates, it will approach 13 billion by 2067 (Rosenberg, 2006). The ever-increasing world population, in combination with the increasing ability of more and more people to consume at accelerated rates (Friedman, 2008), is putting stress on the planet's ability to sustain human beings at the high standards of living expected by a growing number of people in the world. The population of the United States is 300 million people, but that pales in comparison with the populations of China and India, which are 1.5 and 1.1 billion, respectively.

Although world population continues to grow, there is some indication that it is beginning to stabilize. In order to achieve a stable world population, the average number of births per woman should be 2.1. Worldwide, that number now stands at 2.8; in developed countries it is 1.6 and in developing countries 3.1 (World Resources Institute, 2008). In fact some European countries, such as Italy, are so concerned about their declining populations, that they now offer incentives for women to have more children (Shorto, 2008). However, it is clear that the increasing global population, coupled with explosive gains in per capita consumption, is already placing so much stress on the environment that human land uses end up in direct competition with the world's natural habitats and the biodiversity they harbor. Conservation of rain forest biodiversity is not compatible with continued population growth and increased consumption, and the planet may not, in fact, even be capable of sustaining humans at current levels of population and consumption very much further into the future. Mahatma Gandhi recognized this problem when he said "Earth provides enough for every man's need, but not every man's greed."

Why are the plants and animals of tropical forests so vulnerable to increased human population growth and consumption? In the first place, many tropical plants and animals have very limited geographic distributions and, thus, they will disappear if their habitats are destroyed by expanding human populations. The smaller the geographic range of a species, the greater the possibility that its habitat could be destroyed. For example, 53.5% of the trees in the Atlantic coastal forests of Brazil grow nowhere else in the world (Mori et al., 1981), and this area's high degree of endemism holds true for most groups of animals, especially birds and primates. As much as 95% of these forests has already been destroyed by logging, the planting of sugar cane and other crops, the conversion to cattle pasture, the expansion of cities, and the construction of second homes in the *restinga* vegetation that parallels beautiful beaches (Mori, 1989); therefore it is most likely that an undetermined number of species have already been lost to posterity. However, not a single one of these has been documented with certainty because the distributions of neotropical species has not been adequately determined. In addition, plants that depend on animals for pollination and seed dispersal are especially vulnerable to overhunting and habitat destruction. When pollinators and dispersers are eliminated by hunting, the plants that depend upon them for propagation also disappear. For example, when the monkeys that disperse the seeds of trees in the sapodilla family (Sapotaceae) are no longer there, the monkey-dispersed species of Sapotaceae will also disappear unless they adapt to alternative methods of seed dispersal, but this is unlikely if other potential dispersal agents have also been eliminated by hunters.

Biofuels

If the ecosystem services needed to grow corn, rice, soy beans, sugar cane, wheat, and other crops were properly calculated in their pricing, these common staples would be much more expensive. Michael Pollan (2008) pointed out that, in 1940, farmers consumed only one calorie of fossil-fuel energy to produce 2.3 calories of food energy; whereas, today, agribusinesses use 10 calories of fossil-fuel energy to produce a single calorie of food energy. He adds that food production is second only to transportation in the consumption of fossil fuels. Moreover, manufacturing the packaging used to market food in the developed world also consumes energy that is only partially included in food prices. The production of food, especially animal protein, depends on ecosystem services that are not considered in the pricing structure. For example, 21,120 liters (5000 gallons) of water are needed to produce slightly less than one-half kilogram of beef (ca. one pound). According to Pollan (2008), we pay a low price for food because governments subsidize the cost of the energy used to produce it. He argues that "oil is one of the most important ingredients in our food, and people ought to know just how much of it they are eating."

The world's energy demands should not be met by generating bio-fuels from plant biomass harvested from land that was previously used to produce food; moreover, production of biofuels from food plants can have a major negative impact on tropical biodiversity. Corn provides an example of how biofuel production from a major temperate food crop can negatively impact rain forest conservation. Although it is well known that it takes nearly as much energy to grow and distill alcohol from corn as the energy yielded when the alcohol is subsequently burned as fuel, there are still subsidies for growing corn to produce biofuel. Satisfying the transportation needs of the United States would require the country to devote much of its arable—and borderline arable—land to corn production. A great deal of the corn that is currently used, directly or indirectly, to feed humans or animals would have to be used for biofuel production, thereby raising the cost of the astonishing number of corn-derived food products (Pollan, 2006). Using corn to produce biofuel is economically unsound because it would increase the cost of food, and it would have a negative impact on biodiversity conservation in the United States because land that is not currently used for agriculture would be converted into corn fields to meet the demand for fuel.

Although the United States does not have large expanses of tropical forests, using corn to produce biofuel will negatively impact rain forests in places as far away as the Amazon. Growing corn for a source of biofuel has already become so lucrative that U.S. farmers are converting soy bean acreage into corn fields. This in turn causes countries, such as China, to turn to Brazil for soy beans they formerly imported from the United States. The Brazilian Agricultural Research Corporation (EMBRAPA) has been so successful in growing soy beans in the poor soils of the central Brazilian savanna vegetation (*cerrado*) that they have taken over leadership in soy production from the United States. Unfortunately, the only cerrado left in Brazil is that which has already been protected in reserves so Brazil is now turning to the Amazon for expansion of soy bean acreage. As a result, much of the Amazonian rain forest may become soy bean plantations, and the biodiversity it harbors will be lost, along with the ecosystem services it provides.

Another potential source of biofuels is the African oil palm (*Elaeis guineensis* Jacq.), which was introduced into Brazil from West Africa by slaves in the seventeenth century (Medeiros Santos, 2008). This plant has many uses, and is especially valued for food products because it is free from trans-fats. The fruits are used to produce biscuits, margarine, cooking oil, creams for cosmetics, colorants, and detergents, and can also be used to make biofuels. Only seven or eight years after planting, African oil palms produce as much as 25 tons of fruit per year per hectare, from which five tons of oil can be extracted. They rival soy beans in productivity and have an advantage over soy beans because palms are perennial plants and plantations would not incur the cost of planting seeds year after year. The African oil palm is well adapted to the growing conditions of the Amazon because its native habitats have similar rainfall and soil conditions to those throughout much of the Amazon. In addition, because the diseases and predators from its native habitat are left behind, the oil palm (like many other introduced tropical crops) does well when moved from one tropical area into another. It is estimated that in the Amazon 70 million hectares (172.9 million acres) are suitable for African oil palm production. Between 2004 and 2007, the market value of palm oils increased 36% (Medeiros Santos, 2008), so there is an expanding and lucrative market for palm oil. The downside, of course, is that converting a hectare of Amazonian forest to African oil plantation reduces the number of tree species in that hectare from as many as 285 to just one: the African oil palm. An additional, very real danger is that clearing forest for oil palm plantations will release greenhouse gases into the atmosphere, which might counteract any benefit of replacing petroleum with palm-derived biofuels.

The three main options for replacing petroleum are biofuels, solar energy, and hydrogen. Because biofuels (1) are not energy efficient to produce, (2) will raise food prices when farmland is used to grow biomass for biofuel production rather than for food, and (3) will replace natural landscapes rich in biodiversity with monocultures, they will not solve the world's energy problems. From an ecological point-of-view, hydrogen and solar energy lead the field because they would have the smallest environmental impact (they emit no carbon dioxide and the by-products of production are not harmful). Hydrogen can be produced from coal, natural gas, or water. However, hydrogen remains a long-shot to fulfill the world's energy needs due to many challenges, especially with large-scale fuel production, vehicle technology, and infrastructure for the distribution of a hydrogen-derived fuel (Mouawad, 2008). In contrast, 0.3% of the solar energy illuminating the Sahara Desert could meet all of the Europe's energy demands (Meinhold, 2009).

Water and Climate Change

As mentioned earlier, there are many different vegetation types in the Neotropics, and their species compositions are determined to a large extent by water availability and temperature variation. I will use the lowlands of Amazonia to illustrate the impact that climate change could have on the extent of lowland rain forest, but before doing that I need to explain an important ecological principle. Liebig's Law of the Minimum states that plant or animal growth is controlled by the scarcest resource in the environment, which is called the limiting factor. For example, if a soil possesses sufficient nutrients needed for plant growth except potassium, then the paucity of that nutrient will limit the growth of all species except those that can grow in potassium-poor soils. Potassium is therefore a limiting factor in this particular soil for some species.

One of the most important limiting factors for plant growth is water, as evidenced by the fact that rain forest plants need more water than savanna plants and savanna plants need more water than desert plants. Water seasonality, and not just total annual rainfall, is critical in determining the type of vegetation that will occur under a particular rainfall regime. The time of year when the least amount of water is available determines the vegetation type that will develop and whether a particular species will grow in a given area. Thus, two areas with 1500 millimeters of rain a year can have completely different vegetation types; for example, in Bahia, Brazil, an area may be covered by either rain forest or savanna, depending on whether the rain is distributed evenly throughout the year (results in forest) or interrupted by a period of drought (results in savanna) (Mori, 1989; Mori et al., 1983a).

Botanists have a good idea what happens when rainfall patterns change in the Amazon Basin because that experiment has been carried out during the Pleistocene. During that time there were oscillations between greater amounts of the world's water captured as ice in glaciers (colder periods) and lesser amounts in ice (warmer periods). When more water was tied up in glaciers, two things happened to the climate and vegetation: (1) it became cooler, allowing high altitude vegetation, such as cloud forest, to spread into lower elevations, and (2) it became drier, causing rain forests to contract and savannas to expand (Prance, 1982). These contractions and expansions of higher altitude vegetation, lowland rain forest, and savanna separated once continuous populations of species into separate refugia, for example, rain forests split into isolated patches during drier periods and this promoted the evolution of new species following geographic isolation (Prance, 1982).

The difference between what happened in the past, and what is happening today as a result of global warming, is that human-induced temperature increases are occurring over a much shorter period of time, and could potentially reach much higher levels than the increases that occurred during interglacial periods. Furthermore, due to widespread anthropogenic forest fragmentation, natural habitats will not have the same ability to expand and contract. The impacts are difficult to predict because this is the first time in the planet's history that temperature changes have been caused by man's release of greenhouse gases into the atmosphere.

Human populations have reached the current population and consumption levels under agriculturally favorable climatic conditions during the last several hundred years. Rapid changes in climate will undoubtedly have negative impacts on the world's agricultural infrastructure, which developed under the assumption that favorable conditions for growing crops would last indefinitely. Coastal cities will be flooded; agriculturally productive areas will be lost; other presently marginal areas will become more productive for crops; species will become extinct; the atmosphere may not be able to support life as we know it today; and additional, as-yet-unpredicted negative impacts, will almost certainly occur (Friedman, 2008; Gore, 2006a, 2006b). Even should future generations escape ecological collapse, they may well inhabit a planet supporting vastly diminished biodiversity; inhabited only by those resourceful opportunists that already thrive in highly disturbed environments.

I will address the effect that global warming could have on one of Amazonia's most ecologically dominant tree families, the Brazil nut family (Lecythidaceae). These trees are symbols of lowland Amazonia, and their fate will also be the fate of trees in other plant families (Mori & Prance, 1987a). Moreover, because trees form the framework upon which all rain forest plants and animals depend for their survival, the health of trees in the Brazil nut family can serve as indicators of the health of rain forests themselves (Mori, 2001; Mori et al., 2002b).

Trees in the Brazil nut family grow predominantly in non-flooded, lowland rain forests of the Amazon. Although they may occur in savannas, periodically flooded forests, or at higher elevations, relatively few species occur in these sub-optimal habitats (Prance & Mori, 1979, Mori & Prance, 1990). As global warming increases, two scenarios are possible. In the first scenario, glaciers will melt and rainfall will increase, enabling rain forest to expand. In this scenario, species of the Brazil nut family will be able to expand their ranges as more habitats become available. On the other hand, Amazonian temperatures will also rise and if rainfall decreases, this would cause savannas to expand and replace the rain forest habitat where most species of Lecythidaceae grow. In fact, most global climatic models predict substantial reduction of rainfall in the Amazon Basin (Betts et al., 2008; Harris et al., 2008). In this scenario, Amazonian forests, with as many as 24 species of the Brazil nut family per hectare (Mori et al., 2001), will be replaced by savanna that never has more than one or two species of this family per hectare. Moreover, as forest is replaced by savanna there will be even less moisture available to support forest growth, because nearly 50% of the rainfall generated in the Amazon is the result of transpiration of water from the trees themselves (Salati & Vose, 1984). These changes, combined with increased fires caused by logging and swidden (slash-and-burn) agricultural practices, will cause the loss of vast areas of Amazonian forest and lead to the extinction of many dominant tree species now restricted to old growth lowland rain forest throughout the Neotropics (Aragão et al., 2008; Nepstad et al., 2008).

The plants and animals that interact with the Brazil nut family may also be affected by changes in rainfall. In a study of a population of *Gustavia superba* (Kunth) O. Berg in Panama, Mori and Kallunki (1976) found that this species, which usually flowers in the dry season and sets fruit in the wet season, did not have significant fruit set in the 1975/1976 season. Although it flowered normally at the beginning of the dry season, fruits were not produced because of severe dry weather: flower buds failed to open, and developing fruits aborted. When rain began at the beginning of the following wet season, the repressed buds produced flowers, but no fruits were set—probably because the bee pollinators were no longer present. Changes in climate caused by global warming are already throwing temperate plant/animal interactions out-of-sync, and tropical plants that depend on specific pollinators may become extinct if they cannot adapt to increasing aridity as well as evolve new relationships with pollinators that frequent drier habitats.

Some plant/animal interactions are so important that changing them could have major negative impacts on the ecology of tropical forests. Amy Berkov (Chapter 2) studies the interactions of cerambycid beetles with species of the Brazil nut family. These beetles lay their eggs in the dead trunks and branches of Lecythidaceae and the larvae make galleries as they eat their way through the wood. In turn, the galleries allow wood-decaying fungi and bacteria access to the wood, thereby speeding up its decomposition. Without cerambycids, the wood of tropical plants would accumulate on the forest floor, significantly slowing the recycling of nutrients needed to maintain mineral cycles. As Golley et al. (1975) have emphasized, tropical forests are able to grow on nutrient poor soils because they have evolved a system that recycles minerals efficiently, and we now know that cerambycid beetles are an important part of that cycle (Tavakilian et al. 1997).

Berkov and her collaborators demonstrated that, in French Guiana, seasonal changes make a dramatic impact on branch colonization by cerambycids (Berkov & Tavakilian, 1999; Berkov 2002). Canopy branches are colonized by beetles during both the wet and dry seasons, but wood on the forest floor is densely colonized only during the dry season. Contemporary measurements of microclimate and insect niche can help predict responses to the future climate change. These cerambycid species appear to be partitioning resources along a fairly subtle thermal gradient, with small changes in temperature directly linked to and amplified by changes in relative humidity (Berkov, pers. comm.). Thus, it is almost certain that major changes in global temperature, accompanied by changes in rainfall, will impact the recycling of nutrients from wood in tropical forests by cerambycid beetles and have repercussions on other wood-decomposing organisms that depend upon cerambycids.

Is there Hope for the Future?

The future of plant and animal diversity in tropical American rain forests depends on an understanding of the extreme fragility of the plant/animal interactions found in this ecosystem. The relationships between plants and animals in the tropics are so closely co-evolved that man's exploitation of tropical forests always results in loss of biodiversity (Robinson, 1993). Thus, the most important conclusion to be made from this essay is that, although humans have had little to do with the evolution of biodiversity, they are largely responsible for its destruction, especially in the tropics. As Thomas Friedman (2008) summarized "We are the only species in this vast web of life that no animal or plant in nature depends on for its survival—yet we depend on this whole web of life for our survival."

Increasing human population and consumption throughout the world is incompatible with the preservation of biodiversity in any of the world's biomes. The destruction of tropical vegetation for agricultural fields and pasture is particularly alarming because these activities replace biodiversity rich vegetation types. Vast areas of tropical vegetation now grow on nutrient-poor soils that will never support the agricultural demands of increased human populations without massive inputs of fertilizers and pesticides. Thus, the destruction of tropical forests destroys both biodiversity and the ecosystem services that these natural areas provide, in exchange for short-term profits by a limited number of individuals. The costs of the ecosystem services lost, such as sequestration and hydrological regulation, and the costs of externalities, such as polluted waters, outweigh the short-term benefits obtained by attempting to transform tropical habitats with nutrient poor soils into agricultural fields and pastures. If tropical areas are not sufficiently productive to contribute significant support to today's world population of 6.5 billion, what makes us think that these same areas will be able to contribute in any meaningful way to the support of a population of nine to 11 billion humans by 2050? The consumptive power of a contemporary resident of the Amazon is orders of magnitude greater than that of the pre-Colombian inhabitants who based their economy on agricultural systems that did not require inputs of fertilizers and pesticides. In short, residents of the tropics, and of the world in general, will not be able to protect biodiversity at meaningful levels unless both human population and consumption are brought under control. It will be especially important to control our inclination toward excess, and develop a new index of human success to replace the holy grail of economic growth that seems to motivate most of today's human activity. For the planet's biodiversity to survive (including top predators like, you guessed it, ourselves), the earth must no longer be regarded as a supermarket of resources to be exploited any way we choose, merely to satisfy our every want and desire.

The globalization (or flattening) of the world has placed environmental pressures on the world's ecosystems that could not have been foreseen 20 years ago. Hence, all large-scale agriculture, water management, carbon emission, and other resource-use schemes must be weighed against the resulting worldwide impacts. For example, the development of biofuels, which initially seemed like a sound strategy to wean ourselves from petroleum dependence, will have both direct and indirect impacts on tropical forest biodiversity and all of the world's ecosystems. We must evaluate these negative impacts before implementing large-scale biofuel production projects (such as growing corn as a raw material) or, in hindsight, we will lament that which we have thoughtlessly surrendered.

The true cost of tropical forest exploitation must be considered in any scheme to modify tropical ecosystems for human profit. Simply stated, non-tropical countries will have to pay more for products such as bananas or tropical hardwoods, because the loss of ecosystem services, and the costs of negative externalities caused by deforestation, must be offset. Those increased prices should subsidize the cost of protecting large reserves of tropical ecosystems to (1) maintain the ecosystem services they provide, (2) control the cost of externalities caused by their destruction, and (3) protect the biodiversity they harbor. The cost of maintaining the undisturbed tropical forests that are needed to sequester carbon must be shared by the entire world as part of humanity's effort to control climate change. In *The Myth of Progress*, Wessels (2006) argues that if we fail to recognize that continuous economic growth violates fundamental ecological laws, and that blind adherence to this economic model will result in the loss of ecosystem services and biodiversity throughout the entire planet.

Biologists now have enough knowledge about tropical forests, and the conditions under which they grow, to determine which areas are appropriate for sustainable agriculture and cattle grazing; which areas should be managed as extractive reserves; and which areas should be set aside as biological reserves. It is no longer the responsibility of botanists, zoologists, and ecologists to demonstrate that tropical rain forests are valuable; it is the responsibility of those who wish to "develop" the forests to demonstrate that the proposed use justifies the loss of biodiversity and ecosystem services that will follow.

Finally, we need to accept the idea that in order to preserve biodiversity, we need to do more than simply calculate its financial value to humans. Natural vegetation types and the biodiversity they harbor have an inherent right to exist, and that right must be protected. We humans have the capacity to destroy all species, including ourselves. This awesome power brings with it tremendous ecological responsibility: our obligation to protect all life on earth! Thomas Friedman (2008) eloquently summarizes this as follows: "An ethic of conservation declares that maintaining our natural world is a value that is impossible to quantify but also impossible to ignore, because of the sheer beauty, wonder, joy, and magic that nature brings to being alive."

Appendix A
Adopt-a-Forest-Strategy
by Scott A. Mori

The first step in the "Adopt-a-Forest-Strategy," outlined by the Committee on Research Priorities in Tropical Biology (1980) and advocated by Bill Laurance (2008), is the preparation of botanical and faunal inventories. However, a complete botanical inventory of a lowland tropical rain forest is difficult to accomplish because there are so many species to find and collect, species composition usually varies over relatively short distances, sterile plants are often difficult to identify but some species only sporadically flower and set fruit, and large trees—and the plants that grow in their canopies—are difficult to collect. There are published inventories of the vascular plants (Mori et al., 1997, 2003), the mosses (Buck, 2003), and liverworts (Gradstein & Ilkiu-Borges, 2009) of central French Guiana. In addition, an e-Flora project has been initiated for all of French Guiana, including a specimen-based checklist of the flowering plants, a key to the flowering plant families, botanical line drawings and field images, and species pages and keys for Lecythidaceae (Mori et al., 2007c). Although there is still much to be accomplished, in central French Guiana one of the first requirements in an "Adopt-a-Forest-Strategy," an inventory that facilitates the identification of plants, has been accomplished. How do more detailed ecological studies benefit from plant and animal inventories?

Although it is well known that bats play an important role in the pollination and dispersal of tropical plants, the details of these interactions in lowland neotropical forests are poorly known. I initially developed an interest in bat/plant interactions because bats pollinate the flowers of a few species of the Brazil nut family, and probably disperse the seeds of many of its species. This interest remained latent until 1987, when Merlin Tuttle, the founder and long-time Director of Bat Conservation International, joined an expedition to central French Guiana to photograph bats visiting plants. We placed a bat in a black-walled tent with cuttings of plants that we thought might be dispersed by bats. Once a bat was in the tent with a suitable fruiting plant, it quickly located and consumed the fruit. Merlin was ready to document the interaction, and took a number of classic photographs of bats handling and eating fruits. Merlin's trip to French Guiana convinced me to pursue studies of bat/plant interactions, but initially it was difficult to study them because there were no areas in the Neotropics with published inventories of both plants and bats.

With the publication of the *Guide to the Vascular Plants of Central French Guiana* (Mori et al., 1997, 2002) and *Les Chauves-souri de Guyane* (Charles-Dominque et al., 2001), central French Guiana became an ideal place to study bat ecology. When French botanist Fred Blanchard visited me at NYBG in 2000, and told me he was looking for an opportunity to participate in botanical research, we started by developing a database summarizing everything that was known about pollination and seed dispersal by neotropical bats. The results of that effort are available online (Geiselman et al., 2002). The bat team took on another participant when Tatyana Lobova, trained in seed morphology and anatomy at the Komarov Botanical Institute in St. Petersburg, Russia, joined it in 2001. Tatyana's ability to use anatomical and morphological characters to identify plants based on their seeds enabled her to determine which plants bats were feeding on by examining the seeds in their feces. In 2003 Cullen Geiselman, who had gained expertise in bat biology during the five years she worked for Bat Conservation International, joined the bat team. Cullen's knowledge made it possible to identify the captured bats. This collaboration resulted in the publication of *Seeds Dispersed by Bats in the Neotropics* (Lobova et al., 2009), a project that would not have been possible had we not known the identities of the plants and the bats that were the subjects of the study!

This example demonstrates that selecting given tropical areas for detailed study is an effective way of increasing our understanding of tropical biology. It results in inherently valuable biological inventories of selected areas that, in turn, facilitate the identification of focal organisms in ecological studies. In the Neotropics, the research conducted at La Selva in Costa Rica (McDade et al., 1994), Barro Colorado Island in Panama (Leigh et al. 1996), the Biological Dynamics of Forest Fragments Project in central Amazonian Brazil (Bierregaard et al., 2001), Manu in Peru (Gentry, 1990), and the Nouragues Nature Park in French Guiana (Bongers et al., 2001) are well-known examples of the resounding success of the "Adopt a Forest Strategy."

Appendix B
Funding for Systematic Botany
by Scott A. Mori

The projects mentioned in this book are expensive and, thus, one of the most difficult, but essential parts of any research program is acquiring financial support. Even relatively small-scale efforts, such as writing the *Guide to the Vascular Plants of Central French Guiana*, are surprisingly expensive. These studies are not expensive due to high priced equipment needed to undertake them, but because of the many hours botanists spend to: (1) collect specimens; (2) process specimens; (3) extract data based on field observations, the study of herbarium specimens, and laboratory studies; (4) enter information, including images, into databases; (5) keep specimen determinations current on the specimens themselves and in the database records; (6) write the descriptions of species that serve as the basis for Floras and monographs; (7) keep current with research advances and add literature references to databases; (8) prepare the keys used for plant identifications and, if the keys are electronic, maintain concordance between the key matrix and the specimen and species page databases; and (9) raise the money to support the work. Without adequate funding to support this type of research, the time needed to complete Floras and monographs is measured in decades.

My research career has been based at three botanical institutions: the Missouri Botanical Garden, for which I collected plants in Panama from 1974 to1975; the Centro de Pesquisas do Cacau in Bahia, Brazil where I served as a curator of the herbarium from 1978 to 1980; and The New York Botanical Garden (NYBG), where I have been a curator from 1975 to 1978 and then from 1980 to the present. Because my entire fund-raising experience occurred at NYBG, this Appendix is based on my efforts to support my research there.

Why Support Systematic Botany?

"Forget, for the moment, the rationales advanced for saving nature, from wonder drugs yet to be discovered to carbon sequestration. The simple yet awesome truth is that life is precious and deserving of our respect in all of its myriad guises." (Young, 2000)

In the quote above Allen Young, Curator Emeritus of the Milwaukee Public Museum, has captured one of the main reasons for studying the classification of plants (systematic botany); the desire of systematists to enable others to learn the names of plants so that they can understand, appreciate, and protect earth's plant diversity in "all of its myriad guises." For the most part, funding for systematic botany comes from individuals and organizations that appreciate the natural world, recognize the importance of knowing the names of plants and animals, and wish to contribute to conservation. In my various efforts to raise money for studies of tropical plants, it has been apparent that those supporting this kind of research have an innate appreciation for science and the natural world. Individuals who have contributed to my work—two retired physicians; a scientist who worked in the pet food industry; a scientist with several patents who was employed by the food industry; a forester; two botanists with significant publication records; several eco-tour participants; and, most importantly, my botanist wife, Carol Gracie, who raised money and provided support by organizing eco-tours, pressed specimens in the field, contributed countless images to my projects, and proofread my papers—are fascinated by natural history. In addition, the organizations that have supported my work have, as part of their mission, the goal of advancing fields of science that contribute to an understanding and conservation of biodiversity.

Individual donors may be inclined to support work in systematics and make unsolicited contributions based on their knowledge of the research and researcher, but, at the other end of the spectrum, proposals to foundations such as the National Science Foundation of the United States (NSF) are subject to critical review, and may take months to prepare. Funding organizations, administrators, and grant reviewers demand more concrete reasons than merely appreciating the work before they will recommend or award financial support for studying systematics.

A successful grant application will usually demonstrate that research in plant systematics has immediate and lasting value to society. Systematics Agenda 2000 (1994) defined three interrelated missions: one, "to discover, describe, and inventory global species diversity"; two, "to analyze and synthesize the information derived from this global discovery effort into a predictive classification system that reflects the history of life"; and three, "to organize the information derived from this global program in an efficiently retrievable form that best meets the needs of science and society." In short, a competitive proposal must provide arguments that it will better address these goals than other, competing proposals.

The diversity of plants is described in two important publication—Floras (including the lists of plants of a region, known as checklists, which lead to Floras) and monographs. A Flora describes all plants in a given geographic area, and a monograph treats all species of a chosen group of plants throughout its geographic range. Many grant requests in plant systematics seek support for monographic and floristic work as well as for the herbaria that house the collections used to document these studies. Thus, it is important that proposals provide arguments demonstrating the utility of Floras and monographs as well as justifying the importance of the herbarium for which support is requested.

It is difficult to put a monetary value on the products of systematic botany. Consider the scientific background leading to Steve Clemants and Carol Gracie's *Wildflowers in the Field and Forest of the Northeastern United States* (Clemants & Gracie, 2006). Their field guide is based on original research provided by systematic botanists, and publication of their book would not have been possible without the scientific publications that paved its way, such as Gleason and Cronquist's (1991) *Manual of Vascular Plants of the Northeastern United States, the Flora of North America* project (http://hua.huh.harvard.edu/FNA/), online resources such as the United States Department of Agriculture's Plants Database (http://plants.usda.gov/), and the Angiosperm Phylogeny Group's most recent classification of the flowering plants (www.mobot.org/MOBOT/research/APWeb/). Although these resources serve as basic references for professional botanists, they are often difficult for non-specialists to use. On the other hand, amateur naturalists can easily use Clemants and Gracie's field guide to learn the names of plants, and thereby access the information known about them. Their book sells at a reasonable price and generates income, but the scientific studies that made this guide possible are not factored into the price of the book. Moreover, there is no way that a dollar value can be placed on the satisfaction that a user feels when he or she successfully identifies a plant. In this sense, the pleasure derived from using a field guide is comparable to that derived from listening to music, viewing art, or reading classical literature; but these difficult-to-calculate values are usually not considered as justification for awarding a research grant.

On the other hand, there are many direct economic benefits derived from the work of scientists in general, and systematists in particular, and grant proposals that convincingly demonstrate these benefits probably have a better chance of being funded than those that do not. An example of a pure science grant that resulted in immense benefits to mankind was the one leading to the elucidation of the structure of DNA by James Watson and Francis Crick (Watson, 1968). Their work resulted in the creation of a trillions of dollars industry (genetic engineering), new fields of science, the potential to improve human health, and the ability to engineer genetically more productive crops and animals. The returns from this single discovery have easily paid for the original grant investment as well as for all grants supporting scientific research far into the future. There are many less spectacular examples of how research support can pay high dividends on relatively small investments, some of which are listed on the BioNet website (www.bionet-intl.org/opencms/opencms/index1.jsp) under "Why taxonomy matters."

Importance of Grants

Members of the general public may have the impression that there is a "pot" of money available to curators that supports the direct costs of research, but this is not the case. To understand the true cost of research at botanical institutions, one needs to understand the difference between direct and indirect costs. The former refers to the actual expense of doing the research, such as the cost of field work, stipends for research assistants, supplies, mounting specimens, and sequencing genes. The latter includes the costs of maintaining the research facilities (e.g., herbaria, libraries, and labs), offices in which to work, and paying base salaries for curators. The indirect costs are usually higher than the direct costs of research. In return for support received from their home institutions, curators are expected to raise money to pay the direct costs of the research, as well as part of the indirect costs of the home institutions of the researchers.

It is especially difficult to raise money for the indirect costs for systematic botany because funding agencies are reluctant to support long-term projects such as Floras and monographs. An extreme example of how long it takes to produce a Flora is Flora Brasiliensis (Oxford University Herbaria. 2004), which was initiated by Carl Friedrich Philipp von Martius (1794–1864) in 1840. The first volume of this monumental Flora appeared in 1845 and the last volume was published in 1906. Seventy-five botanists from around the world contributed to Flora Brasiliensis, which comprises 40 volumes, 20,773 pages, 3,811 pen and ink plates, and 1,071 lithographs. Nearly 23,000 species are treated, of which 5,939 were described as new to science. This expensive and time-consuming undertaking would not have been possible without the financial support of Ferdinand I, Emperor of Austria; Ludovic I, King of Bavaria; and Dom Pedro II, Emperor of Brazil. Although computer technology facilitates the preparation of modern Floras and monographs, they still take a great deal of time to complete.

Fortunately, there are botanical gardens and universities with botanists on their staffs specifically hired to pursue the long-term studies needed for the preparation of Floras and monographs, but finding the funds needed to conduct this research is probably the biggest challenge most of them will face. The diverse sources of research funding are the topic of the remainder of this Appendix.

The National Science Foundation (NSF)

Among funding sources, NSF meets both the needs of the scientist, by covering direct costs of research, and the home institution, by paying part of the indirect costs. For this reason, curators are encouraged to hold an NSF grant or to be in the process of preparing an application. In 2010, a successful grant would provide a researcher with project money, and the home institution (NYBG) with overhead amounting to 46.6% of onsite and offsite direct costs. The overhead figures vary from year-to-year and from institution to institution, but indirect costs always provide a significant source of income when institutions are awarded grants. Because NSF grants meet the needs for both direct and indirect project support, they are highly competitive and currently only 10 to 30 percent of the applications are successful. The failure to obtain NSF funding for research makes it extremely difficult for free-standing botanical gardens to maintain strong research and educational programs.

NSF has programs designed to stimulate research in all aspects of biodiversity research. There are pre-doctoral and post-doctoral fellowships; fellowships supporting collaborations with scientists from other countries; research support for students at all levels; programs specifically designed to support research in systematics and biodiversity exploration; opportunities for mid- to late-career scientists to synthesize the findings of their research; and efforts to promote large-scale, multidisciplinary studies. NSF also has programs to support museum infrastructure, e.g., the purchase of herbarium cabinets, reorganization of herbaria, databasing and imaging of specimens, and many programs that provide support for developing educational tools for teaching botany. These programs are described on NSF's Environmental Biology (DEB) home page at http://nsf.

gov/funding/pgm_summ.jsp?pims_id=503618&org=DEB&from=home. This page opens the door to all aspects of the environmental biology program of NSF and all those planning to submit proposals should periodically visit the site to learn about what is new in the DEB program.

The funding requested in an NSF proposal is important to the institutions receiving grants because the larger the budget, the more money the home institution receives in indirect costs. As an example, recent awards for REVSYS have ranged from $100,000 to $600,000, while those of the PBI ranged from approximately 2–4.4 million dollars. Traditional systematic biology research centered on describing species using morphological data is less expensive than research based on combined morphological and molecular data. Therefore, curators can not only enhance their research, but also obtain larger grants, by including molecular techniques as part of their methodology.

The following tips, provided to help others write better proposals, are based mostly on what I have learned from having my share of NSF proposals declined.

Tip 1. Before starting to write a proposal, the grant guidelines should be read carefully. For example, the NSF guidelines clearly indicate that the intellectual merits of the proposal must be stated, the broader impacts of the research addressed, and the role of students in the research clearly outlined. Failure to follow these guidelines will reduce an applicant's chance for NSF funding. Other useful information is available from the NSF website, which provides descriptions of the kinds of grants that have been previously funded and gives an idea of the normal grant amount for a particular program. This allows the researcher to assess the suitability of a particular research project for a given panel, and helps avoid budget exaggerations.

Tip 2. Applicants should have an ongoing dialogue with their scientific administration about submitting NSF proposals, and should seek the advice of those most successful in obtaining grants to get their perspectives on writing proposals. NSF has a stringent peer review process, so having colleagues review a proposal critically before its submission increases the chance of success. The first draft of the proposal should be written at least one month before the submission deadline, so that colleagues have enough time to read it, and the PI has adequate time to incorporate suggestions.

Tip 3. Submit only proposals for research that can be completed within the specified time frame. Novices often propose to accomplish more than will be possible in the time stipulated. For example, many floristic projects take decades to complete, so if a research team proposes that it will finish a Flora of, for example, 2000 species in three years, it is almost certain that this will not be accomplished. Our *Guide to the Vascular Plants of Central French Guiana* serves as an example. NSF support for this Flora was awarded in 1991, the first volume was produced in 1997, and the second and last volume appeared in 2002. The NSF award provided support for three years, but the project continued for seven years after the money was spent. Because the time needed to complete the Flora was underestimated, a subsequent proposal submitted to NSF to fund additional work on the same project was not successful. A much better strategy would have been to write three separate proposals, of three years each, to (1) gather and database the collections needed to prepare an online, specimen-based check list of the flora, (2) publish the treatments of the ferns and monocots in a first volume, and (3) finish with the publication of the dicots in a second volume. If a promised product is not delivered at the end of a grant period, it will be extremely difficult, if not impossible, to obtain subsequent funding for the same project.

Tip 4. Do not get discouraged by a first or even second denial of funding, but rather learn from what the reviewers have to say, address the issues they raise, and submit at least a third time before giving up on a particular project. NSF grants are awarded through a stringent review process involving outside "ad hoc" reviewers, who receive electronic versions of the proposals; panels of reviewers assembled in Washington, who evaluate the outside reviews and discuss the merits of the proposals; and the staffs of the respective NSF programs. Nearly all of the proposals have merit, especially if they have been critically reviewed by colleagues prior to submission. Nevertheless, relatively few proposals are funded the first-time they are submitted because reviewers sometimes discover flaws as part of the review process. Subsequent submissions that have addressed these flaws stand a much better chance of funding. Declined proposals can be submitted numerous times, especially if the original review does not discourage resubmission.

Tip 5. Clearly state the hypotheses to be tested. Although systematic botany is based on hypothesis testing, that is not always apparent to reviewers, who remark that research in systematics is too descriptive and, thus, not as worthy for funding as proposals in which testable hypothesis are articulated (see Hypothesis Conundrum below). Even programs that support the work of biological inventory, such as Biotic Surveys and Inventories, can include hypothesis testing but they should be a natural outcome of the inventory.

Tip 6. Include preliminary data that demonstrate the ability to address the hypotheses stated using the techniques proposed. Successful proposals make it very clear that the applicant has the qualifications to carry out the research, and that the methodology described in the proposal will provide the data needed to address the hypotheses. For example, if a goal of a proposal is to test hypotheses about the monophyly of genera in a given plant family by employing morphological and molecular data, then preliminary trees demonstrating the ability of the applicant to do the work should be included in the proposal. It is especially important to convince reviewers that the selected genes can be successfully sequenced, that the resulting sequences are sufficiently variable to test the proposed hypotheses, and that the data analysis methods are adequate.

Tip 7. Plant systematics is based to a large extent on the collection and archiving of specimens and, thus, these aspects need to be covered in proposal. In a floristic study, the collection protocol needs to be established and justified. The plan may be as basic as visiting the area encompassed by the Flora at different seasons of the year, walking the trails through all vegetation types, and collecting specimens from every plant found in flower or fruit. If large numbers of specimens have already been entered in a database for the area to be studied, then more sophisticated analyses can be used to justify the need for additional collecting. For example, additional collecting can be justified if a graph demonstrates that the species accumulation curve has not yet begun to level off. A graph showing the number of collections made in the different months of the year might demonstrate that fewer collections have been collected at a particular time of the year, for example the rainy season, making it important to gather more collections then. In the case of monographic studies, the plan should state what areas have not been adequately collected, and what species need to be collected in order to document all stages of their life cycles. In addition, the proposal needs to make clear that the proper permits can be obtained, and where specimens resulting from the exploration will be deposited. Most important, do not assume that all reviewers will understand concepts and methods that are obvious to you. For this reason, the proposal should make complicated concepts as simple as possible, and be written without the jargon of your field of expertise. Proposals should always include recommendations for potential reviewers with expertise in the field for which support is requested; these recommendations are considered when reviewers are selected for writing the "ad hoc" reviews.

Tip 8. Make sure that the roles of all participants are spelled out. For the most part, proposals that include training of students, especially local ones, are favored over proposals that do not include such training. In this regard, the role of the student should indicate that he or she will gain experience doing more than just entering data into a database or scanning specimens. Although these tasks are important aspects of any project, the reviewer needs to be convinced that the student will be trained in such a way that he or she will gain skills necessary to build a successful career in systematics.

Tip 9. Describe the products that will result from the research. Will a website be produced, and, if so where will it be hosted, what database will be used, and how will it be maintained? If species pages will be created, what program will be used to create them, and what program will be used to prepare the electronic keys? Where will the results be published? Will a Flora or a monograph be prepared, and will it be in hard copy, electronic format, or both? In other words, what will be the legacy that justifies an NSF award to carry out the proposed research?

Tip 10. Because NSF proposals are so critically reviewed, a convincing NSF proposal is difficult to write without already having accumulated considerable background data. This dictates that researchers plan for future projects while working on current projects. An advantage of long-term research planning is that field trips planned for current projects can also be used to collect preliminary data for future projects. An efficient way to respond to calls for research proposals is to have boilerplate grant proposals ready for submission to NSF and other funding agencies as opportunities arise.

Hypothesis conundrum. Proposals must clearly state the hypotheses to be tested, because some critics of systematic botany claim that it is not a science, but merely "stamp collecting" or an enumeration of facts about the living world. This criticism fails to recognize how important inventories are to the study of biodiversity, and that systematics, just like any other science, is based on hypothesis testing. Bernd Heinrich (2007) addresses the issue as follows:

"Science rests on empiricism, and the 'enumeration of objects' (or phenomena) is the first and necessary step in any science. Only afterward can one proceed to try to compare and search for patterns and regularities, and only then can one try to make sense of it all by formulating hypotheses, testing them, and ultimately searching for the causes and effects of given phenomena. One can't do one before the other, any more than one can cross the finish line of a race before taking all the steps leading up to it. And in any organismal biology, before any enumeration can take place, the observer must first name. Naming before knowing."

Critics fail to recognize that the overwhelming amount of descriptive work in systematics, which sometimes masks the hypothesis testing, is an integral part of the discipline. But each and every taxon (family, genus, species) defined by a systematist is a hypothesis that is open to testing by the addition of new collections, and the accumulation of new information. In short, there are as many hypotheses in systematic biology as there are taxa on the planet!

One of the most important contributions of systematic botany is that it leaves a legacy (Floras and monographs) that enables others to identify plants and subsequently both enjoy and learn more about them. Fortunately, NSF recognizes the value of plant collection and description in its Biotic Surveys and Inventories, Planetary Biodiversity Inventory, and the Revisionary Synthesis in Systematics programs, which allow for both descriptive activities and hypothesis testing.

National Geographic Society (NGS)

Funding needed to carry out botanical expeditions to remote and botanically unexplored regions of the planet is generally easier to obtain than funds for other direct and indirect project costs. Foundations and nongovernmental organizations are often willing to take on the financial commitment needed to pay for plane fares, field expenses, and equipment needed to make new discoveries, and even the part-time help required to organize collections after they have been gathered. The results of an expedition are tangible: specimens are collected, new species are described, and conservation recommendations are made based on the information acquired. For example, researchers at Chicago's Field Museum of Natural History mounted an expedition into the Andes of northern Peru for three weeks in August 2000, and found an amazing diversity of vegetation types from which they collected 28 species previously unknown to science. Working with local conservation groups, the Field Museum played an instrumental role in convincing (then) President Valentín Pantiagua to establish the Cordillera Azul as one of the world's largest national parks... about the size of the state of Connecticut (Kinzer, 2002).

One of the most important sources of funds for botanical exploration is the National Geographic Society (NGS). The NGS Committee for Research and Exploration has a website (www.nationalgeographic.com/field/grants-programs/cre.html) that can be consulted for details about its grant programs. The application procedure includes a short pre-proposal, which prevents researchers from spending a great deal of time writing a detailed proposal for a project that might not be consistent with the NGS funding mission.

NGS does not provide indirect institutional support or salary support for a PI, thus home institutions put no restrictions on proposals to NGS. This simplifies the grant application process and makes NGS one of the leading sources of funding for biodiversity exploration.

Foundations

Financial support from foundations is critical to non-governmental organizations such as botanical gardens and natural history museums. The development departments of these organizations routinely raise money from foundations to support the costs of new facility construction, staff salaries, field work and research (direct and indirect costs), endowments for chairs and programs, and even general budgetary relief. Many foundations have the same goals as the organizations they support and, thus, their interactions can be viewed as partnerships.

The Conservation and Environment Program of the Andrew W. Mellon Foundation (Mellon Foundation, 2003 onward) is, for instance, a foundation program with goals that are consistent to those of botanical institutions: to promote an understanding of plants and advocate for the conservation of biodiversity. On a broad scale, the Mellon Foundation promotes the preservation of natural areas and the support of "organizations concerned with increasing man's understanding of his natural environment, his relation to it, and the effects of his activity upon it." This foundation generally focuses on grants for botany and terrestrial ecosystems because of the importance of plants to the world's ecosystems, and "because other funding sources paid the least attention to them." One of its guidelines is to build, strengthen, and sustain botanical institutions through long-term collaborations. As a result, the Andrew W. Mellon foundation has made significant contributions to botanical science. One of the most recent is the Latin American Plant Initiative (LAPI), started in 2007. This project aspires to provide high-resolution images of Latin American type specimens archived in at least 140 herbaria located in 50 countries. Botanists designate type specimens to represent new species, and the types must be consulted if there are questions about the identity of a particular plant. As one of the first steps in their research, almost all students of systematic botany are required to study type specimens. Until recently, scientists wishing to study types had to request loans in a time-consuming and expensive process. In addition, shipping specimens to other parts of the world frequently resulted in damage, or sometimes even their loss. When LAPI is completed, high-resolution images of all Latin American type specimens will be available for immediate consultation online. Financial support from the Mellon Foundation has helped herbaria throughout the world accomplish a common goal that would not otherwise have been possible.

All of my projects, in one way or the other, have benefited from the generosity of foundations. They have provided support for the direct costs of many of my projects or, for the 35 years that I have been associated with NYBG, been the source of much of my salary and provided the funds needed to maintain the facilities where I work. The Beneficia Foundation has been especially supportive of my work, and has contributed to nearly all of my major projects over many years. This foundation was especially helpful when I was not able to complete the *Guide to the Vascular Plants of Central French Guiana* within the three year window funded by NSF. It also contributed in major ways to the book *Flowering Plants of the Neotropics*, and was instrumental in facilitating every aspect of the bat-plant studies that Lobova, Geiselman, and I conducted.

In some cases I collaborated on institutional efforts to raise money from foundations, but most of the time funds were raised to support my work without my knowledge, as part of larger fund-raising efforts initiated by NYBG's Development Department (under the leadership of Gregory Long since 1989). For example, I have been the Nathaniel Lord Britton Curator of Botany since 1998 and that chair was established by NYBG through a generous donation from the Harriet Ford Dickinson Foundation.

Contracts

Research contracts also contribute to the missions of both the contracting organization and the researcher's home institution. For example, the Plants and Lichens of Saba website (Mori et al., 2007a) was supported by a contract from Conservation International, and the French Guianan e-Flora Project (Mori et al., 2007c) was possible because of a contract from the Centre National de la Recherche Scientifique of France. Although detailed formal proposals are not usually required for contract work, written contracts need to be agreed upon by the PI, his or her home institution, and the granting organization. The contractor has to be assured that the work will be delivered at the end of the project, and the project director has to make sure that there is enough funding to cover the cost of the work.

Fellowships

Fellowships give researchers the opportunity to join the staff of another scientific institution for a limited period of time. For example, for six months in 1991 I was an Andrew W. Mellon Fellow at the Department of Botany at the Smithsonian Institution in Washington, D.C. This fellowship allowed me to dedicate most of my research to studying the species of the Brazil nut family found in a 100-hectare plot at Reserve 1501 (also called Km 41) in central Amazonian Brazil. This plot was part of the Biological Dynamics of Forest Fragments Project (BDFFP), a joint effort between the Smithsonian Institution and the Instituto Nacional de Pesquisas da Amazônia. The fellowship covered my salary, living expenses while in Washington, and the cost of the preparation of botanical illustrations. In addition, the BDFFP awarded a small grant to defray the publication expenses of a booklet based on the research (Mori & Lepsch-Cunha, 1995). The fellowship allowed me to interact with Smithsonian staff, study the collections of Lecythidaceae at the National Herbarium, work on research that contributed to understanding central Amazonian species of Lecythidaceae, and publish the booklet. The Smithsonian benefited by seeing the completion of one of their BDFFP projects, and NYBG was relieved of paying my salary for six months while I worked on research consistent with its mission.

Ecotours

Ecotourism has seen remarkable growth in the last 30 years. Pimentel et al. (1997) calculated that travel to enjoy scenic landscapes, and to observe plants and birds and other animals in their natural habitats, generated 500 billion dollars a year; Munn (1992) estimated that an individual macaw is worth between $22,500 and $165,000 a year in ecotourism dollars.

With the support of NYBG, my wife Carol Gracie and I organized ecotours from 1991 through 2007. Approximately 350 ecotourists (many of whom participated in multiple tours) travelled with us on 34 trips to destinations in Brazil, Costa Rica, Ecuador, France, French Guiana (Mori et al.1998, 1999), Greece, Hawaii, Italy, Spain, Trinidad, and Venezuela. Our goal was to introduce participants to the plants, animals, and cultures of the areas we visited; promote an appreciation for natural environments and an awareness of the need to protect them; and generate income to support botanical research. We offered one or two tours a year, which, to a large extent, were successful because Carol organized them—a task that took considerable time and effort. The ecotours were possible because NYBG included our trips under its insurance coverage.

Our most successful trips were to the Brazilian Amazon with Amazonia Expeditions, a company owned by legendary guide Moacir (Mo) Fortes (Figs. B-1, B-2). Our standard itinerary was to fly to Manaus; spend two to three days on the Amazon River to learn how Amazonian *caboclos* (= natives) make their living by fishing and growing cassava; return to Manaus for a day that included visits to the Teatro Amazonas and the Japanese Natural History Museum, lunch at a typical Brazilian barbecue restaurant called a *churrascaria*, and shopping in the markets; and then proceed upriver to explore the pristine environments of the Rio Negro above Manaus.

Occasionally we would deviate from this itinerary, but part of the success of the standard tour was that Mo and his crew knew exactly where to go to show our group the many interesting things that Amazonia offers. For example, when we arrived at the mouth of the Rio Cuieiras, a tributary of the Rio Negro above Manaus, we knew where to find a night-blooming cactus, *Selenicereus wittei* (K. Schum.) G. D. Rowley, growing as an epiphyte on tree trunks; when we took certain channels through the Anavilhanas Islands at dusk, we were confident that we would see spectacular evening flights of large flocks of scarlet and blue-and-yellow macaws; and we could always impress our travelers by visiting the giant ceiba tree (*Ceiba pentandra* [L.] Gaertn.) growing in flooded forest across from the mouth of the Rio Branco. Unfortunately, this tree crashed to the ground

Fig. B-1. Amazon ecotours. Top: Tour group in 1995. The income from these and other tours partially supported the author's research for nearly 18 years. Bottom: Two eco-tourists (Katie Lee, right, and Charissa Baker, left) photographing an emerald tree boa. Photos by Carol A. Gracie.

several years ago and no longer astounds tourists with its majesty. Flight availability dictated the length of our trips, but we tried to make them for approximately 11 days because our travelers ended the trip still wanting to see more, and that provided us with repeat customers.

The Amazon trip was also our most lucrative, because we dealt directly with Mo rather than using a commercial travel company to make the arrangements (as we did for other trips). Direct contact with local tour operators can nearly double the income, and researchers interested in rais-

Fig. B-2. Amazon ecotours. Top: The Victoria Amazonica, one of the Amazon river boats owned by Moacir Fortes. Photo by Carol A. Gracie. Bottom: Moacir Fortes (left) showing a second day flower of the Amazon water lily (Victoria amazonica Poeppig & J. C. Sowerby) and Jennifer McGee, a student from South Hampton College, showing a first day flower. Photo by Scott A. Mori.

ing funds through ecotours are in a particularly good position to make these contacts. Before establishing business relations with local tour operators, however, it is essential to establish their reliability. In the case of Mo, I knew him even before he owned a boat large enough to accommodate our groups, and had been a fellow guide with him on several Amazonian cruises on large ships. We also received feedback from ecotourists who had made successful trips with him before we began running tours with his company. My confidence in Mo was confirmed every time we exited customs in the Manaus airport, bleary-eyed after our night flights, to be greeted by Mo with a large *abraço* (= hug), a slap on the back, and a cold *cerveja* (= beer)!

Our most successful trips were those in which we had enough passengers to fill two boats. On those occasions we either held the tours sequentially or contracted additional guides and ran two boats simultaneously. Once we ran a tour with one boat for botanical artists and second boat for non-artists. Our friend Katie Lee led the art boat and offered classes on how to draw the flowers, scenes, and animals observed on the daily excursions. The artists, all avid birders and plant lovers, went on the normal excursions and then spent the evening or travel hours preparing their paintings. Periodically they treated the non-artists to artwork displays during evening social hours convened on their boat.

Individual Giving

When curators receive supplemental income for giving lectures, consulting, guiding ecotours, or from book royalties, etc. they may chose to donate the money to a research fund established at the home institution. For the most part, work done outside of normal working hours, such as giving lectures, is considered part of the personal income of the curator. More substantial work, however, needs approval by the department administrator.

Members of the general public are usually surprised to learn that many curators, and research scientists in other fields, are obliged to raise funds for their own projects (Cohen, 1998). I met a number of individuals (while teaching adult education courses early in my career, on ecotours, after presenting public lectures, or referred to me from third parties) who made donations to support my research at NYBG. These were unsolicited contributions that usually amounted to one-time donations of a few hundred dollars, but a few individuals contributed substantial amounts of money over multiple years. In particular, John D. Mitchell has been a major contributor to most of my projects, especially the studies of plants of central French Guiana and the bat work, and Chris Davidson supported several years of the collection program on the Osa Peninsula, as well as the production of a website based on those collections (Aguilar et al., 2008). The Blue Moon Fund supported the second year of the Osa project.

Additional Institutional Support

The New York Botanical Garden, as mentioned earlier, has been the biggest supporter of my research. In addition to providing me with a place to work and most of my salary, it has made it possible for me to receive a great deal of support that would not have been possible working outside of botanical garden dedicated to research, education, and horticulture.

Several volunteers, recruited through the NYBG volunteer program, have donated their time as research assistants. My most helpful volunteers were John Brown and Edmund Hecklau, both retired physicians. Their volunteer work led to quite a few publications (Brown & Mori, 2002; Fritsch et al., 2002; Hecklau et al., 2005; Lohman et al., 2002; Mori & Brown, 1994, 1998, 2002a, 2002b; Mori et al., 2002b, 2005), and Hecklau is one of the editors of this book. Another volunteer who worked with me for a shorter time was Frederick Blanchard, who designed our bat/plant database and entered a great deal of information into it, participated in a field trip, and was a co-author of a paper on the seed dispersal of *Cecropia*. My wife, Carol Gracie, has been instrumental in nearly all of my projects, participated in most of my fieldwork, and taken most of the color photographs that appear in our publications. Another example is Li Gao, a summer intern who made contributions to the ecosystem services section of the Vascular Plants of the Osa Peninsula project.

Over most of my career I have had support from the staff of the Herbarium of The New York Botanical Garden; for example, for years Eileen Whalen has managed my collections, which included label preparation and distributing duplicates to other herbaria. Eileen is a very careful worker, so when she worked on my collections I had no worries about their quality. Personnel of the IT Department and the information managers of the Herbarium staff have been another source of "free" help. Mellisa Tulig, for example, designed and managed all of my web sites, and has always been willing to help me better understand KE EMu, the database system at NYBG.

While I was the Director of the Institute of Systematic Botany (1995 to 2001), I had the advantage of having two research assistants, Scott Heald (1996–1998) and Nathan P. Smith (1998–2001). Scott helped complete the *Guide to the Vascular Plants of Central French Guiana* (Mori et al., 1997, 2002a) and Nate played the lead role in the production of the Flowering Plants of the Neotropics (Smith et al., 2004). Nate now studies the Lecythidaceae, and I am learning a great deal more about this family because of my interaction with him. Another research assistant, Xavier Cornejo, was financed by grants to study the flora of the Osa Peninsula,

Costa Rica. Xavier made outstanding contributions to this project and is an outstanding botanist with a strong work ethic. Even on his vacation time to his home country of Ecuador, Xavier gathered many collections of Lecythidaceae that were embellished with magnificent images, and supplemented with DNA samples.

My students have been another source of help. Brian Boom (Ph. D. in 1983) contributed to my studies of the ecology of Lecythidaceae in central French Guiana (Boom & Mori, 1982; Mori & collaborators, 1987) and projects in eastern Brazil (Mori et al., 1981; Mori et al., 1983a, 1983b). Four of my students added to my knowledge of neotropical Lecythidaceae— Chih-Hua Tsou (Ph. D. in 1990; Mori et al., 1989, 2007a; Tsou & Mori, 2002, 2007), Amy Berkov (Ph. D. in 1999; Mori et al., 2005; Tavakilian et al., 1997), Ya-Yi Huang (Ph. D. in 2010, Huang, 2010; Huang et al., 2008; unpublished data), and Carolina Potascheff (Potascheff, 2010). My other students have also taught me about tropical systematics and ecology because of my interactions with them on their projects: Pedro Acevedo-Rodriguez (Ph. D. in 1989, Sapindaceae), Cullen Geiselman (Ph.D. in 2010, nectar bats and the plants they utilize for nectar), Vanessa Hecquet (M. S. in 2003, plants pollinated by bats), Maria Lúcia Kawasaki (Ph. D. in 1992, Vochysiaceae), Samuel Kisseadoo (Ph. D. in 1992, forest ecology), Amy Litt (Ph., D. in 1999, Vochysiaceae), and John Pipoly (Ph. D. in 1986, Myrsinaceae). In addition, my two post-docs made considerable contributions to my career. Tatyana Lobova (2000 to 2007), working in collaboration with my student Cullen Geiselman, was instrumental in producing our book on bat seed dispersal in the Neotropics (Lobova et al., 2008) and enabled me to learn a great deal about the plants bats depend on for their survival. The other post-doc, John Janovec (2001), brought me into the computer age, published with me (Janovec et al. 2003), helped teach a course in Amazonian botany, and has been an inspiration to me because of his love for botany and his strong work ethic. Without these interactions with volunteers, research assistants, students, and post-docs, my career would have been less interesting and much less productive.

The study of systematic botany requires a number of tedious but important jobs. Without help, curators would become so preoccupied making labels, entering data and managing databases, filing species, distributing collections, etc. that there would be little time for studying plants, writing papers, or obtaining grants. Thus, I have concluded that the most important factor in curatorial productivity is the ability to find research help.

Miscellaneous Fundraising

I have tried several experimental fund-raising activities, in addition to the more traditional funding strategies discussed above. Botanical art is an important part of systematic botany; botanists rely on it to communicate information about the plants they study and, to a lesser extent, to raise money for research. Bobbi Angell prepared nearly all the botanical line drawings that appear in my publications, and Michael Rothman was commissioned to prepare twenty paintings depicting studies of plant/animal interactions, species of various plant families, and habitat types such as the canopy and understory of lowland rain forest in French Guiana. Most of Rothman's paintings, many of which depict the work of ISB curators, can be seen online at: www.nybg.org/botany/mori/mori/Curator_paintings/paintings_home.htm.

NYBG used Rothman's paintings for invitations to several events; in addition, the NYBG gift shop sold a color scarf with the design based on his painting of an Amazon flooded forest, and *potpourri* with packaging based on his painting of the French Guianan rain forest understory. Rothman's paintings have also been used for the covers of the *Flowering Plants of the Neotropics* (Smith et al., 2004), *Seed Dispersal by Bats in the Neotropics* (Lobova et al., 2009), and for this book. We made several efforts to market products based on Rothman's art; for example, we produced note cards from the Amazon flooded forest painting and sold enough of them to pay for their production, but once we reached that point we gave the rest of them away to NYBG visitors. We had more success in selling a few of the original paintings, or giclée prints of them.

Carol prepared note cards and U.S. postage stamps based on her photographs and those of our Costa Rican collaborator, Reinaldo Aguilar. These products are marketed at www.zazzle.com/cgracie and have been available on the Internet for several years, but there have been few purchasers. For the most part, our attempts to market products or promote our ecotours via Internet have not been successful.

Appendix C
Personal Field Supplies
by Scott A. Mori

A. Sleeping Gear

- Hammock
- Hammock ropes (3 meters long and light but strong)
- Mosquito net (see text for design)
- Clothes pins or binder clips for closing bottom of mosquito net
- Small pillow
- Sleeping bag or sleeping bag liner
- Two small head lamps
- Batteries for head lamps

B. Clothing
(the wetter the area and season, the more clothes needed)

B-1. Field

- Rain gear
- 2(3) field pants
- Long running pants (for sleeping and for evening wear to protect against insects)
- 3 field t-shirts
- 2 camp t-shirts
- 2 long-sleeve shirts
- Light jacket (for cool evenings and for sleeping)
- 3 pairs of underpants
- 1 pair sleeping shorts
- 1 pair bathing shorts
- Hat
- Field boots (Muckboots® or hiking boots depending on the season)
- Extra shoe laces
- 2(3) pairs of socks
- Sandals
- Belt
- 2 gear hammocks to organize clothing in camp

text continues on next page

Appendix C (Personal Field Supplies) continued

B-2. City

- Sandals (same as used in the field)
- 1 pair dress socks for wearing with sandals
- 2 light weight shirts
- 2 pairs of convertible pants
- 1 light jacket (same as used in field)

C. Toilet Gear

- Toothbrush
- Dental floss
- Toothpaste
- Soap (127 gram bar of Ivory lasts two weeks)
- Stringed soap holder
- Scrubber mitt (good for dislodging ticks while bathing)
- Razor and one blade for each three weeks
- Nail clipper or scissors
- Cigarette lighters for burning toilet paper (2)
- Spare eyeglasses
- Sewing kit

D. First aid kit
(needs to be tailored for each participant, this list assumes that the expedition will also have a communal first aid kit)

- Epipens (for bee stings)
- Prednisone (for allergic reactions)
- Thiobendazole (for cutaneous lava migrans)
- Band-Aids
- Antibiotic ointment
- Mosquito repellant
- Sunscreen
- Water purification tablets
- Medicated powder for rashes caused by long hikes in wet weather
- Fine point tweezers

E. Back Packs and Contents

- Large back pack
- Small back pack for daily collecting excursions
- Plastic bags to protect contents from getting wet
- Parachute cord for tying items to packs
- Tupperware for lunch
- 1.5 liter canteens (2)
- Cigarette lighter for starting fire if lost
- Small (5 × 5 m) tarp for shelter against rain while collecting

F. Documents

- Passport and visas
- Copy of birth certificate stored separate from passport
- Immunization record
- Insurance documents

Appendix D

Essential Collecting Equipment

by Scott A. Mori

A. Climbing Gear

- Safety belt
- Two lanyards with cams and carabineers
- Small climbing spikes
- Large climbing spikes

B. Collecting Gear

- Inter-nesting clipper poles (see text for details)
- Two clipper pole heads (one as a spare)
- 12 meter rope for clipper pole (plus spare rope)
- Spare buttons for connecting the poles
- Two pairs of Velcro straps for bundling poles (extra pair in case one is lost)
- Hand clipper and sheath
- Machete and sheath
- Plastic collecting bags for temporarily holding plants before pressing (same as those used for specimen storage, see below)
- File for sharpening machete
- Leatherman® tool
- Field notebook
- Waterproof bag for field notebook
- Field Press
- Newspaper
- Fruit bags
- 4 × 4 m tarp with 2 m long parachute cord in each grommet for rain shelter
- Pencils
- Indelible markers for writing collection numbers on newspaper
- Binoculars
- Hand lens
- Maps of the area
- Compass and GPS (the compass as a backup to the GPS)
- Silica gel for DNA samples
- Small Ziplock® bags for DNA samples and pickled material

text continues on next page

Appendix D (Essential Collecting Equipment) continued

- FAA or 70% ETOH for pickled flowers and fruits
- 70% ETOH if specimens are to be preserved in the field
- Spray bottle for moistening specimens preserved in the field
- Plastic bags for herbarium specimen storage (thick gauge, 92.5 cm long × 47 cm wide)
- Twine for tying specimen bundles
- Mothballs for protecting specimens from insect damage

C. Camera Gear

- Waterproof bags
- Packets of desiccant for camera bag
- Camera
- Two 1-gigabyte CF cards
- Close-up lens or extension tubes
- Flash
- Rechargeable batteries for camera and flash
- Battery chargers
- Ziplock plastic bags to protect camera gear
- Computer to manage images (if practical)
- AC adapter for computer
- Currency converter if necessary
- Plug adapters
- Card reader (and cord if applicable)
- External backup device
- Plant holders to secure specimens during photography (Box 4-2)
- Tripod
- Black velvet cloth for background
- Light blue graph paper for photos of seeds, etc.
- Paper rulers with room to write collector and number on

D. Plant Drying Supplies

- Drying frame (Boxes 4-3, 4-4)
- Aluminum and/or cardboard corrugates (see text)
- Felts (blotters) (see text)
- End boards for plant press
- Straps for plant press
- Heat source (see text)

E. Documents

- Passport
- Permit to collect in host country
- Permit to import plants if required by country into which specimens are imported
- Copy of CITES permit if CITES plants are to be collected

Literature Cited

Academia Brasileira de Ciências. 2009 accessed. Paulo de Tarso Alvim. http://www.abc. org.br/~alvimpt.

Aguiar, A. M. & M. C. Gaglianone. 2008. Comportamento de abelhas visitantes florais de *Lecythis lurida* (Lecythidaceae) no norte do estado do Rio de Janeiro. Revista Brasileira de Entomologia 57: 277–282.

Aguilar, R., X. Cornejo, C. Bainbridge, M. Tulig & S. A. Mori. 2008 onward. Flowering Plants of the Osa Peninsula, Costa Rica (http:sweetgum.nybg.org/osa/). The New York Botanical Garden, Bronx, New York.

Alexiades, M. N (ed.). 1996. Selected guidelines for ethnobotanical research: a field manual. Advances in Economic Botany 10: 1–306.

Anderson, R. C. & S. A. Mori. 1967. A preliminary investigation of *Raphia* swamps, Puerto Viejo, Costa Rica. Turrialba 17(2): 221–224.

Anderson, W. R. 1996. The importance of duplicate specimens in herbaria. Pages 239–248 *in* T. F. Steussy & S. H. Sohmer (eds.), Sampling the green world. Columbia University Press, New York.

Aragão, L. E. O. C., Y. Jalhi, N. Barbier, A. Lim, Y. Shimabukuro, L. Anderson & S. Saatchi. 2008. Interactions between rainfall, deforestation and fires during recent years in the Brazilian Amazonia. Philosophical Transactions of the Royal Society of London. Series B 363: 1779–1785.

Asner, G. P., M. Keller & J. N. M. Silva. 2004. Spatial and temporal dynamics of forest canopy gaps following selective logging in the eastern Amazon. Global Change Biology 10: 765–783.

Aublet, F. 1775. Histoire des Plantes de la Guiane Françoise, 4 vols. Paris: Didot.

Balée, W. 1994. Footprints of the Forest. Columbia University Press, New York. 396 pp.

Balick, M. 1985. Useful plants of Amazonia: a resource of global importance. Pages 339–368 *in* G. T. Prance & T. E. Lovejoy (eds.), Amazonia. Pergamon Press.

——— & P. A. Cox. 1996. Plants, people, and culture. The science of ethnobotany. Scientific American Library. 228 pp.

Barlow, J., L. A. M. Mestre, T. A. Gardner & C. A. Peres. 2008a. The value of primary, secondary and plantation forests for Amazonian birds. Biological Conservation 136: 212–231.

Barlow, J., W. L. Overal, I. S. Araujo, T. A. Gardner & C. A. Peres. 2008b. The value of primary, secondary and plantation forests for fruit-feeding butterflies in the Brazilian Amazon. Journal of Applied Ecology 44: 1001–1012.

Barret, J. 2001. Atlas Illustré de la Guyane. Laboratoire de Cartographie de la Guyane & Institut d'Enseignement Supérieur de Guyane. 215 pp.

Bassini, F. & P. Becker. 1990. Charcoal's occurrence in soil depends on topography in terra firme forest near Manaus, Brazil. Biotropica 22: 420–422.

Belbenoit, P., O. Poncy, D. Sabatier, M.-F. Prévost, B. Riera, P. Blanc, D. Larpin & C. Sarthou. 2001. Floristic checklist of the Nouragues area. Pages 301–341 *in* Bongers F., P. Charles-Dominique, P.-M. Forget & M. Théry (eds.), Nouragues: Dynamics and Plant-animal Interactions in a Neotropical Rainforest. Kluwer Academic Publisher, Dordrecht, The Netherlands, Biological Monographs series.

Berkov, A. 2002. The impact of redefined species limits in *Palame* (Coleoptera, Cerambycidae, Lamiinae, Acanthocinini) on assessments of host, seasonal, and stratum specificity. Biological Journal of the Linnean Society 76: 195–209.

———, J. Feinstein, P. Centeno, J. Small & M. Nkamany. 2007. Yeasts isolated from Neotropical wood-boring beetles in SE Peru. Biotropica 39: 530–538.

———, B. Meurer-Grimes & K. Purzycki K. 2000. Do Lecythidaceae specialists (Coleoptera: Cerambycidae) shun fetid tree species? Biotropica 32(3): 440–551.

——— & M. A. Monné. 2010. A new species of *Neobaryssinus* Monné & Martins, and two new species of *Baryssiniella* new genus (Coleoptera: Cerambycidae), reared from trees in the Brazil nut family (Lecythidaceae). Zootaxa 2538: 47–59.

——— & G. Tavakilian. 1999. Host utilization of the Brazil nut family (Lecythidaceae) by sympatric wood-boring species of Palame (Coleoptera, Cerambycidae: Lamiinae, Acanthocinini). Biological Journal of the Linnaean Society 67: 181–198.

Betts, R. A., Y. Malhi & J. Timmons Roberts. 2008. The future of the Amazon: new perspectives from climate, ecosystem and social sciences. Philosophical Transactions of the Royal Society of London. Series B 363: 1729–1733.

Bierregaard, R. O. Jr., C. Gascon, T. E. Lovejoy & R. C. G. Mesquita (eds.). 2001a. Lessons from Amazonia. The ecology and conservation of a fragmented forest. Yale University Press, New Haven & London. 544 pp.

———, W. F. Laurance, C. Gascon, J. Benitez-Malvido, P. F. Fearnside, C. R. Fonseca, G. Ganade, J. R. Malcom, M. B. Martins, S. A. Mori, M. Oliveira, J. Rankin-de Mérona, A. Scariot, W. Spironello & B. Williamson. 2001b. Principles of forest fragmentation and conservation in the Amazon. Pages 371–385 *in* R. O. Bierregaard, Jr., C. Gascon, T. E. Lovejoy & R. C. G. Mesquita (eds.), Lessons from Amazonia. The ecology and conservation of a fragmented forest. Yale University Press, New Haven & London.

Biodiversity Heritage Library. 2010 accessed. Biodiversity Heritage Library. http://www.biodiversitylibrary.org/.

Biodiversity Information Facility (TDWG). 2010 accessed. DarwinCore Group (DwC). http://www.tdwg.org/.

Biodiversity Research Center (University of Kansas). 2009 accessed. Specify 6. http://specifysoftware.org/.

Bongers F., P. Charles-Dominique, P.-M. Forget & M. Théry (eds.). 2001. Nouragues: Dynamics and Plant-animal Interactions in a Neotropical Rainforest. Kluwer Academic Publisher, Dordrecht, The Netherlands, Biological Monographs series. 421 pp.

Bonpland, A. 1807. *Bertholletia*. Pages 122–127, pl. 36 *in* A. von Humboldt & A. Bonpland, Plantae Aequinoxiales. Lutetiae Parisiorum, Paris.

Boom, B. M. 1996. Societal and scientific information needs from plant collectors. Pages 16–27 *in* T. F. Steussy & S. H. Sohmer (eds.), Sampling the Green World. Columbia

University Press, New York.

———— & S. A. Mori. 1982. Falsification of two hypotheses on liana exculsion from tropical trees possessing buttresses and smooth bark. Bulletin of the Torrey Botanical Club 109: 447–450.

Bortolus, A. 2008. Error cascades in the biological sciences: the unwanted consequences of using bad taxonomy in ecology. Ambio 37(2): 114–118.

Botanical Research Institute of Texas. 2009 accessed. Atrium. Biodiversity Information System. http://www.atrium-biodiversity.org/.

Bramwell, D. 2002. How many plant species are there? Plant Talk 28: 32–34.

Bridson, D. & L. Forman (eds.). 1992. The Herbarium Handbook. Royal Botanic Gardens, Kew. 303 pp.

Brown, D. 2000. Chronicle of the Guayaki Indians (book review). Ethnobotany 47(3–4): 818–821.

Brown, J. L. & S. A. Mori. 2002. Canellaceae. *In* S. A. Mori, G. Cremers, C. Gracie, J.-J. de Granville, S. V. Heald, M. Hoff & J. D. Mitchell. 2002. Guide to the Vascular Plants of Central French Guiana. Part 2. Dicotyledons. Memoirs of The New York Botanical Garden 76(2): 184–186.

Brown, K. S. 1979. Ecologia, Geografia e Evolução na Florestas Neotropicais. Tese de Livre Docéncia. Universidade Estadual de Campinas, São Paulo, Brazil. 485 pp.

Brummit, R. K. & C. E. Powell. 1992. Authors of Plant Names. Royal Botanic Gardens, Kew. 732 pp.

Buchmann, S. L. 1983. Buzz pollination in angiosperms. Pages 73–113 *in* C. E. Jones & R. J. Little (eds.), Handbook of Experimental Pollination Biology, S&AE Scientific and Academic Editions, New York.

Buck, W.R. 2003. Guide to the Plants of Central French Guiana. Part 3. Mosses. Memoirs of The New York Botanical Garden 76(3): 1–167.

———— & B. M. Thiers. 1996. Guidelines for collecting bryophytes. Pages 143–146 *in* M. N. Alexiades (ed.), Selected Guidelines for Ethnobotanical Research: A Field Manual. The New York Botanical Garden.

Burger, W. C. 2006. Flowers: how they changed the world. Prometheus Books, Amherst, New York. 337 pp.

Butler, D. 2006. Murders halt rainforest research. Nature 44: 555.

Campbell, D. G., D. C. Daly, G. T. Prance, & U. N. Maciel. 1986. Quantitative ecological inventory of terra firme and várzea forest on the Rio Xingu, Brazilian Amazon. Brittonia 38(4): 369–393.

Cannon, P. F. & B. C. Sutton. 2004. Microfungi on wood and plant debris. Chapter 11, pages 217–239 *in* Mueller, G. M., G. Bills, and M. S. Foster (eds), Biodiversity of Fungi. Inventory and monitoring methods. Elsevier Academic Press, San Diego, CA. 777 pp.

Carrasquilla R., L. G. 2005. Árboles y arbustos de Panamá. Trees and Shrubs of Panama. Universidad de Panamá and the Autoridad Nacional del Ambiente. 478 pp.

Carrillo, E., G. Wong & A. D. Cuarón. 2000. Monitoring mammal populations in Costa Rican protected areas under different hunting restrictions. Conservation Biology 14(6): 1580–1591.

Castilho, C. V., W. Magnusson, R. N. O. Araújo & E. C. da Pereira. 2006. The use of French spikes to collect botanical vouchers in permanent plots: evaluation of potential impacts. Biotropica 38(4): 555–557.

Charles-Dominique, P. Brosset, A. & Jouard, S. 2001. Les Chauves Souris de Guyane. Museum d'Histoire Naturelle, Paris, France. 172 pp.

Chave, J. 2006 onward. Bridge Project. http://ecofog.cirad.fr/bridge/index.html. Accessed 2009.

Chazdon, R. L., S. G. Letcher, M. van Breugel, M. Martínez-Ramos, F. Bongerx & B. Finegan. 2007. Rates of change in tree communities of secondary Neotropical forests following major disturbances. Philosophical Transactions of the Royal Society of London. Series B. 362: 273–289.

Chippaux, J. P., L. Sanite & D. Heuclin. 1988. Serpents de Guyane. Nature Guyanaise. Sepanguy – Cayenne, French Guiana. 55 pp.

Chivian, E. & A. Bernstein. 2008. Sustaining Life. How Human Health Depends on Biodiversity. Oxford University Press, New York. 542 pp.

Claborn, D. M. The biology and control of leishmaniasis vectors. Journal of Global Infectious Disease. 2010 onward. http://www.jgid.org/text.asp?2010/2/2/127/62866.

Clastres, P. 1998. Chronicle of the Guayaki Indians. Translated and with a foreword by P. Auster. Zone Books, New York. 349 pp.

Clemants, S. & C. Gracie. 2006. Wildflowers in the Field and Forest. A Field Guide to the Northeastern United States. Oxford University Press. 445 pp.

Clement, C. R. 2008. À espera de mercados. Pages 37–43 *in* Amazônia. A Floresta e o Futuro. Scientific American Brasil, Suplementos.

CNRS (Centre National de la Recherche Scientifique). 2009 accessed. Nouragues. Gateway to European Rainforest. http://www.nouragues.cnrs.fr/indexenglish.html.

Cohen, J. 1998. Scientists who fund themselves. Science 279: 178–181.

Cohen, N. 2009. Wikipedia to limit changes to articles on people. The New York Times. 24 August 2009.

Collins, J. O. 2009 accessed. Hotel Tivoli. Ancon, Panama, Canal Zone. http://panamaliving.com/tivoli.html.

Colwell, R. K. 1985. A bite to remember: one strike and you might be out, if you accidently meet a viper in a rain forest. Natural History 94(4): 2, 4, 6, 8.

Committee on Research Priorities in Tropical Biology. 1980. Research Priorities in Tropical Biology. The National Research Council, National Academy of Sciences, Washington, DC. 116 pp.

Costanza, R., R. d'Arge, R. de Groot, S. Farber, M. Grasso, B. Hannon, K. Limburg, S. Naeem, R. V. O'Neill, J. Paruelo, R. G. Raskin, P. Sutton & M. van den Belt. 1997. The value of the world's ecosystem services and natural capital. Nature 387: 253–259.

Cottam, G. & J. T. Curtis. 1962. Plant Ecology Workbook. Burgess Publishing Company, Minneapolis, Minnesota. 193 pp.

Colwell, R. K. 1985. A bite to remember. Natural History. April, pp. 2, 4, 6, 8.

Croat, T. B. 1971. Summit Garden, Canal Zone: Its role in botanical research. Taxon 20: 769–772.

———. 1978. Flora of Barro Colorado Island. Stanford University Press. Stanford, California. 956 pp.

——— & S. A. Mori. 1974. A new *Gustavia* from Panama and Colombia. Brittonia 26: 22–26.

Cunha, O. Rodrigues da. 1989. Walter Alberto Egler. Pages 150–158 *in* Talento e Atitude: Estudos Biográficos do Museu Emílio Goeldi, I. Museu Paraense Emílio Goeldi. Coleção Alexandre Rodrigues Ferreira. Belém, Pará, Brazil.

Dallwitz, M. J., T. A. Paine & E. J. Zurcher. 1999. User's guide to the DELTA editor. http://delta-intkey.com/. Accessed 2009.

Davis, T. A. W. & P. W. Richards. 1933. The vegetation of Moraballi Creek, British Guiana: An ecological study of a limited area of tropical rain forest. Part I. Journal of Ecology 21: 350–384.

——— W. & P. W. Richards. 1934. The vegetation of Moraballi Creek, British Guiana: An ecological study of a limited area of tropical rain forest. Part II. Journal of Ecology 22:

106–155.

Denslow, J. 1980. Gap partitioning among tropical rain forest trees. Biotropica 12(suppl.): 47–55.

———. 1987. Tropical rain forest gaps and tree species diversity. Annual Review of Ecology and Systematics 18: 431–451.

Department of Plant Sciences (University of Oxford). 2009 accessed. Brahms. Botanical Research and Herbarium Management System. http://dps.plants.ox.ac.uk/bol/BRAHMS/Home/Contact/.

Dirzo, R. & P. H. Raven. 2003. Global state of biodiversity and loss. Annual Review of Environmental Resources 28: 137–167.

Dmitriev, D. A. 2003 onward. 31 Interactive Keys and Taxonomic Databases. http://ctap.inhs.uiuc.edu/dmitriev/index.asp. 2009 accessed.

Dransfield, J. 1986. A guide to collecting palms. Annals of the Missouri Botanical Garden 73: 166–176.

Dressler, R. L. 1982. Biology of orchid bees (Euglossini). Annual Review of Ecology and Systematics 13: 373–394.

Dubernat, P.-F. 1982. Les Solitaires Crèvent Seuls. Editions de la Compté, Cayenne, French Guiana. 322 pp.

Dugand, A. 1947. Observaciones taxonomicas sobre las *Lecythis* del norte de Colombia. Caldasia 4:411–426.

Dwyer, J. D. 1965. Notes on the Lecythidaceae of Panama. Annals of the Missouri Botanical Garden 56(2): 351–363.

———. 1967. A new herbarium in the Canal Zone. Taxon 16: 158–159.

———. 1980. The history of plant collecting in Panama 1700–1981. Annals of the Missouri Botanical Garden 67 (4): ix–xv.

Ek, R. C. 1991. Index of Suriname plant collectors. Pages 1–9 *in* A.R.A. Görts van Rijn (ed.), Flora of the Guianas Supplementary Series, Fascicle 2. FRG, Koelz Scientific Books.

Elias, T. S. 1980. The Complete Trees of North America. Outdoor Life/Nature Books, Van Nostrand Reinhold Company, New York. 948 pp.

Ellis, B., D. C. Daly, L. J. Hickey, K. R. Johnson, J. D. Mitchell, P. Wilf & S. L. Wing. 2009. Manual of Leaf Architecture. Comstock Publishing Associates, a division of Cornell University Press. 190 pp.

Emmons, L. H. 1989. Tropical rain forests: why they have so many species, and how we may lose this biodiversity without cutting a single tree. Orion 3: 8–14.

Emmons, L. H. 1990. Neotropical rain forest mammals. A field guide. The University of Chicago Press, Chicago and London. 281 pp.

Encyclopedia of Life. 2009 accessed. Encylopedia of Life (EOL). http://www.eol.org.

Enserink, M. 2010. Malaria's drug miracle in danger. Science 328: 844–846.

Erlich, P. 1968. The Population Bomb. E. P. Dutton & Co. 201 pp.

Erwin, T. R. 1982. Tropical forests: their richness in Coleoptera and other arthropod species. The Coleopterists Bulletin 36: 74–75.

———. 1997. Biodiversity at its utmost: tropical forest beetles. Pages 27–40 *in* M. L. Reaka-Cudla, D. E. Wilson & E. O. Wilson. Biodiversity II, Joseph Henry Press, Washington, D.C.

ETI Bioinformatics. 2004. Linnaeus II. http://www.eti.uva.nl/Products/linnaeus/info.php. Accessed 2009.

Ewel, J. J. 1986. Designing agricultural ecosystems for the humid tropics. Annual Review of Ecology and Systematics 17: 245–271.

Farr, D. E. 2006. Online keys: more than just paper on the web. Taxon 55 (3): 589–596.

Feinstein, J., S. A. Mori & A.Berkov. 2007. Saproflorivory: a diverse insect community in fallen flowers of Lecythidaceae in French Guiana. Biotropica 39(4): 549–554.

————, K. L. Purzycki, S. Mori, V. Hequet & A.Berkov. 2008. Neotropical soldier flies (Stratiomyidae) reared from *Lecythis poiteaui* in French Guiana: Do bat-pollinated flowers attract saprophiles? Journal of the Torrey Botanical Society 135: 200–207.

Ferri, M. G. & N. L. de Menezes. 1990. Glossário Ilustrado de Botânica. Nobel, São Paulo. 197 pp.

Fidalgo, O. & V. L. Ramos Bononi. 1989. Técnicas de coleta, preservação e herborização de material botânico. Série Documentos, da Secretaria do Meio Ambiente (Estado de São Paulo, Brazil). 62 pp.

Foley, J, A., G. P. Asner, M. H. Costa, M. T. Coe, R. DeFries, H. K. Gibbs, E. A. Howard, S. Olson, J. Patz, N. Ramankutty & P. Snyder. 2007. Amazonia revealed: forest degradation and loss of ecosystem goods and services in the Amazon Basin. Frontiers in Ecology and the Environment 5(1): 25–32.

Font Quer, P. 1985. Diccionario de Botánica. Editorial Labor, S. A. Barcelona etc. 1244 pp.

Forero, E. & S. Mori. 1995. The Organization for Flora Neotropica. Brittonia 47(4): 379–293.

Forsyth, A. and K. Miyata. 1984. Tropical Nature. Charles Scribner's Sons, New York. 249 pp.

Fosberg, F. R.& M.-H. Sachet. 1965. Manual for tropical herbaria. International Bureau for Plant Taxonomy and Nomenclature. Utrecht, Netherlands. 132 pp.

Friedman, T. L. 2008. Hot, Flat, and Crowded. Farrar, Straus, and Giroux, New York. 438 pp.

Friends of the Osa. 2010 accessed. Friends of the Osa. http://www.osaconservation.org/.

Fritsch, P. W., J. L. Brown & S. A. Mori. 2002. Styracaceae. *In* S. A. Mori, G. Cremers, C. Gracie, J.-J. de Granville, S. V. Heald, M. Hoff & J. D. Mitchell. 2002. Guide to the Vascular Plants of Central French Guiana. Part 2. Dicotyledons. Memoirs of The New York Botanical Garden 76(2): 706–708.

Frodin, D. G. 2001. Guide to the Standard Floras of the World (ed. 2). Cambridge University Press, Cambridge, U. K. 1100 pp.

Galindo-Leal, C. & I. de Gusmão Câmara. 2003. Atlantic forest hotspot status: An overview. Pages 3–11 *in* C. Galindo-Leal & I. de Gusmão Câmara (eds.), The Atlantic Forest of South America: Biodiversity Status, Threats, and Outlook. Island Press, Washington, D.C.

Gamboa-Gaitán, M. A. 1997. Biología reproductiva de *Eschweilera bogotensis* (Lecythidaceae), en la Cordillera Occidental de Colombia. Caldasia 19(3): 479–485.

Geiselman, C. K, S. A. Mori & F. Blanchard. 2002 onward. Database of Neotropical bat/plant interactions. The Virtual Herbarium of The New York Botanical Garden (http·// www.nybg.org/botany/dubuva/mori/batsplants/database/dbase_frameset.htm). Accessed 2009.

Gentry, A. 1975a. Dr. Gentry returns from Latin America, reports on Cerro Tacarcuna expedition. Missouri Botanical Garden Bulletin 43(13): 6–7.

————. 1975b. Dr. Gentry continues report on Cerro Tacarcuna expedition. Missouri Botanical Garden Bulletin 43(14): 6.

————. 1975c. End of the expedition: Dr. Gentry evaluates the exploration of Cerro Tacarcuna expedition. Missouri Botanical Garden Bulletin 43(15): 6–7.

————. 1977. Botanical exploration of Cerro Tacarcuna. Explorers Journal 55(1): 40–45.

————. 1982. Neotropical floristic diversity: phytogeographical connections between Central and South America, Pleistocene climatic fluctuations or an accident of the Andean orogeny. Annals of the Missouri Botanical Garden 69: 557–593.

————. 1988. Tree species richness of upper Amazonian forests. Proceedings of the National Academy of Sciences of the United States of America 85: 156–159.

——— (ed.). 1990. Four Neotropical Rainforests. Yale University Press, New Haven, CT. 627 pp.

Ghazoul, J. & D. Sheil. 2010. Tropical Rain Forest Ecology, Diversity, and Conservation. Oxford University Press.

Gill, G. E., Jr., R. T. Fowler & S. A. Mori. 1998. Pollination biology of Symphonia globulifera (Clusiaceae) in central French Guiana. Biotropica 30(1): 139–144.

Gleason, H. A. & A. Cronquist. 1991. Manual of Vascular Plants of Northeastern United States and Adjacent Canada. The New York Botanical Garden, Bronx, New York. 910 pp.

Global Biodiversity Information Facility (GBIF). 2010 accessed. GBIF Data Portal. http://data.gbif.org/welcome.htm.

Goldblatt, P., P. C. Hoch, & L. M. McCook. 1992. Documenting scientific data: the need for voucher specimens. Annals of the Missouri Botanical Garden. 79: 969–970.

Golley, F. B., J. T. McGinnis, R. G. Clements, G. I. Child & M. J.Duever. 1975. Mineral Cycling in a Tropical Moist Forest Ecosystem. University of Georgia Press, Athens. 248 pp.

Gore, A. 2006a. An Inconvenient Truth: The Planetary Emergency of Global Warming and What We can Do About It. Rodale, Inc. New York. 327 pp.

———. 2006b. Earth in the Balance: Ecology and the Human Spirit. Rodale, Inc., New York. 408 pp.

Goulding, M., N. J. H. Smith & D. J. Mahar. 1996. Floods of fortune. Ecology & economy along the Amazon. Columbia University Press, New York. 193 pp.

Govaerts, R. 2001. How many species of seed plants are there? Taxon 50:1085–1090.

Gradstein, S. R. & A. L. Ilkiu-Borges. 2009.Guide to the Plants of Central French Guiana. Part 4. Liverworts and Hornworts. Memoirs of The New York Botanical Garden. 76(4): 1–144.

Grann, D. 2010 (3rd edition). The Lost City of Z. A Tale of Deadly Obsession in the Amazon. Vintage Departures, Vintage Books. A Division of Random House, Inc., New York. 400 pp.

Granville, J.-J. de. 1975. Projets de reserves botaniques et forestières en Guyane. Office de la Recherche Scientifique et Technique Outre-Mer. Centre ORSTOM de Cayenne.

Granville, J.-J. de. 1992. Un cas de distribution particulier: les espèces forestières peri-amazoniennes. Compte Rendus de la Societé de Biogéographie 68: 1–33.

——— & Sastre, C. 1991. Remarks on the montane flora and vegetation types in the Guianas. Willdenowia 2: 201–213.

Grayum, M., B. E. Hammel & N. Zamora. 2004. El ambiente físico/The physical environment. Manual de Plantas de Costa Rica 1: 51–90. Missouri Botanical Garden/INBio/Museo Nacional de Costa Rica.

Greenhall, A. M. 1965. Sapucaia nut dispersal by greater spear-nosed bats in Trinidad. Caribbean Journal of Science 5: 167–171.

Grenand, P. C. Moretti, H. Jacquemin & M.-F. Prevost. 2004. Pharmacopées Traditionnelles en Guyane. IRD Éditions, Institut de Recherche pour le Développement, Paris. 816 pp.

Guala, G. F. 2006. SLIKS: Stinger's Lightweight Interactive Key Software. Computer program distributed by the G. F. Guala. http://www.stingersplace.com/SLIKS/. Accessed 2009. Guariguata, M. R., R. L. Chazdon, J. S. Denslow, J. M. Dupuy & L. Anderson. 1997. Structure and florsitics of secondary and old-growth forest stands in lowland Costa Rica. Plant Ecology 132: 107–120.

Guariguata, M. R., R. L. Chazdon, J. S. Denslow, J. M. Dupuy & L. Anderson. 1997. Structure and floristics of secondary and old-growth forest stands in Costa Rica. Plant Ecology 132(1): 107–120.

Gusson, E., A. M. Sebben & P. Y. Kageyama. 2006. Sistema de reprodução em populações de *Eschweilera ovata* (Cambess.) Miers. Revista Arvore 30: 491–502.

Haemig, P. D. 2002 onward. Sympatric white-lipped peccary and collared peccary. ECOL-OGY. INFO #10 (http://www.ecology.info/ecology-peccaries.htm). Accessed 2009.

Haffer, J. 1974. Avian Speciation in Tropical South America. Publications of the Nuttall Ornithological Club 14: 1–390.

Hagedorn, G. 2002 onward. DeltaAccess. http://www.diversityworkbench.net/OldModels/Descriptions/index.html. Accessed 2009.

Halling, R. 1996. Recommendations for collecting mushrooms. Pages 135–141 *in* M. N. Alexiades (ed.), Selected Guidelines for Ethnobotanical Research: A Field Manual. The New York Botanical Garden. 306 pp.

Hamilton, M. 1998. Successful seed dispersal measured with chloroplast DNA polymorphism is highly localized in a Brazilian canopy tree, *Corythophora alta* (Lecythidaceae). Biotropica 30: 10–11.

Hammel, B. E., M. H. Grayum, C. Herrera & N. Zamora (eds.). 2004. Introducción. Manual de Plantas de Costa Rica, Vol. 1: 1–299. Missouri Botanical Garden, St. Louis.

——— & N. A. Zamora. 2005. *Pleodendron costaricense* (Canellaceae), a new species for Costa Rica. Lankesteriana 5(3): 211–218.

Harris, J. G. & M. W. Harris. 1994. Plant Identification Terminology. An Illustrated Glossary. Spring Lake Publishing, Payson, UT. 188 pp.

Harris, P. P., C. Huntingford & P. M. Cox. 2008. Amazon Basin climate under global warmings: the role of sea surface temperature. Philosophical Transactions of the. Royal Society of London. Series B 363: 1753–1759.

Hart, R. D. 1980. A natural ecosystem analog approach to the design of a successional crop system for tropical forest environments. Biotropica 12(supplement): 73–82.

Hartshorn, G. 1980. Tropical succession: manifold routes to maturity. Biotropica 12(supplement): 23–30.

———. 1989. Gap-phase dynamics and tropical tree species richness. Pages 66–73 *in* L. B. Holm-Nielson, I. C. Neilsen & H. Balslev (eds.), Tropical forests. Academic Press, New York.

Harvard University Herbaria. 2009 accessed. SPNHC. Society for the Preservation of Natural History Collections. http://140.247.98.87/.

Hecht, S. & A. Cockburn. 1989. Fate of the Forest. Routledge, Chapman & Hall, New York. 266 pp.

Heckenberger, M. J., J. Christian Russel, C. Fausto, J. R. Toney, M. J. Schmidt, E. Pereira, B. Franchetto & A. Kuikuro. 2008. Pre–Columbian urbanization, anthropogenic landscapes and the future of the Amazon. Science 321(5893): 1214–1217.

Hecklau, E. F., S. A. Mori & J. L. Brown. 2005. Specific epithets of the flowering plants of central French Guiana. Brittonia 57(1): 68–87.

Heinrich, B. 2007. The Snoring Bird. Harper Collins Publishers, New York. 461 pp.

Herrera, W. 1985. Clima de Costa Rica. *In* L. D. Gómez (ed.). Vegetación y clima de Costa Rica. Vol. 2. UNED, San José, CR. 118 pp.

Herwaldt, B. L., S. L. Stokes & D. D. Juranek. 1993. American cutaneous leishmaniasis in U.S. travelers. Annals of Internal Medicine 118(10): 779–784.

Hickey, M. & C. King. 2000. The Cambridge Illustrated Glossary of Botanical Terms. Cambridge University Press. 208 pp.

Hiepko, P. 1987. The collections of the Botanical Museum Berlin–Dahlem (B) and their history. Englera 7: 219–252.

Hilty, S. L. & W. L. Brown. 1986. Birds of Colombia. Princeton University Press. 836 pp.

Hogue, C. L. 1983. *Eutrombicula* (coloradillas, chiggers). Pages 723–724 *in* D. H. Janzen (ed.), Costa Rican natural history. The University of Chicago Press, Chicago and London.

———. 1993. Latin American insects and entomology. University of California Press, Berkeley, Los Angeles, and Oxford. 535 pp.

Holdridge, L. R. 1967. Life Zone Ecology. Tropical Science Center, San José, Costa Rica. 206 pp.

Howard, A. R. 1983. The plates of Aublet's Histoire des Plantes de la Guiane Françoise. Journal of the Arnold Arboretum 64: 255–292.

Howell, D.J. & D. Burch. 1974. Food habits of some Costa Rican bats. Revista de Biología Tropical 21: 281–294.

Huang, Y.-Y. 2010. Systematics of Lecythidoideae (Lecythidaceae): With an Emphasis on *Bertholletia, Corythophora, Eschweilera,* and *Lecythis*. Ph.D. Program in Biology, Lehman College, City University of New York. 70 pp.

———, S. A. Mori & G. T. Prance. 2008. A phylogeny of *Cariniana* (Lecythidaceae) based on morphological and anatomical data. Brittonia 60(1): 69–81.

Hurtley, S., C. Ash & L. Roberts. 2010. Landscapes of infection. Science 328: 841.

Iltis, H. H. 1988. Serendipity in the exploration of biodiversity. What good are weedy tomatoes? Pages 98–105 *in* E. O. Wilson & F. M. Peter (eds.), Biodiversity. National Academy Press, Washington, D.C.

———. 2002. The impossible race. Population growth and the fallacies of agricultural hope. Pages 35–39 *in* A. Kimbrell (ed.), Fatal Harvest. The Tragedy of Industrial Agriculture. Island Press, Washington, DC.

IRD. 2009 accessed. Herbier de Guyane (CAY). Institut de Recherche pour le Développment. http://www.cayenne.ird.fr/aublet2/.

Iwokrama International Centre for Rain forest Conservation and Development. 2008. Iwokrama. The Green Heart of Guayan. http://www.iwokrama.org/home.htm. Accessed 10 Oct 2008.

Jackson, B. D. 1965. A Glossary of Botanic Terms. Hafner Publishing Company, New York. 481 pp.

Jackson, G. C. & J. B. Salas. 1965. Insect visitors of *Lecythis elliptica* H. B. K. Journal of Agriculture of the University of Puerto Rico 49: 133–140.

Janovec, J. P., L. G. Clark & S. A. Mori. 2003. Is the Neotropical flora ready for the Phylo-Code. Botanical Review (Lancaster) 69(1): 22–43.

Janzen, D. H. 1971. Seed predation by animals. Annual Review of Ecology and Systematics 2: 465–492.

———. 1990. An abandoned field is not a tree fall gap. Vida Silvestre Neotropical 2(2): 64–67.

Jørgensen, P. M., J. E. Lawesson & L. B. Holm-Nielsen. 1984. A guide to collecting passionflowers. Annals of the Missouri Botanical Garden 71: 1172–1174.

Katz, M., D. D. Despommier & R. Gwadz. 1982. Parasitic diseases. Springer-Verlag, New York, Heidelberg, Berlin. 264 pp.

KE Software. 2009 accessed. KE EMu. The World's Number One Solution in Collections Management. http://www.kesoftware.com/component/option,com_frontpage/Itemid,1/lang,en/.

Kiger, R. W. & D. M. Porter. 2001. Categorical Glossary for the Flora of North America Project. Hunt Institute for Botanical Documentation Carnegie Mellon University, Pittsburgh.

Kincaid, D. T., P. J. Anderson & S. A. Mori. 1998. Leaf variation in a tree of *Pourouma tomentosa* (Cecropiaceae) in French Guiana. Brittonia 50(3): 324–338.

Kinzer, S. 2002. Museum's goal: save the world's wild places. The New York Times. 5 November 2002, pages E1, E6).

Kinzey, W. G. 1981. Distribution of primates and forest refuges. Pages 455–482 *in* G. T. Prance (ed.), Biological Diversification in the Tropics. Columbia University Press, New York.

Knudsen, J. & S. A. Mori. 1996. Floral scents and pollination in neotropical Lecythidaceae. Biotropica 28(1): 42–60.

Kricher, John 1997. A Neotropical Companion: A Guide to the Animals, Plants, and Ecosystems of the New World Tropics. Second Edition, revised and expanded. Princeton University Press, Princeton, NJ. 436 pp.

Lafaix, P. 2003. La Loi de la Jungle. Chronique d'une Zone de Non-droit: la Guyane Française. http://www.dailymotion.com/video/xg051_doc-la-loi-de-la-jungle.

Langmead, C. 1995. A Passion for Plants: From the Rainforest of Brazil to Kew Gardens. A Lion Book. 201 pp.

Laurance, W. F. 2008. Adopt a forest. Bitotropica 40(1): 3–6.

Lawrence, A. & W. Hawthorne. 2006. Plant Identification. Creating User-friendly Field Guides for Biodiversity Management. Earthscan, London & Sterling, VA. 268 pp.

Le Parc amazonien de Guyane-Parc national. 2010 accessed. Le Parc de la forêt. http://www.parc-guyane.gf/site.php?id=1.

Leigh, E. G., Jr., A. S. Rand & D. M. Windsor. 1982. The Ecology of a Tropical Forest: Seasonal Rhythms and Long-term Changes. Smithsonian Institution Press, Washington, DC. 468 pp.

———., Jr., A. S. Rand & D. M. Windsor. 1996. The Ecology of a Tropical Forest: Seasonal Rhythms and Long-term Changes, 2nd ed. Smithsonian Institution Press, Washington, DC. 503 pp.

Leite, E. J. 2007. State-of-knowledge on *Cariniana estrellensis* (Raddi) Kuntze (Lecythidaceae) for genetic conservation in Brazil. Research Journal of Botany 2(3): 138–160.

Lepsch da Cunha, N. M., P. Y. Kageyama & R. Vencovsky. 1999. Genetic diversity of *Couratari multiflora* and *Couratari guianensis* (Lecythidaceae): Consequences of two types of rarity in central Amazonia. Biodiversity and Conservation 8: 1205–1218.

Lepsch da Cunha, N. M. & S. A. Mori. 1999. Reproductive phenology and mating potential in a low density tree population of *Couratari multiflora* (Lecythidaceae) in central Amazonia. Journal of Tropical Ecology 15: 97–121.

Lindeman, J. C. & S. A. Mori. 1989. The Guianas. Pages 375–390 *in* Campbell, D. G. & H. D. Hammond (eds.), Floristic Inventory of Tropical Countries. The New York Botanical Garden.

Linnaeus, C. 1775. Plantae Surinamensis 17. Uppsala.

Lobova, T. L., C. K, Geiselman & S. A. Mori. 2009. Seed Dispersal by Bats in the Neotropics Memoirs of The New York Botanical Garden. 101: 1–470.

Lobova, T. A., S.A. Mori, F. Blanchard, H. Peckham & P. Charles-Dominique. 2003. *Cecropia* as a food resource for bats in French Guiana and the significance of fruit structure in seed dispersal and longevity. American Journal of Botany 90(3): 388–403.

Lohman, L. G., J. L. Brown & S. A. Mori. 2002. Bignoniaceae. *In* S. A. Mori, G. Cremers, C. Gracie, J.-J. de Granville, S. V. Heald, M. Hoff & J. D. Mitchell. 2002. Guide to the vascular plants of central French Guiana. Part 2. Dicotyledons. Memoirs of The New York Botanical Garden. 76(2): 118–139.

Longman, K. A. & J. Jenik. 1987. Tropical Forest and its Environment. Longman Scientific and Technical, 2nd edition, New York. 347 pp.

Lopes, M. A. & S. F. Ferrari. 1994. Differential recruitment of *Eschweilera albiflora* (Lecythidaceae) seedlings at two sites in western Brazilian Amazonia. Tropical Ecology. 35: 25–34.

Lot, A. & F. Chiang. 1986. Manual de herbario. Consejo Nacional de La Flora de México. México. 142 pp.

Lucid. 2009 accessed. Lucidcentral.org. http://www.lucidcentral.com/.

Luckow, M. 1995. Species concepts: assumptions, methods, and applications. Systematic Botany. 20(4): 589–605.

Luja, V. H., S. Herrando-Pérez & D. González-Solis. 2008. Secondary forests are not havens for reptile species in tropical Mexico. Biotropica 40(6): 747–757.

Maia, L. M de Alencar. 2008. Frutos, alimentos para peixes. Pages 72–75 *in* Amazônia. A Floresta e o Futuro. Scientific American Brasil, Suplementos.

Mabberley, D. J. 1997. The Plant Book. Cambridge University Press, New York. 858 pp.

Manbar, U. 2007. Encouraging people to contribute knowledge. The Official Google Blog. http://googleblog.blogspot.com/2007/12/encouraging-people-to-contribute.html. Accessed 2009.

Mares, M. A. 1991. How scientists can impede the development of their discipline: egocentrism, small pool size, and the evolution of "sapismo." Pages 57–75 *in* M. A. Mares & D. J. Schmidly (eds.), Latin American Mammalogy: History, Biodiversity, and Conservation. University of Oklahoma Press, Norman and London.

Maués, M. M. 2006. Reproductive phenology and pollination of the Brazil nut tree (*Bertholletia excelsa* Humb. & Bonpl. Lecythidaceae) in eastern Amazonia. Pages 267–277 *in* P. G. Kevan & V. L. Imperatriz-Fonseca (eds.), Pollinating Bees: The Conservation Link Between Agriculture and Nature, 2nd edition. Ministério do Meio Ambiente, Brasilia, Brasil.

McDade, L. A., K. S. Bawa, H. A. Hespenheide & G. S. Hartshorn (eds.). 1994. La Selva: Ecology and Natural History of a Neotropical Rainforest. The University of Chicago Press, Chicago, IL. 486 pp.

McNeill, J., F., R. Barrie, H. M. Burdet, V. Demoulin, D. C. Hawksworth, K. Marhold, D. H. Nicolson, J. Prado, P. C. Silva, J. E. Skog, J. H. Wiersema & N. J. Turland. 2006. International code of botanical nomenclature (Vienna Code). Regnum Vegetabile 146: 1–568.

Medeiros Santos, A. 2008. Riqueza em cachos. Pages 52–57 *in* Amazônia. A Floresta e o Futuro. Scientific American Brasil, Suplementos.

Meinhold, B. 2009. Worlds Largest Solar Project Planned for Saharan Desert. Inhabitat. (www. http://inhabitat.com/worlds-largest-solar-project-sahara-desert/).

Mellon Foundation. 2003 onward. The Andrew W. Mellon Foundation. http://www.mellon.org/. Accessed 2010.

Melo, M. C., A. Berkov & M. Coscoran. 2005. Redescription of *Manicocoris rufipes* (Fabricius, 1787), including nymphs I, II, III and V (Reduviidae: Harpactorinae: Apiomerini), and its association with *Clusia* fruits. Studies on Neotropical Fauna and Environment 40: 55–64.

Metsger, D. A. & S. C. Byers (eds.). 1999. Managing the Modern Herbarium. Elton-Wolf Publishing, Vancouver, Canada. 384 pp.

Meyke, E. 2004. TAXIS. http://www.bio-tools.net/. Accessed 2009.

Miller, J. S., T. M. Barkley, H. H. Iltis, W. L. Lewis, E. Forero, M. Plotkin, O. Phillips, R. Rueda & P. Raven. 1996. Alwyn Howard Gentry, 1945–1993. A tribute. The life and work of Al Gentry. Annals of the Missouri Botanical Garden. 83(4): 433–460.

Ministère de l'Ecologie, de l'Energie, du Développement Durable et de la Mer, 2009 accessed. http://www.ecologie.gouv.fr/Le-parc-amazonien-de-Guyane-est.html.

Missouri Botanical Garden. 2009 accessed. Tropicos. http://www.tropicos.org/Home.aspx.

Mittermeir, R. A., P. Robles Gil, M. Hoffman, J. Pilgrim, T. Brooks, C. Goettsch Mittermeir, J. Lamoureux & G. A. B. da Fonseca. 2005. Hotspots Revisited. Earth's Biologically Richest and Most Endandgered Terrestrial Ecosystems. University of Chicago Press,

Chicago. 392 pp.

Modde, M. 1981. Autrefois, Saül. La Semaine 30: 4–7.

Morellato, L. P. & C. F. B. Haddad. 2000. Introduction: The Brazilian Atlantic Forest. Biotropica 32(4b): 786–792.

Mori, S. A. 1970. The ecology and uses of the species of *Lecythis* in Central America. Turrialba 20(3): 344–350.

———. 1971. A new species of *Lecythis* from Panama. Annals of the Missouri Botanical Garden. 57: 386–388.

———. 1976. Taxonomic and Anatomic Studies in *Gustavia*. Ph.D. Dissertation, University of Wisconsin, Madison, Wisconsin. 415 pp.

———. 1984. Use of "Swiss Tree Grippers" for making botanical collections of tropical trees. Biotropica 16(1): 79–80.

———. 1987a. A guide to collecting Lecythidaceae. Annals of the Missouri Botanical Garden. 74: 321–330.

———. 1987b. Chapter I. Introduction. *In* S. A. Mori & collaborators, The Lecythidaceae of a Lowland Neotropical forest: La Fumée Mountain, French Guiana. Memoirs of The New York Botanical Garden. 44: 3–8.

———. 1989. Eastern, extra-Amazonian Brazil. Pages 427–454 *in* D. G. Campbell and H. D. Hammond (eds.), Floristic inventory of tropical countries. The New York Botanical Garden, New York.

———. 1991. The Guayana lowland floristic province. Compte Rendus de la Societé de Biogéographie 67(2): 67–75.

———. 1992a. The Brazil nut industry—past, present, and future. Pages 241–251 in M. Plotkin and L. Famolare (eds.), Sustainable harvest and marketing of rain forest products. Island Press, Washington, D.C.

———. 1992b. *Eschweilera pseudodecolorans* Lecythidaceae), a new species from central Amazonian Brazil. Brittonia 44(2): 244–246.

———. 1992c. Neotropical floristics and inventory: Who will do the work? Brittonia 44(3): 372–375.

———. 1995a. Exploring for plant diversity in the canopy of a French Guianan forest. Selbyana 16(1): 94–98.

———. 1995b. Observações sobre as espécies de Lecythidaceae do leste do Brasil. Boletim de Botânica, Departamento de Botânica, Instituto de Biociencias, Univsidade de São Paulo 14: 1–31.

———. 1995c. Lecythidaceae. Pages 66–73 *in* M. Nee, Flora preliminar do Projeto Dinâmica Biológica de Fragmentos Florestais (PDBFF). New York Botanical Garden and INPA/ Smithsonian Projeto Dinâmica Biológica de Fragmentos Florestais.

———. 1998. Botanical vouchers: seldom discussed problems and recommendations for inventories. Mesoamericana 3(2): 37–38.

———. 1999. Ghillean T. Prance—recipient of the 1998 Asa Gray Award. Systematic Botany 24(1): 1–4.

———. 2001. A família da castanha-do-Pará: Símbolo do Rio Negro. Pages 119–141 *in* A. A. de Oliveira and D. C. Daly (eds.), Florestas do Rio Negro. Companhia das Letras, Universidade Paulista and The New York Botanical Garden, São Paulo and New York.

———. 2004. Tropical Forests: Lecythidaceae. Pages 1745–1752 *in* J. Burley, J. Evans & J. A. Youngquist (eds.), Encyclopedia of Forest Sciences, vol. 4. Elsevier Ltd., Oxford.

———. 2010. Soybean agriculture threatens biodiversity in Brazil. Blog in Plant Talk, The New York Botanical Garden.

———, P. Becker & D. Kincaid. 2001. Lecythidaceae of a central Amazonian lowland forest. Implications for conservation. Pages 54–67 *in* R. O. Bierregaard, Jr., C. Gascon, T. E.

Lovejoy & R. C. G. Mesquita (eds.), Lessons from Amazonia. The ecology and conservation of a fragmented forest. Yale University Press, New Haven & London.

—— & J. D. Boeke. 1987. Chapter XII. Pollination. *In* S. A. Mori & Collaborators, The Lecythidaceae of La Fumée Mountain, French Guiana. Memoirs of The New York Botanical Garden. 44: 137–155.

—— & B. M. Boom. 1987. Chapter II. The forest. *In* S. A. Mori & collaborators, The Lecythidaceae of a Lowland Neotropical forest: La Fumée Mountain, French Guiana. Memoirs of The New York Botanical Garden. 44: 113–123.

——, ——, A. M. de Carvalho & T. S. dos Santos. 1983a. Southern Bahian moist forests. Botanical Review (Lancaster) 49(2): 155–232.

——, ——, —— & ——. 1983b. Ecological importance of Myrtaceae in an eastern Brazilian wet forest. Biotropica 15: 68–70.

——, —— & G. T. Prance. 1981. Distribution patterns and conservation of eastern Brazilian coastal forest tree species. Brittonia 33: 233–245.

—— & J. L. Brown. 1994. Report on wind dispersal in central French Guiana. Brittonia 46(2): 105–125.

—— & ——. 1998. Epizoochorous dispersal by barbs, hooks, and spines in a lowland moist forest in central French Guiana. Brittonia 50(2): 165–173.

—— & ——. 2002a. Symplocaceae. *In* S. A. Mori, G. Cremers, C. Gracie, J.-J. de Granville, S. V. Heald, M. Hoff & J. D. Mitchell, Guide to the Vascular Plants of Central French Guiana. Part 2. Dicotyledons. Memoirs of The New York Botanical Garden. 76(2): 708–709.

—— & ——. 2002b. Hernandiaceae. *In* S. A. Mori, G. Cremers, C. Gracie, J.-J. de Granville, S. V. Heald, M. Hoff & J. D. Mitchell, Guide to the Vascular Plants of Central French Guiana. Part 2. Dicotyledons. Memoirs of The New York Botanical Garden. 76(2): 344–347.

——, W. R. Buck, C. A. Gracie & M. Tulig. 2007a onward. Plants and Lichens of Saba (http://sweetgum.nybg.org/saba/). Virtual Herbarium of The New York Botanical Garden.

—— & Collaborators. 1987. The Lecythidaceae of La Fumée Mountain, French Guiana. Memoirs of The New York Botanical Garden. 44: 1–290.

——, G. Cremers, C. Gracie, J.-J. de Granville, S. V. Heald, M. Hoff & J. D. Mitchell. 2002a. Guide to the vascular plants of central French Guiana. Part 2. Dicotyledons. Memoirs of The New York Botanical Garden. 76(2): 1–776, pls. 1–128.

——, ——, ——, ——, M. Hoff & J. D. Mitchell. 1997. Guide to the vascular plants of central French Guiana. Part 1. Pteridophytes, Gymnosperms, and Monocotyledons. Memoirs of The New York Botanical Garden. 76(1): 1–422, pls. 1–72.

—— & H. García-Barriga. 1975. A new species of *Gustavia* (Lecythidaceae) endemic to the Magdalena Valley of Colombia. Caldasia 11: 51–53.

——, Gracie, C. A., E. F. Hecklau, T. A. Lobova, A. Berkov & J.-J de Granville. 2005. Documenting plant diversity in Central French Guiana: The first step toward understanding biocomplexity. *In* Friis, I. & H. Baslev (eds.), Plant Diversity and Complexity Patterns. Local, Regional and Global Dimensions. Biologiske Skrifter. 55:11–24.

——, C. A. Gracie & J. D. Mitchell. 1998. Écotourisme et protection du patrimoine naturel dans le centre de la Guyane Française. Journal d'Agriculture Tropicale et de Botanique Appliquée, Revue d'Ethnobiologie 40(1–2): 299–310.

——, —— & ——. 1999. Guyana Francesa central una unica y ruda experiencia turística. Plumeria 7: 29–44.

—— & S. V. Heald. 1997. John J. Wurdack's contribution to South American floristics. BioLlania Edición Especial 6: 127-132.

——, E. F. Hecklau & T. Kirchgessner. 2002b. Life form, habitat, and nutritional mode of the flowering plants of central French Guiana. Journal of the Torrey Botanical Society. 129(4): 331–345.

—— & L. B. Holm-Nielsen. 1981. Recommendations for botanists visiting Neotropical countries. Taxon 30(1): 87–89.

—— & J. A. Kallunki. 1976. Phenology and floral biology of *Gustavia superba* (Lecythidaceae) in Central Panama. Biotropica 8(3): 143–165.

—— & N. Lepsch-Cunha. 1995. The Lecythidaceae of a central Amazonian moist forest. Memoirs of The New York Botanical Garden. 75: 1–55.

—— & L. A. Mattos Silva. 1979a. The herbarium of the "Centro de Pesquisas do Cacau" at Itabuna, Brazil. Brittonia 31(2): 177–196.

—— & ——. 1979b. O herbario do Centro de Pesquisas do Cacau. Ciência e Cultura 31(7): 808–809.

—— & ——. 1979c. Flora de Região Cacaueira da Bahia Plano geral para sua elaboração. Anais da Sociedade Botânica de Brasil 30: 102–104.

——, ——, G. Lisboa, R. C. Pereira & T. S. dos Santos. 1980a. Subsídios para estudos de plantas invasoras no sul da Bahia. I. Produtividade e fenologia. Commisão Executiva do Plano da Lavoura Cacaueira. Boletim Técnico 73: 1–18.

——, ——, —— & L. Coradin. 1985. Manual de manejo do herbario fanerogâmico. Centro de Pesquisas do Cacau, Ilhéus, Bahia. 97 pages.

——, ——, —— & ——. 1989. Manual de manejo do herbario fanerogamico (2nd. Ed.). Centro de Pesquisas do Cacau, Ilhéus, Bahia, Brasil.

——, —— & T. S. dos Santos. 1980b. Observações sobre a fenologia e biologia floral de *Lecythis pisonis* Cambess. (Lecythidaceae). Revista Theobroma 10: 103–111.

——, J. E. Orchard & G. T. Prance. 1980c. Intrafloral pollen differentiation in the New World Lecythidaceae, subfamily Lecythidoideae. Science 209: 400–403.

—— & G. T. Prance. 1987a. Species diversity, phenology, plant animal interactions, and their correlation with climate as illustrated by the Brazil nut family (Lecythidaceae). Pages 69–89 *in* R. E. Dickinson (ed.), The geophysiology of Amazonia. John Wiley & Sons, New York.

—— & ——. 1987b. Chapter XI. Phenology. *In* S. A. Mori & Collaborators, The Lecythidaceae of La Fumée Mountain, French Guiana. Memoirs of The New York Botanical Garden. 44: 124–136.

—— & ——. 1987c. A guide to collecting Lecythidaceae. Annals of the Missouri Botanical Garden 74(2): 321–330.

—— & ——. 1990. Lecythidaceae – Part II: The zygomorphic-flowered New World genera (*Couroupita, Corythophora, Bertholletia, Couratari, Eschweilera, & Lecythis*) With a study of secondary xylem of Neotropical Lecythidaceae by Carl de Zeeuw. Flora Neotropica Monograph 21(II): 1–376.

——, —— & A. B. Bolten. 1978. Additional notes on the floral biology of Neotropical Lecythidaceae. Brittonia 30: 113–130.

——, B. V. Rabelo, Chih Hua Tsou & Douglas Daly. 1989. Composition and structure of an eastern Amazonian forest at Camaipi, Amapá, Brazil. Bol. Mus. Paraense Hist. Nat. 5(1): 3 18.

——, N. P. Smith, X. Cornejo & G. T. Prance. 18 March 2010 onward. The Lecythidaceae Pages (http://sweetgum.nybg.org/lp/index.php). The New York Botanical Garden, Bronx, New York.

—— & D. Swarthout. 2007 onward. Brazil nut family (Lecythidaceae) in the New World. *In* C. J. Cleveland (ed.), Encyclopedia of Earth. Accessed 2009. Washington, D.C.: Environmental Information Coalition, National Council for Science and the Environment. http://www.eoearth.org/article/Brazil_nut_family_(Lecythidaceae)_in_the-New_World. Accessed 2009.

——— & ———. 2008a onward. Brazil nut (*Bertholletia excelsa*)." *In* C. J. Cleveland (ed.), Encyclopedia of Earth. Washington, D.C.: Environmental Information Coalition, National Council for Science and the Environment. Accessed 2009. http://www.eoearth. org/article/Brazil_nut_(Bertholletia_excelsa). Accessed 2009.

——— & ———. 2008b onward. Cannon ball tree (*Couroupita guianensis*). *In* C. J. Cleveland, Encyclopedia of Earth. Washington, D.C.: Environmental Information Coalition, National Council for Science and the Environment. http://www.eoearth.org/article/ Cannon_ball_tree_(Couroupita_guianensis). Accessed 2009.

———, C.-H. Tsou, C.C.Wu, B. Cronholm & A. A. Anderberg. 2007b. Evolution of Lecythidaceae with an emphasis on the circumscription of neotropical genera: information from combined *ndh*F and *trn*L-F sequence data. American Journal of Botany. 94(3): 289–301.

———, M. Tulig, J.-J. de Granville, S. Gonzalez & V. Guerin. 2007c onward. French Guianan e-Flora Project. The New York Botanical Garden and the Institut de Recherche pour le Développement. http://sweetgum.nybg.org/fg/cite.html. Accessed 2009.

Moritz, A. 1984. Estudo biológicos de floração e da frutificação da castanha-do-Brasil (*Bertholletia excelsa* H. B. K.). EMBRAPA-CPATU Documentos 29: 1–84.

Morton, C. M., S. A. Mori, G. T. Prance, K. G. Karol & M. W. Chase. 1997. Phylogenetic relationships of Lecythidaceae: A cladistic analysis using rbcL sequence and morphological data. American Journal of Botany 84(4): 530–540.

Mouawad, J. 2008. Pumping hydrogen. The New York Times. Business Section. September 24, 2008. Pages 1, 8.

Müller, P. 1973. The Dispersal Centres of Terrestrial Vertebrates in the Neotropical Realm. Junk, The Hague. 244 pp.

Munn, C. A. 1992. Macaw biology and ecotourism, or "When a bird in the bush is worth two in the hand." Pages 47–72 *in* S. R. Beissinger & N. F. Snyder (eds.), New World Parrots in Crisis. Smithsonian Institution Press, Washington and London.

Myers, N. 1988. Threatened biotas: 'hotspots' in tropical forests. Environmentalist 8: 187–208.

———, R. A. Mittermeier, C. G. Mittermeier, G. A. B. da Fonseca & J. Kent. 2000. Biodiversity hotspots for conservation priorities. Nature 403: 853–858.

Myster, R. W. 2004. Post-agricultural invasion, establishment, and growth of neotropical trees. Botanical Review (Lancaster) 70(4): 381–402.

Nelson, B. W., M. L. Absy, E. M. Barbosa & G. T. Prance. 1985. Observations on flower visitors to *Bertholletia excelsa* H. B. K. and *Couratari tenuicarpa* A. C. Sm. (Lecythidaceae). Acta Amazonica, Supl. 15(1–2): 225–234.

Nepstad, D. C., C. M. Stickler, B. Soares-Filho & F. Merry. 2008. Interactions among Amazon land use, forests and climate: prospects for a near-term forest tipping point. Philosophical Transactions of the Royal Society of London, Series B 363: 1737–1746.

Nichols, E., S. Spector, J. Louzada, T. Larsen, S. Amezquita, & M. E. Favila. 2008. Ecological functions and ecosystem services by Scarabaeinae dung beetles. Biological Conservation 141: 1461–1474.

Nixon, K. 2009 accessed. Cladistics.com. http://www.cladistics.com/.

Norconk, M. A. 2010 accessed. Feeding ecology and mechanical properties of fruit and seeds ingested by white-faced and bearded sakis. http://www.personal.kent.edu/~mnorconk/saki-feeding-ecology.html

Nugent, M. 2001. La reserve naturelle de Nouragues. Pages 382–385 *in* C. Richard-Hansen & R. Le Guen (eds.), Guyane ou le Voyage Écologique. Éditions Roger Le Guen, Garies,

France.

Oldeman, R. A. A. 1974. L'architecture de la forêt guyanaise. Mémoirs Office de la Recherche Scientifique et Technique Outre-Mer (O.R.S.T.O.M.) 73: 1–204.

Oliveira, A. A. de & S. A. Mori. 1999. A central Amazonian terra firme forest. I. High tree species richness on poor soils. Biodiversity and Conservation 8: 1219–1244.

Oliveiro-Filho, A. T. & M. A. L. Fontes. 2000. Patterns of floristic differentiation among Atlantic forests in southeastern Brazil and the influence of climate. Biotropica 32(4b): 793–810.

Ortiz, E. G. 1995. Survival in a nutshell. Americas. September– October, pp. 7–17.

Oxford University Herbaria. 2004. Flora Brasiliensis. http://herbaria.plants.ox.ac.uk/flora brasiliensis.htm. Accessed 1 January 2010.

Pennington, T. D. Sapotaceae. 1990. Flora Neotropica Monograph 52:1–770.

————— & S. A. Mori. 1993. *Guarea michel-moddei*, a new species from central French Guiana. Brittonia 45(3): 231–234.

Peres, C. A., C. Baider, P. A. Zuidema, L. H. O. Wadt, K. A. Kainer, D. A. P. Gomes-Silva, R. P. Salomão, L. L. Simões, E. R. N. Franciosi, F. Cornejo Valverde, R. Gribel, G. H. Shepard, Jr., M. Kanashiro, P. Coventry, D. W. Yu, A. R. Watkinson & R. P. Freckleton. 2003. Demographic threats to the sustainability of Brazil nut exploitation. Science 302: 2112–2114.

Pimentel, D., C. Wilson, C. McCullum, R. Huang, P. Dwen, J. Flack, Q. Tran, T. Saltman & B. Cliff. 1997. Economic and environmental benefits of biodiversity. BioScience 47(11): 747–757.

Piperno, D. R. & P. Becker. 1996. Vegetational history of a site in central Amazon Basin derived from phytolith and charcoal records from natural soils. Quaternary Research 45: 202–209.

Poiteau, M. A. 1825. Mémoire sur les Lecythidées. Mémoires du Muséum d'Histoire Naturell. Paris 13: 141–165.

Pollan, M. 2006. Ominvore's Dilemma: A Natural History of Four Meals. Penguin Press. 450 pp.

————— 2008. Farmer in chief. What the next president can and should do to remake the way we grow and eat our food. The New York Times Magazine. October 12, 2008. Pages 62–71, 92.

Potascheff, C. de Moraes. 2010. Ecologia da polinização de *Eschweilera nana* Miers, uma Lecythidaceae do cerrado. Undergraduate thesis. Universidade Estadual Paulista "Julio de Mesquita Filho" Instituto de Biociências – Rio Claro, Brazil. 53 pp.

Prance, G. T. 1972a. Chrysobalanaceae. Flora Neotropica Monograph. 9: 1–410.

—————. 1972b. Dichapetalaceae. Flora Neotropica Monograph. 10; 1–84.

—————. 1972c. Rhabdodendraceae. Flora Neotropica Monograph. 11: 1–22.

—————. 1976. Pollination and androphore structure of some Amazonian Lecythidaceae. Biotropica 8: 235–241.

—————. 1979. *Cariniana*. Flora Neotropica Monograph. 21: 218–244.

—————. 1982. Biological Diversification in the Tropics. Columbia University Press, New York. 714 pp.

—————. 1989. Chrysobalanaceae. Flora Neotropica Monograph. 9S: 1–267.

—————. 2001. Discovering the plant world. Taxon 50: 345–359.

—————. 2007. The protection of the Yabotí Biosphere Reserve, Misiones, Argentina and its Guaraní people. EnviroNews 13: 1–3.

—————. 2008. A revision of *Foetidia* (Lecythidaceae subfam. Foetidioideae). Brittonia 60(4): 336–348.

—————, H. Beentje, J. Dransfield & R. Johns. 2000. The tropical flora remains undercollected. Annals of the Missouri Botanical Garden 87: 67–71.

——— and D. G. Campbell. 1988. The present state of tropical floristics. Taxon 37(3): 519–548.

——— & M. Freitas da Silva. 1973. Caryocaraceae. Flora Neotropica Monograph. 12: 1–75.

——— & S. A. Mori. 1979. Lecythidaceae-Part I. The actinomorphic-flowered New World Lecythidaceae (*Asteranthos, Gustavia, Grias, Allantoma,* & *Cariniana*). Flora Neotropica Monograph. 21: 1–270.

——— & S. A. Mori. 1998. Pollination and dispersal of neotropical Lecythidaceae. Pages 13–27 *in* H.C.F. Hopkins, C.R. Huxley, C.M. Pannell, G.T. Prance & F. White (eds.), The Biological Monograph. Royal Botanic Gardens, Kew.

———, V. Plana, K. S. Edwards & R. T. Pennington. 2007. Proteaceae. Flora Neotropica Monograph 100:1–218.

——— & F. White. 1988. The genera of Chrysobalanaceae: a study in practical and theoretical taxonomy and its relevance to evolutionary biology. Philosophical Transactions of the Royal Society of London, Series B 320: 1–184.

Procópio, L. C. & R. de S. Secco. 2008. A importância da indentificação botânica nos inventários florestais: o exemplo do "tauari" (*Couratari* spp. e *Cariniana* spp. - Lecythidaceae) em duas áreas manejadas no estado do Pará. Acta Amazonica 38: 31–44.

Putz, F. E. , P. A. Zuidema, M. A. Pinard, R. G. A. Boot, J. A. Sayer, D. Sheil, P. Sist, Elias & J. K. Vanclay. 2008. Improved tropical forest management for carbon retention, PLoS Biology 6(7): e166 an online journal available at http://dx.doi.org/10.1371/journal.pbio.0060166.

Raven, P. 1988. Tropical floristics tomorrow. Taxon 37(3): 549–560.

Redford, K. H. 1991. The ecologically noble savage. Cultural Survival Quarterly 15(1): 46-48.

———. 1992. The empty forest. BioScience 42(6) 412–422.

Reis, A. M. M., A. C. Braga, M. R. Lemes, R. Gribel & R. G. Collevatti. 2009. Development and characterization of microsatellite markers for the Brazil nut tree *Bertholletia excelsa* Humb. & Bonpl. (Lecythidaceae). Molecular Ecology Resources 9(3): 920–923.

Ribeiro, J. E. L. da, M. J. G. Hopkins, A. Vicentini, C. A. Sothers, M. A. da S. Costa, J. M. de Brito, M. A. D. de Souza, L. H. P. Martins, L. G. Lohmann, P. A. C. L. Assunção, E. da C. Pereira, C. F. da Silva, M. R. Mesquita & L. C. Procópio. 1999. Flora da Reserva Ducke. Guia de Identificação das Plantas Vasculares de uma Floresta de Terra-Firme na Amazônia Central. INPA, Manaus. 816 pp.

Robbins, R. K., A. Aiello, J. Feinstein, A. Berkov, A. Caldas, R. C. Busby & M. Duarte. 2010. A tale of two species: detritivory, parapatry, and sexual dimorphism in *Lamprospilus collucia* and *L. orcidia* (Lycaenidae: Theclinae: Eumaeini). Journal of Research on the Lepidoptera 42: 64–73.

Roberts, L. Shrinking the malaria mpa from the outside in. Science 828: 849–851

Robinson, J. G. 1993. The limits to caring: sustainable living and the loss of biodiversity. Conservation Biology 7(1): 20–28.

———. & K. H. Redford (eds.). 1991. Neotropical Wildlife Use and Conservation. The University of Chicago Press. 520 pp.

Roe, K. E. 1967. A revision of *Solanum* sect. *Brevantherum* (Solanaceae) in North and Central America. Brittonia 19(4): 353–373.

Roosevelt, A. C. 1992. Secrets of the forest.The Sciences. November/December pp 23–28.

———, M. Lima da Costa, D. Lopes Machado, M. Michab, N. Mercier, H. Valladas, J. Feathers, W. Barnett, M. Imazio da Silveira, A. Henderson, J. Sliva, B. Chernoff, D. S. Reese, J. A. Holman, N. Toth & K. Schick. 1996. Paleoindian cave dwellers in the Amazon: the peopling of the Americas. Science 272: 373–384.

Roosmalen, M. G. M. van. 1985. Fruits of the Guianan Flora. Institute of Systematic Botany, Utrecht University and Silvicultural Department of Wageningen Agricultural University, The Netherlands. 483 pp.

Rosenberg, M. 2006. Population growth rates and doubling time. http://geography.about.com/od/populationgeography/a/populationgrow.htm. Accessed 2009.

Rovira, I., A. Berkov, A. Parkinson, G. Tavakilian, S. Mori & B. Meurer-Grimes. 1999. Antimicrobial activity of Neotropical wood and bark extracts. Pharmaceutical Biology 37(3): 208–215.

Royal Botanic Gardens Kew. 2008. Helping the Planet Breathe. Royal Botanic Gardens Kew.

Salati, E. & P. B. Vose. 1984. Amazon Basin: A system in equilibrium. Science 225: 125–138.

Saldarriaga, J. G. 1986. Recovery following shifting cultivation. Pp. 24–33. In C. F. Jordan (ed.), Amazonian Rain Forests. Springer-Verlag, New York.

——— & C. Uhl. 1991. Recovery of forest vegetation following slash-and-burn agriculture in the upper Rio Negro. In A. Gómez-Pompa, T. C. Whitmore & M. Hadley (eds.), Rain forest regeneration and management. Man and the Biosphere Series 6: 303–312.

——— & D. C. West. 1986. Holocene fires in the northern Amazon Basin. Quaternary Research 26(3): 358-366.

Sánchez-Azofeifa, G. A., B. Rivard, J. Calvo & I. Moorthy. 2002. Dynamics of tropical deforestation around national parks: Remote sensing of forest change on the Osa Peninsula of Costa Rica. Mountain Research and Development 22(4): 352–358.

Sanford, R. L., Jr.,]. Saldarriaga, K. E. Clark, C. Uhl & R. Herrera. 1985. Amazon rain-forest fires. Science 227: 53–55.

Santos, G. M. dos, E. J. G. Ferreira & A. L. Val. 2008. Pages 65–71 in Amazônia, o universo dos peixes. Scientific American Brasil, Suplementos..

Scotland, R. W. & A. H. Wortley. 2003. How many species of seed plants? Taxon 52:101–104.

Scratchpads. 2010 accessed. Scratchpads. http://scratchpads.eu/.

Shorto, R. 2008. ¿No hay bebes? Keine Kinder? Nessun bambino? The New York Times Magazine. June 28, 2008. Pages 34–41, 68–69, 70–71.

Sibley, D. A. 2000. The Sibley guide to birds. Alfred A. Knopf, New York. 544 pp.

Simmons, N. B. & R. S. Voss. 1998. The mammals of Paracou, French Guiana: a neotropical lowland rain forest fauna part 1. bats. Bulletin of the American Museum of Natural History 237: 1–219.

Smith, N., S. A. Mori, A. Henderson, D. Wm. Stevenson & S. V. Heald. 2004. Flowering Plants of the Neotropics. Princeton University Press, Princeton and Oxford. 594 pp.

Smith, N., R. Vásquez & W. H. Wust. 2007. Amazon River Fruits. Flavors for Conservation. Missouri Botanical Garden Press, St. Louis, Missouri. 274 pp.

Sork, V. L. 1987. Effects of predation and light on seedling establishment in Gustavia superba. Ecology 68(5): 1347–1350.

Sowls, L. K. 1984. The Peccaries. The University of Arizona Press, Tucson, Arizona.

Stap, D. 1990. A Parrot Without a Name: The Search for the Last Unknown Birds on Earth. University of Texas Press. 255 pp.

Steege, H. ter, N. C. A. Pitman, O. L. Phillips, J. Chave, D. Sabatier, A. Duque, J.-F. Molino, M.-F. Prévost, R. Spicher, H. Castellanos, P. von Hildebrand & R. Vásquez. 2006. Continental-scale patterns of canopy tree composition and function across Amazonia. Nature 443: 444–447.

Stern, W. T. 1992. Botanical Latin, ed. 4. David & Charles, Newton Abbot, England. 546 pp.

Stewart, R. R. 1984. How did they die? Taxon 33(1): 48–52.

Stokstad, E. 2008. A second chance for rainforest biodiversity. Science 320: 1436–1439.

Stouffer, P. C. & R. O. Bierregaard, Jr. 1995. Use of Amazonian forest fragments by understory frugivorous birds. Ecology 76(8): 2429-2445.

Systematics Agenda 2000. 1994. Systematics Agenda 2000. Charting the Biosphere. A Global Initiative to Discover, Describe and Classify the Word's Species. Department of Ornithology, American Museum of Natural History, New York. 20 pp.

Tavakilian, G., A. Berkov, B. Meurer-Grimes & S. A. Mori. 1997. Neotropical tree species and their faunas of xylophagous longicorns (Coleoptera: Cerambycidae) in French Guiana. Botanical Review (Lancaster): 63(4): 3–55.

Terborgh, J. 1999. Requiem for nature. Island Press/Shearwater Books, Washington, D.C. & Corvelo, California. 234 pp.

———, L. Lopez & J. Tello S. 1997. Bird communities in transition: the Lago Guri Islands. Ecology 78(5): 1494–1501.

Thiers, B. 1997 onward. Index Herbariorum: A global directory of public herbaria and associated staff. New York Botanical Garden's Virtual Herbarium. http://sweetgum.nybg.org/ih/.

Thomas, W. W. 1999. Conservation and monographic research on the flora of tropical America. Biodiversity and Conservation 8: 1007–1015.

———. 2005 onward. Organization for Flora Neotropica. http://www.nybg.org/botany/ofn/OFN.html.

———. 2008. The Atlantic Coastal Forest of Northeastern Brazil. Memoirs of The New York Botanical Garden 100: 1–586.

Tiroler Landsmuseen. 2009 accessed. BioOffice. http://www.biooffice.at/index.php/home__page.html.

Tsou, C.-H. 1994. The Embryology, Reproductive Morphology, and Systematics of Lecythidaceae. Memoirs of The New York Botanical Garden 71: 1–110.

———. & S. A. Mori. 2002. Seed coat anatomy and its relationship to seed dispersal in subfamily Lecythidoideae of the Lecythidaceae (the Brazil nut family). Botanical Bulletin of Academia Sinica. 43: 37–56.

——— & S. A. Mori. 2007. Floral organogenesis and floral evolution of the Lecythidoideae (Lecythidaceae). American Journal of Botany. 94(5): 716–736.

Tuttle, M. D. 1988. America's Neighborhood Bats. University of Texas Press. 36 pp.

U. S. Department of State. 2010a accessed. Background note: Guyana. http://www.state.gov/r/pa/ei/bgn/1984.htm.

———. 2010b accessed. Background note: Suriname. http://www.state.gov/r/pa/ei/bgn/1893.htm.

Vázquez G., J. A., R. Cuevas G., T. S. Cochrane, H. H. Iltis, F. J. Santana M. & L. Guzmán H. 1995. Flora de Manantlán. Sida, Botanical Miscellany 12: 1–312.

Vogel, G. 2010. The 'do unto others' malaria vaccine. 329: 847–848.

Watson, J. 1968. The Double Helix: A Personal Account of the Discovery of the Structure of DNA. Atheneum, New York. 226 pp.

Wessels, T. 2006. The Myth of Progress. Toward a Sustainable Future. University of Vermont Press, Burlington, Vermont. 131 pp.

Whigham A. E., Baxt A. & Berkov A. In prep. Healthy Choices: preferential foraging of insects within senescent, but nutrient-rich, Lecythidaceae flowers. To be submitted to Journal of Tropical Ecology.

Wilson, E. O. 1991. Ants. Wings. Fall, 1991, pages 4–13.

———. 2001. Foreword. Page xiv in R. O, Bierregaard, Jr., C. Gascon, T. E. Lovejoy & R. C. G. Mesquita (eds.). 2001. Lessons from Amazonia. The ecology and conservation of a fragmented forest. Yale University Press, New Haven & London.

World Resources Institute. 2008. Population growth: stabilization. http://www.wri.org/publication/content/8599. Accessed 13 Oct 2008.

Wortley, A. H. & R. W. Scotland. 2004. Synonymy, sampling and seed plant numbers. Taxon 53: 478–480.

Wright, J. & H. Muller-Landau. 2006. The future of tropical forest species. Biotropica 38(3): 287–301.

Yoder, M. J., K. Dole, K. Seltmann & A. Deans. 2006 onward. Mx, a collaborative web based content management for biological systematists. http://hymenoptera.tamu.edu/wiki/index.php/Main_Page. Accessed 2009.

Young, A. M. 2000. Small Creatures & Ordinary Places. Essays on nature. The University of Wisconsin Press. 217 pp.

Zamora, N., B. E. Hammel & M. H. Grayum. 2004. Vegetación. Manual de Plantas de Costa Rica. Volume I. Monographs in Systematic Botany from the Missouri Botanical Garden 97: 91–216.

ZooKeys. 2009 onward. ZooKeys. http://pensoftonline.net/zookeys/index.php/journal/index

Index to Scientific Names, Common Names, and Place Names

This index includes the names of people; scientific names of plants, animals, and microorganisms; common names of plants, animals, and microorganisms; names of organizations; geographic localities; and names of important books. Entries with illustrations are marked with an * (asterisk). The names of scientific and common names of well known animals, plants, and microorganisms are grouped together under the common name of the group; for example, common and scientific names of birds are placed under birds. The same is true for the names of geographic localities which are grouped under the country where they occur; for example, all names in Brazil are grouped under Brazil. The Table of Contents should be consulted for subject categories.

Cavanillesia arborea, 29*
chocolate, 9, 27, 67, 71, 245, 259
Ceiba pentandra, 288
cocoa, 27, 28
Quararibea, 22
    asterolepis, 23,
Pseudobombax munguba, 97
Theobroma, 245
    cacao, 9, 27
Mammals, 41, 49, 51, 71, 94, 95, 96, 102,
    106, 110, 118, 120, 124, 236, 244,
    252, 256
    agoutis, 26, 120, 244, 256
    Bats, 17, 60, 61, 63, 71, 84, 104, 121,
        162, 163, 242, 243, 245, 251, 252,
        254, 257, 261, 273, 274, 286, 291,
        292, 293, 294
            Desmodus rotundus, 84
            Sturnira lilium, 251*
            vampire, 84
    Dasyprocta, 244
    Felis concolor, 119
    giant anteater, 71, 72
    four-eyed opossum, 71
    jaguar, 40*, 56, 71, 72, 83, 118, 119, 258
        black, 119
    monkeys, 26, 109, 118, 123, 124, 132,
        245, 256, 258, 264
            Aloutta, 123
            Ateles, 123
                paniscus, 123
            Cacajao calvus, 132
            capuchin, 123, 258
                white-faced, 123, 124*
            Cebus, 123
                capuchinus, 123, 124*
            howler, 82, 123, 162
            spider, 123, 162
            white uacari, 132
    Myrmecophaga tridactyla, 71
    Nasua nasua, 41*
    opossums, 71
        Philander opossum, 71
        four-eyed, 71
    Panthera onca, 40*, 71, 118
    peccaries, 39, 71, 118, 120, 121, 122*,
    123, 124, 162, 245, 256
        Catagonus wagneri, 120
        Chacoan, 120
        collared, 120, 122*, 123

Tayassu pecari, 120, 121, 122*
Pecari tajacu, 120, 122*
white-lipped, 22, 26, 120, 121, 122*,
    123
puma, 118, 119, 120, 258
South American coati, 41*
Suidae, 120
tapir, 39, 71, 256
Ursus maritimus, 235
Manhattan Graphics Center, 59
Manual de Plantas de Costa Rica, 227
*Manual of Vascular Plants of the Northeast-
ern United States,* 277
Marie, M., 68
Martinique, 182
Martius, C. F. P. von, 278
Mayacaceae
    Mayaca sellowiana, 35
McBryde Fellow, 24
McGee, J., 290*
Melastomataceae
    Maieta, 97
    melastome, 11
    Tococa, 97
Meliaceae, 29
    Guarea michel-moddei, 36, 40*
Mellon Foundation
    Andrew W. Mellon Foundation, 285,
        286
    Conservation and Environment Pro
        gram, 285
    LAPI, 286
    Latin American Plant Initiative, 286
Menispermaceae, 142, 246
Mesoamerica, 56, 218, 236
Mexico, 1, 6, 7, 55
    Jalisco, 6
    Oaxaca, 97
        Puerto Escondido, 7*
        Rio Verde, 7*
    Tabasco, 7
        Cárdenas, 7
        Mal Paso, 8
MG, see Museu Goeldi
midges
    Culicoides, 84
    no-see-um midges, 84
Miller, J., 20, 22
Milton, 3
Milton Union High School Conservation

Club, 3
Milwaukee Public Museum, 276
Mimi's Guest House, 151
Minimum Critical Size of Ecosystems
    Project, 43, 44
Missouri Botanical Garden, 15, 17, 20, 134,
    206, 275
    MO, 196, 204, 206
    Annals of, 20
Mitchell, J. D., 17*, 197, 291
MO, see Missouri Botanical Garden
Modde, M., 35, 36*, 39, 40
monkey pot, see Lecythidaceae
Moonen
    Bernie, 34*
    Joep, 33, 34*
    Marijke, 34*
Moonseed family, 246
Moraceae, 195
    Cecropia, 97, 251*, 252, 292
        obtusa, 254*,
    Pourouma, 149
Moran, R. C., 53
Mori, S. A., 1, 2, 7, 10*, 13, 14, 15, 16, 18*,
    20, 25*, 28, 29*, 30, 36, 38*, 39, 44, 46,
    47, 50*, 51, 52, 53, 54*, 55, 60, 61, 62*,
    64, 77, 81, 90, 91*, 95, 98*, 103, 107*,
    121, 126, 131, 136, 137, 141, 144, 151,
    153*, 154, 156*, 158, 161, 164*, 165*,
    166*, 168, 170, 173, 176, 179, 183*, 191,
    193, 197, 203, 204, 209, 212, 215, 221,
    223, 227, 228, 237, 239, 244, 269, 273,
    275, 290, 295, 297
mosquitoes, 9, 64, 72, 75, 80, 82, 83, 84, 85,
    104, 106, 108, 109, 110, 112, 113,
    119, 129, 295, 296
    Aedes aegypti, 108
    Anopheles, 112
mosses, see bryophytes
moths, 103
    Hylesia metabus, 103
    papillionite, 104
Museu Goeldi
    Department of Botany, 93
    MG, 139, 193
Museum National d'Histoire Naturelle, 49,
    192, 194
mushrooms, see fungi
mycota, 31
Myrtaceae, 53

Nathaniel Lord Britton Curator of Botany,
    286
National Geographic Society, 20, 198, 284,
    285
    Committee for Research and Explora-
        tion, 285
National Herbarium La Botanica, 212
National Science Foundation (NSF), 24,
    26, 32, 33, 42, 276, 279, 280, 281,
    282, 283, 284, 286
    Biotic Systems and Inventories (BS&I),
        284
    Dimensions in Biodiversity, 280
    Opportunitites for Promoting Under
        standing through Synthesis (OPUS),
        280
    Partnerships for Enhancing Expertise
        in Taxonomy (PEET), 279
    Planetary Biodiversity Inventory (PBI),
        279
    Revisionary Synthesis in Systematics
        (REVSYS), 279, 280, 284
National Tropical Botanical Garden, 24
Nature Conservancy, 5
Nouragues, see French Guiana
NCB, see Netherlands
Nee, M., 13, 14*, 17, 55, 87, 97, 125, 126
nematodes, see parasites
Neotropics, 2, 17, 26, 31, 43, 77, 87, 89,
    100, 104, 108, 115, 116, 119, 125, 194,
    240, 241, 242, 243, 255, 266, 269. 274,
    286, 292, 293, 294
Netherlands, 32, 192, 195
    Centre for Biodiversity (NCB), 195
    Leiden, 195
    Utrecht, 195
Netherlands Antilles, 215
    Saba, 215
    St. Maarten, 52
New York Botanical Garden (The), 24, 26,
    60, 108, 126, 139, 186, 192, 197, 275,
    292
    Development Department, 286
    Herbarium, 140, 141, 192, 229, 292
    Institute of Systematic Botany, 2, 31,
        292
    NY, 138, 141, 142, 144, 192, 193, 204,
        209, 219, 229
    NYBG Press, 26
    Virtual Herbarium, 33, 147, 228, 229

About the Editors

Scott A. Mori. Dr. Scott A. Mori attended the University of Wisconsin-Stevens Point where he obtained his B.S. degree in 1964 in Biology and Conservation, and the University of Wisconsin-Madison where he was awarded his Ph.D. in botany in 1974. He is now the Nathaniel Lord Britton Curator of Botany at The New York Botanical Garden (NYBG). Dr. Mori is a former Executive Director of Flora Neotropica, a former Director of the Institute of Systematic Botany at NYBG, and an adjunct professor at the City University of New York, the Center for Environmental Research and Conservation at Columbia University, and the Yale School of Forestry and Environmental Studies. Dr. Mori has been awarded the David Fairchild Medal for Plant Exploration and the Asa Gray award by the American Society of Plant Taxonomists for life-time achievement based on his studies of the classification, ecology, and conservation of New World tropical plants.

Amy Berkov. Dr. Amy Berkov attended the University of Colorado-Denver, where she obtained her B.F.A degree in 1977 in fine art, and the City University of New York-Lehman College, where she was awarded her Ph.D. in biology in 1999. She is currently an Assistant Professor in the Department of Biology at the City College of New York (CCNY, CUNY), an Honorary Research Associate at NYBG, and an Associate in the Division of Invertebrate Zoology at the American Museum of Natural History (AMNH). Research in her lab focuses on the evolutionary and community ecology of neotropical wood-boring beetles, particularly those associated with the Brazil nut family.

Carol A. Gracie. Carol A. Gracie has a B.S. in Plant Studies from the City University of New York, Lehman College. She is retired from NYBG, where she served as Senior Administrator of Children's Education and Director of Foreign Tours, among other positions. She subsequently worked with her husband, Scott Mori, on tropical research projects, including the preparation of a flora of central French Guiana. Ms. Gracie's current interests include the temperate flora of northeastern North America. She has co-authored a field guide to the wildflowers of that region (*Wildflowers in the Field and Forest: A Field Guide to the Northeastern United States*). Her current book, *Spring Wildflowers of the Northeast: A Natural History*, is in press.

Edmund F Hecklau. Dr. Edmund F. Hecklau attended Wagner College, Staten Island, New York, where he obtained his B.S. degree in Biology in 1950, and NYU College of Medicine where he was awarded his M.D. degree in 1954. Following a residency at Buffalo Children's Hospital, he was in the private practice of General Pediatrics in Greenwich, CT from 1959–1986, and then served as Vice-President of Medical Services at the Greenwich (CT) Hospital from 1986–1991. In his 20 years in retirement, he has been able to refine his some 70-year interest in horticulture and botany, culminating in a volunteer position at NYBG under the mentorship of Dr. Scott Mori, with whom he has co-authored several published papers relating to the flora of Central French Guiana. As a self-described field botanist, from 2006–2010, he conducted educational sessions in field botany and plant name etymology for the naturalist staff of the Natural History Museum of the Adirondacks. He initiated and made available to visitors at that museum an educational herbarium of some 150+ species, designed for hands-on public education.